T0188625

Introduction to Mixed Modelling

Introduction to Mixed Modelling

Beyond Regression and Analysis of Variance

Second Edition

N. W. Galwey

Statistical Consulting Group, GlaxoSmithKline, UK

WILEY

This edition first published 2014
© 2014 John Wiley & Sons, Ltd

Registered office
John Wiley & Sons Ltd, The Atrium, Southern Gate, Chichester, West Sussex, PO19 8SQ, United Kingdom

For details of our global editorial offices, for customer services and for information about how to apply for permission to reuse the copyright material in this book please see our website at www.wiley.com.

Library of Congress Cataloging-in-Publication Data

Galwey, Nicholas
  Introduction to mixed modelling : beyond regression and analysis of variance / N. W. Galwey. – Second edition.
       pages cm
  Includes bibliographical references and index.
  ISBN 978-1-119-94549-9 (cloth)
  1. Multilevel models (Statistics) 2. Experimental design. 3. Regression analysis. 4. Analysis of variance. I. Title.
  QA276.G33 2014
  519.5–dc23

                                                                          2014021670

A catalogue record for this book is available from the British Library.

ISBN: 978-1-119-94549-9

Set in 10/11.5pt TimesLTstd-Roman by Laserwords Private Limited, Chennai, India.
Printed and bound in Malaysia by Vivar Printing Sdn Bhd

1   2014

Contents

Preface

This book is intended for research workers and students who have made some use of the statistical techniques of regression analysis and analysis of variance (anova), but who are unfamiliar with *mixed models* and the criterion for fitting them called *REsidual Maximum Likelihood* (*REML*, also known as *REstricted Maximum Likelihood*). Such readers will know that, broadly speaking, regression analysis seeks to account for the variation in a response variable by relating it to one or more explanatory variables, whereas anova seeks to detect variation among the mean values of groups of observations. In regression analysis, the statistical significance of each explanatory variable is tested using the same estimate of residual variance, namely the residual mean square, and this estimate is also used to calculate the standard error of the effect of each explanatory variable. However, this choice is not always appropriate. Sometimes, one or more of the terms in the regression model (in addition to the residual term) represents *random variation*, and such a term will contribute to the observed variation in other terms. It should therefore contribute to the significance tests and standard errors of these terms: but in an ordinary regression analysis, it does not do so. Anova, on the other hand, does allow the construction of models with additional random-effect terms, known as *block terms*. However, it does so only in the limited context of balanced experimental designs.

The capabilities of regression analysis can be combined with those of anova by fitting to the data a mixed model, so called because it contains both fixed-effect and random-effect terms. A mixed model allows the presence of additional random-effects terms to be recognized in the full range of regression models, not only in balanced designs. Any statistical analysis that can be specified by a general linear model (the broadest form of linear regression model) or by anova can also be specified by a mixed model. However, the specification of a mixed model requires an additional step. The researcher must decide, for each term in the model, whether effects of that term (e.g. the deviations of group means from the grand mean) can be regarded as values of a random variable – usually taken to mean that they are a random sample from some much larger population – or whether they are a fixed set. In some cases, this decision is straightforward: in others, the distinction is subtle and the decision difficult. However, provided that an appropriate decision is made (see Section 6.3), the mixed model specifies a statistical analysis which is of broader validity than regression analysis or anova, and which is nearly or fully equivalent to those methods in the special cases where they are applicable.

It is fairly straightforward to specify the calculations required for regression analysis and anova, and this is done in many standard textbooks. For example, Draper and Smith (1998) give a clear, thorough and extensive account of the methods of regression analysis, and Mead (1988) does the same for the analysis of variance. To solve the equations that specify a mixed model is much less straightforward. The model is fitted – that is, the best estimates of its parameters are obtained – using the REML criterion, but the fitting process requires recursive numerical methods. It is largely because of this burden of calculation that mixed models are less familiar than regression analysis and anova: it is only in about the past three decades that the development of computer power and user-friendly statistical software has allowed them to be used routinely in research. This book aims to provide a guide to the use of mixed models that is accessible to the broad community of research scientists. It focuses not on the details of calculation, but on the specification of mixed models and the interpretation of the results.

The numerical examples in this book are presented and analysed using three statistical software systems, namely

- GenStat, distributed by VSN International Ltd, Hemel Hempstead, via the website https://www.vsni.co.uk/.

- R, from The R Project for Statistical Computing. This software can be downloaded free of charge from the website http://www.r-project.org/.

- SAS, available from the SAS Institute via the website http://www.sas.com/technologies /analytics/statistics/stat/.

GenStat is a natural choice of software to illustrate the concepts and methods employed in mixed modelling because its facilities for this purpose are straightforward to use, extensive and well integrated with the rest of the system and because their output is clearly laid out and easy to interpret. Above all, the recognition of random terms in statistical models lies at the heart of GenStat. GenStat's method of specifying anova models requires the distinction of random-effect (block) and fixed-effect (treatment) terms, which makes the interpretation of designed experiments uniquely reliable and straightforward. This approach extends naturally to mixed models and provides a firm framework within which the researcher can think and plan. Despite these merits, GenStat is not among the most widely used statistical software systems, and the numerical examples are therefore also analysed using the increasingly popular software R, as well as SAS, which is long-established and widely used in the clinical and agricultural research communities.

The book's website, http://www.wiley.com/go/beyond_regression, provides solutions to the end-of-chapter exercises, as well as data files, and programs in GenStat, R and SAS, for many of the examples in this book.

This second edition incorporates many additions and changes, as well as some corrections. The most substantial are

- the addition of SAS to the software systems used;

- a new chapter on meta-analysis and the multiple testing problem;

- recognition of situations in which it is appropriate to specify the interaction between two factors as a random-effect term, even though both of the corresponding main effects are fixed-effect terms;

- an account of the Bayesian interpretation of mixed models, an alternative to the random-sample (frequentist) interpretation mentioned above;

- a fuller account of the 'great mixed model muddle';

- the random coefficient regression model.

I am grateful to the following individuals for their valuable comments and suggestions on the manuscript of this book, and/or for introducing me to mixed-modelling concepts and techniques in the three software systems: David Balding, Aruna Bansal, Caroline Galwey, Toby Johnson, Peter Lane, Roger Payne, James Roger and David Willé. I am also grateful to the participants in the GenStat Discussion List for their helpful responses to many enquiries. (Access to this lively forum can be obtained via the website https://www.jiscmail.ac.uk/cgi-bin/webadmin?A0=GENSTAT.) Any errors or omissions of fact or interpretation that remain are my sole responsibility. I would also like to express my gratitude to the many individuals and organizations who have given permission for the reproduction of data in the numerical examples presented. They are acknowledged individually in their respective places, but the high level of support that they have given me deserves to be recognized here.

References

Draper, N.R. and Smith, H. (1998) *Applied Regression Analysis*, 3rd edn, John Wiley & Sons, Inc., New York, 706 pp.

Mead, R. (1988) *The Design of Experiments: Statistical Principles for Practical Application*, Cambridge University Press, Cambridge, 620 pp.

The need for more than one random-effect term when fitting a regression line

1.1 A data set with several observations of variable *Y* at each value of variable *X*

One of the commonest, and simplest, uses of statistical analysis is the fitting of a straight line, known for historical reasons as a *regression line*, to describe the relationship between an *explanatory variable*, *X* and a *response variable*, *Y*. The departure of the values of *Y* from this line is called the *residual variation*, and is regarded as random. It is natural to ask whether the part of the variation in *Y* that is explained by the relationship with *X* is more than could reasonably be expected by chance: or more formally, whether it is *significant* relative to the residual variation. This is a simple *regression analysis*, and for many data sets it is all that is required. However, in some cases, several observations of *Y* are taken at each value of *X*. The data then form natural groups, and it may no longer be appropriate to analyse them as though every observation were independent. observations of *Y* at the same value of *X* may lie at a similar distance from the line. We may then be able to recognize two sources of random variation, namely

- variation among groups

- variation among observations within each group.

This is one of the simplest situations in which it is necessary to consider the possibility that there may be more than a single *stratum* of random variation – or, in the language of mixed modelling, that a model with more than one *random-effect term* may be required. In this chapter, we will examine a data set of this type and explore how the usual regression analysis is modified by the fact that the data form natural groups.

Introduction to Mixed Modelling: Beyond Regression and Analysis of Variance, Second Edition. N. W. Galwey.
© 2014 John Wiley & Sons, Ltd. Published 2014 by John Wiley & Sons, Ltd.
Companion website: http://www.wiley.com/go/beyond_regression

We will explore this question in a data set that relates the prices of houses in England to their latitude. There is no doubt that houses cost more in the south of England than in the north: these data will not lead to any new conclusions, but they will illustrate this trend, and the methods used to explore it. The data are displayed in a spreadsheet in Table 1.1. The first cell in each column contains the name of the variable held in that column. The variables 'latitude' and 'price_pounds' are *variates* – lists of observations that can take any numerical value, the commonest kind of data for statistical analysis. However, the observations of the variable 'town' can take only certain permitted values – in this case, the names of the 11 towns under consideration. A variable of this type is called a *factor*, and the exclamation mark (!) after its name indicates that 'town' is a factor. The towns are the groups of observations: within each town, all the houses are at nearly the same latitude, and the latitude of the town is represented by a single value in this data set. In contrast, the price of each house is potentially unique. The conventions introduced here apply to all other spreadsheets displayed in this book.

Before commencing a formal analysis of this data set, we should note its limitations. A thorough investigation of the relationship between latitude and house price would take into account many factors besides those recorded here – the number of bedrooms in each house, its state of repair and improvement, other observable indicators of the desirability of its location, and so on. To some extent such sources of variation have been eliminated from the present sample by the choice of houses that are broadly similar: they are all 'ordinary' houses (no flats, maisonettes, etc.) and all have three, four or five bedrooms. The remaining sources of variation in price will contribute to the residual variation among houses in each town, and will be treated accordingly. We should also consider in what sense we can think of the latitude of houses as 'explaining' the variation in their prices. The easily measurable variable 'latitude' is associated with many other variables, such as climate and distance from London, and it is probably some of these, rather than latitude *per se*, that have a causal connection with price. However, an explanatory variable does not have to be causally connected to the response in order to serve as a useful predictor. Finally, we should consider the value of studying the relationship between latitude and price in such a small sample. The data on this topic are extensive and widely interpreted. However, this small data set, illustrating a simple, familiar example, is highly suitable for a study of the *methods* by which we judge the significance of a trend, estimate its magnitude and place confidence limits on the estimate. Above all, this example will show that in order to do these things reliably, we must recognize that our observations – the houses – are not mutually independent, but form natural groups – the towns.

1.2 Simple regression analysis: Use of the software GenStat to perform the analysis

We will begin by performing a simple regression analysis on these data, before considering how this should be modified to take account of the grouping into towns. The standard linear *regression model* (Model 1.1) is

$$y_{ij} = \beta_0 + \beta_1 x_i + \varepsilon_{ij} \tag{1.1}$$

where

x_i = value of X (latitude) for the ith town,

y_{ij} = observed value of Y ($\log_{10}$(house price in pounds)) for the jth house in the ith town,

β_0, β_1 = constants to be estimated, defining the relationship between X and Y and

ε_{ij} = the residual effect, that is, the deviation of y_{ij} from the value predicted on the basis of x_i, β_0 and β_1.

Table 1.1   Prices of houses in a sample of English towns and their latitudes.

	A	B	C		A	B	C
1	town!	latitude	price_pounds	34	Crewe	53.0998	84950
2	Bradford	53.7947	39950	35	Crewe	53.0998	112500
3	Bradford	53.7947	59950	36	Crewe	53.0998	140000
4	Bradford	53.7947	79950	37	Durham	54.7762	127950
5	Bradford	53.7947	79995	38	Durham	54.7762	157000
6	Bradford	53.7947	82500	39	Durham	54.7762	169950
7	Bradford	53.7947	105000	40	Newbury	51.4037	172950
8	Bradford	53.7947	125000	41	Newbury	51.4037	185000
9	Bradford	53.7947	139950	42	Newbury	51.4037	189995
10	Bradford	53.7947	145000	43	Newbury	51.4037	195000
11	Buxton	53.2591	120000	44	Newbury	51.4037	295000
12	Buxton	53.2591	139950	45	Newbury	51.4037	375000
13	Buxton	53.2591	149950	46	Newbury	51.4037	400000
14	Buxton	53.2591	154950	47	Newbury	51.4037	475000
15	Buxton	53.2591	159950	48	Ripon	54.1356	140000
16	Buxton	53.2591	159950	49	Ripon	54.1356	152000
17	Buxton	53.2591	175950	50	Ripon	54.1356	187950
18	Buxton	53.2591	399950	51	Ripon	54.1356	210000
19	Carlisle	54.8923	85000	52	Royal Leamington Spa	52.2876	147000
20	Carlisle	54.8923	89950	53	Royal Leamington Spa	52.2876	159950
21	Carlisle	54.8923	90000	54	Royal Leamington Spa	52.2876	182500
22	Carlisle	54.8923	103000	55	Royal Leamington Spa	52.2876	199950
23	Carlisle	54.8923	124950	56	Stoke-On-Trent	53.0041	69950
24	Carlisle	54.8923	128500	57	Stoke-On Trent	53.0041	69950
25	Carlisle	54.8923	132500	58	Stoke-On-Trent	53.0041	75950
26	Carlisle	54.8923	135000	59	Stoke-On-Trent	53.0041	77500
27	Carlisle	54.8923	155000	60	Stoke-On-Trent	53.0041	87950
28	Carlisle	54.8923	158000	61	Stoke-On-Trent	53.0041	92000
29	Carlisle	54.8923	175000	62	Stoke-On-Trent	53.0041	94950
30	Chichester	50.8377	199950	63	Witney	51.7871	179950
31	Chichester	50.8377	299250	64	Witney	51.7871	189950
32	Chichester	50.8377	350000	65	Witney	51.7871	220000
33	Crewe	53.0998	77500				

Source: Data obtained from an estate agent's website in October 2004.

Note that in this model the house prices are transformed to logarithms, because preliminary exploration has shown that this gives a more linear relationship between latitude and price and more uniform residual variation. The model is illustrated graphically in Figure 1.1.

The model specifies that a sloping straight line is to be used to describe the relationship between latitude and log(house price). The *parameters* β_0 and β_1 specify, respectively, the intercept and slope of this line. *Estimates* of these parameters, $\widehat{\beta}_0$ and $\widehat{\beta}_1$, respectively, are to be obtained from the data, and these estimates will define the *line of best fit* through the data. An estimate of each of the ε_{ij}, designated $\widehat{\varepsilon}_{ij}$, will be given by the deviation of the *i*th data point, in a vertical direction, from the line of best fit. The parameter estimates chosen are those that minimize the sum of squares (s.s. or SS) of the $\widehat{\varepsilon}_{ij}$. It is assumed that the *true* values ε_{ij} are independent values of a variable E, which is *normally distributed with mean zero and variance* σ^2. The meaning of this statement, which can be written in symbolic shorthand as

$$E \sim N(0, \sigma^2), \tag{1.2}$$

is illustrated in Figure 1.2. The area under this curve between any two values of E gives the probability that a value of E will lie between these two values. For example, the probability that a value of E will lie between σ and 2σ, represented by the hatched area in the figure, is 0.1359. Hence the total area under the curve is 1, as any value of E must lie between minus infinity and plus infinity. The variable plotted on the vertical axis, $f(E)$, is referred to as the *probability density*. It must be integrated over a range of values of E in order to give a value of probability, just as human population density must be integrated over an area of land in order to give a value of population. For the reader unfamiliar with such regression models, a fuller account is given by Draper and Smith (1998).

The calculations required in order to fit a regression model to data (i.e. to estimate the parameters of the model) can be performed by many computer software systems, and one of these,

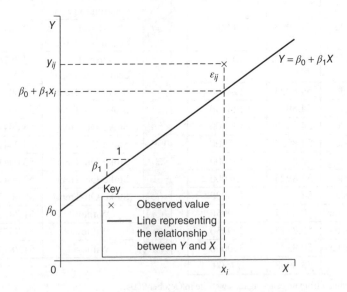

Figure 1.1 Linear relationship between an explanatory variable X and a response variable Y, with residual variation in the response variable.

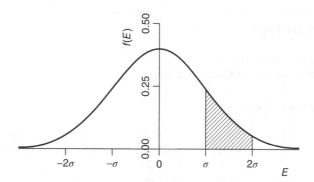

Figure 1.2   A normal distribution with mean zero and variance σ^2. $f(E) =$ probability density of E. Total area under curve $= 1$. Hatched area $=$ probability that $\sigma \le E \le 2\sigma$.

GenStat, will be referred to throughout this book. Information on obtaining access to Gen-Stat is given in the preface of this book. The GenStat command language, used to specify the models to be fitted, provides a valuable tool for thinking clearly about these models, and the GenStat statements required will therefore be presented and discussed here. However, the details of a computer language should not be allowed to distract from the statistical concepts that are our central topic. We will therefore note only a few key points about these statements: a full introduction to the GenStat command language is given by Payne, Murray and Harding (2013): this manual is available via GenStat's Graphical User Interface (GUI). The analyses presented in this book were performed using GenStat Release 16.1.

The following statements, in the GenStat command language, import the data into the Gen-Stat environment and fit Model 1.1:

```
IMPORT 'IMM Edn 2\\Ch 1\\house price, latitude.xlsx'
CALCULATE logprice = log10(price_pounds)
MODEL logprice
FIT [FPROB = yes; TPROB = yes] latitude
```

The IMPORT statement specifies the file that contains the data and makes the data available to GenStat. The CALCULATE statement performs the transformation to logarithms and stores the results in the variate 'logprice'. The MODEL statement specifies the response variable in the regression model (Y, logprice) and the FIT statement specifies the explanatory variable (X, latitude). The *option setting* 'FPROB = yes' indicates that when an F statistic is presented, the associated probability is also to be given (see below). The option setting 'TPROB = yes' indicates that this will also be done in the case of a t statistic. The same operations could be specified – perhaps more easily – using the menus and windows of GenStat's GUI: the use of these facilities is briefly illustrated in Section 1.13 and fully explained by Payne *et al.* (2013).

A researcher almost always receives the results of statistical analysis in the form of a computer output, and the interpretation of this, the extraction of key pieces of information and their synthesis in a report, is an important statistical skill. The output produced by GenStat is therefore presented and interpreted here. That from the FIT statement is shown in the box below.

Regression analysis

Response variate: logprice
Fitted terms: Constant, latitude

Summary of analysis

Source	d.f.	s.s.	m.s.	v.r.	F pr.
Regression	1	0.710	0.70955	20.55	<0.001
Residual	62	2.141	0.03453		
Total	63	2.850	0.04524		

Percentage variance accounted for 23.7.
Standard error of observations is estimated to be 0.186.

Message: the following units have large standardized residuals.

Unit	Response	Residual
1	4.602	−2.72
17	5.602	2.46

Message: the residuals do not appear to be random; for example, fitted values in the range 5.009–5.074 are consistently smaller than observed values and fitted values in the range 5.162–5.231 are consistently larger than observed values.

Estimates of parameters

Parameter	Estimate	s.e.	t(62)	t pr.
Constant	9.68	1.00	9.68	<0.001
Latitude	−0.0852	0.0188	−4.53	<0.001

This output begins with a specification of the model fitted. Note that the fitted terms include not only the explanatory variable, latitude, but also a constant, although none was specified: any regression model includes a constant by default (β_0 in this case). Next comes an analysis of variance (anova) table, which partitions the variation in log(house price) between the *terms* in Model 1.1, namely

- the effect of latitude (represented in the row labelled Regression in the anova table) and

- the residual effects (represented in the row labelled Residual).

After the names of the terms, the next two columns of the anova table hold the degrees of freedom (abbreviated to d.f. or DF) and the SS for each term. The methods for calculating these will not be given here (for an account, see Draper and Smith, 1998, Section 1.3, pp. 28–34), but it should be noted that the DF for each term represents the number of independent pieces of information to which that term is equivalent. Thus the effect of latitude is a single piece of information, and $DF_{latitude} = 1$. There are 64 houses in the sample, each of which gives a value of $\hat{\varepsilon}_{ij}$, so it might be thought that the residual term would comprise 64 pieces of information. However, two pieces of information have been 'used up' by the estimation of the intercept and the effect of latitude, so

$$DF_{Residual} = 64-2 = 62.$$

This reduction in the residual DF is equivalent to the fact that a line of best fit based on only two observations passes exactly through the data points – such a data set provides no information on residual variation.

The mean square for each term (m.s. or MS) is given by SS/DF and is a measure of the part of the variation in log(house price) that is accounted for by the term. If there is no real effect of latitude on log(house price), the expected values of $MS_{latitude}$ and $MS_{Residual}$ are the same. Hence on this *null hypothesis* (H_0), the expected value of $MS_{latitude}/MS_{Residual}$ is 1, though the actual value will vary from one data set to another. This ratio, called the *variance ratio* (abbreviated to *v.r.*), thus provides a test of H_0. Provided that the residual variation is normally distributed (see Section 1.10), v.r. is also known as the *F statistic*. If H_0 is true, the distribution of F over an infinite population of samples (data sets) has a definite mathematical form. The precise shape of this distribution depends on the DF in the numerator and denominator of the ratio: hence the variable F is referred to more precisely as $F_{DF_{numerator}, DF_{denominator}}$. The distribution in the present case (i.e. the distribution of the variable $F_{1,62}$) is illustrated in Figure 1.3. This curve is interpreted in the same way as the normal distribution illustrated earlier. The area under the curve between any two values of F gives the probability that an observation of F will lie between these values. Hence again the total area under the curve is 1, as any observation of F must have some value between 0 and infinity. Again the variable plotted on the vertical axis is the probability density. This F distribution can be used to determine the probability p of obtaining by chance a value of F larger than that actually observed, as shown in the figure. For example, if $F_{1,62} > 4.00$, then $p < 0.05$, and it is said that the effect under consideration is significant at the 5% level. Similarly, if $F_{1,62} > 7.06$, then $p < 0.01$, and it is said that the effect is significant at the 1% level, that is, highly significant. In the present case

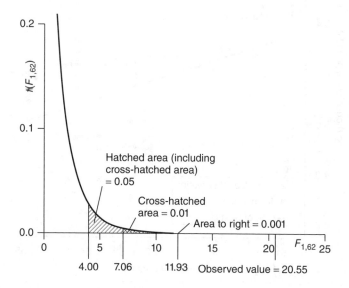

Figure 1.3 Distribution of the variable $F_{1,62}$, showing critical values for significance tests and the corresponding critical regions.

$F_{1,62} = 20.55$, and both the anova table and the figure show that p (called F pr. in the table) is less than 0.001: that is, the relationship between latitude and log(house price) in this data set is highly significant – provided that the model specified is correct. However, there is a diagnostic message that indicates that this may not be the case: GenStat has detected that the residuals do not appear to be random.

The appropriateness of using such a *significance test* to decide whether a relationship between two variables is real and/or important has been hotly debated by statisticians. An alternative way of using the concept of probability to help in the interpretation of data, the *Bayesian* approach, is introduced in Section 5.6, and a lively critique of over-reliance on significance testing is given by Royall (1997, Chapter 3, pp. 61–82). However, for most of this book, we will follow the common practice of using significance tests as a key tool to assess the explanatory power of statistical models.

The next item in the output is the parameter estimates, which give the intercept ($\widehat{\beta}_0$, the constant term) and slope ($\widehat{\beta}_1$) of the line of best fit, with their standard errors (SEs). The negative slope indicates that house prices are higher in the south of England than in the north. The SE is a measure of the precision of the corresponding estimate, a low SE indicating high precision. The t statistic for the effect of latitude is given by estimate/$SE_{estimate} = -0.0852/0.0188 = -4.53$. Note that $t^2 = (-4.53)^2 = 20.55 = F$, and that for both these statistics the p-value (when calculated to a greater degree of precision than is given by the GenStat output) is 0.0000271 – that is, the t test for the significance of the slope is equivalent to the F test in the analysis of variance.

The line of best fit is displayed, together with the data and the mean value for each town, in Figure 1.4. This figure shows that, overall, the regression line fits the data reasonably well.

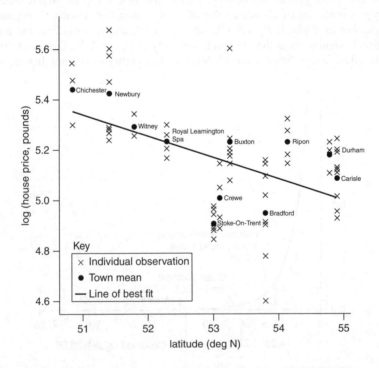

Figure 1.4 Relationship between latitude and house price in a sample of English towns, showing individual observations, town means and line of best fit.

However, observations from the same town generally lie on the same side of the line – that is, the residual values within each town are not mutually independent. For example, as noted in GenStat's diagnostic message, the observations from Ripon, Durham and Carlisle generally lie above the line (GenStat identifies these by the fact that their fitted values of log(house price) lie in the range 5.009–5.074), whereas those from Stoke-On-Trent and Crewe lie below the line (their fitted values lie in the range 5.162–5.231).

1.3   Regression analysis on the group means

Because the residual values are not mutually independent, the analysis presented above cannot be relied upon, even though the regression line appears reasonable. In particular, the residual degrees of freedom (DF$_{Residual}$ = 62) are an overestimate: we deceive ourselves if we believe that the data comprise 64 independent observations. The simplest way to overcome this problem is to fit the regression model using the mean value of log(house price) for each town. These means are displayed in a spreadsheet in Table 1.2.

The following statements will perform a regression analysis on these mean values:

```
IMPORT 'IMM Edn 2\\Ch 1\\price, lat, town means.xlsx'
MODEL meanlogprice
FIT [FPROB = yes; TPROB = yes] meanlatitude
```

The output of the FIT statement is as follows:

Regression analysis

Response variate: meanlogprice
Fitted terms: Constant, meanlatitude

Summary of analysis

Source	d.f.	s.s.	m.s.	v.r.	F pr.
Regression	1	0.1160	0.11601	5.19	0.049
Residual	9	0.2010	0.02234		
Total	10	0.3170	0.03170		

Percentage variance accounted for 29.5.
Standard error of observations is estimated to be 0.149.

Estimates of parameters

Parameter	Estimate	s.e.	t(9)	t pr.
Constant	9.44	1.87	5.04	<0.001
Meanlatitude	−0.0804	0.0353	−2.28	0.049

The residual DF are much fewer (DF$_{Residual}$ = 9), reflecting the more reasonable assumption that each town can be considered as an independent observation. Consequently, the relationship

Table 1.2   Mean values of $\log_{10}$(price) of houses in a sample of English towns, and their latitudes.

	A	B	C	D
1	town_unique	n_houses	meanlatitude	meanlogprice
2	Bradford	9	53.7947	4.94745
3	Buxton	8	53.2591	5.23083
4	Carlisle	11	54.8923	5.08552
5	Chichester	3	50.8377	5.44034
6	Crewe	4	53.0998	5.00394
7	Durham	3	54.7762	5.17775
8	Newbury	8	51.4037	5.42456
9	Ripon	4	54.1356	5.23106
10	Royal Leamington Spa	4	52.2876	5.23337
11	Stoke-On-Trent	7	53.0041	4.90642
12	Witney	3	51.7871	5.29207

between latitude and house price is much less significant: $p = 0.049$, compared with $p < 0.001$ in the previous analysis. However, the estimates of the intercept and slope of the regression line are not much altered.

1.4   A regression model with a term for the groups

The analysis of the town means, presented in the previous section, gives no account of the variation within the towns. Nor does it take account of the variation in the number of houses sampled, and hence in the precision of the mean, from one town to another. An alternative approach, which overcomes these deficiencies, is to add a term to the regression model to take account of the variation among towns that is not accounted for by latitude, so the model becomes:

$$y_{ij} = \beta_0 + \beta_1 x_i + \tau_i + \varepsilon_{ij} \tag{1.3}$$

where

τ_i = mean deviation from the regression line of observations from the ith town.

The following statements will fit this model (Model 1.3) using the ordinary methods of regression analysis:

```
IMPORT 'IMM Edn 2\\Ch 1\\house price, latitude.xlsx'
CALCULATE logprice = log10(price_pounds)
MODEL logprice
FIT [FPROB = yes; TPROB = yes; \
    PRINT = model, estimates, accumulated] \
    latitude + town
```

The output of the FIT statement is as follows. It is quite voluminous, and each part will be discussed before presenting the next.

> *Message: term town cannot be fully included in the model because one parameter is aliased with the terms already in the model.*
>
> (town Witney) = 26.80 − (latitude)*0.4981 − (town Buxton)*0.2668 + (town Carlisle)*0.5467 − (town Chichester)*1.473 − (town Crewe)*0.3461 + (town Durham)*0.4889 − (town Newbury) *1.191 + (town Ripon)*0.1698 − (town Royal Leamington Spa)*0.7507 − (town Stoke-On-Trent)*0.3938

First comes a message noting that because the variation among the town means is partly accounted for by latitude, the term 'town' cannot be fully included in the regression model. It is said to be *partially aliased* with latitude. (The technical consequence of this partial aliasing is that the effect of one of the towns − Witney, arbitrarily chosen because it comes last when the towns are arranged in the alphabetical order − is a function of the effects of the other towns and of latitude, but the numerical details of this relationship, given in the message, need not concern us.)

Next come the statement of the regression model and the estimates of its parameters:

Regression analysis

Response variate: logprice
Fitted terms: Constant + latitude + town

Estimates of parameters

Parameter	Estimate	s.e.	$t(53)$	t pr.
Constant	14.18	2.32	6.12	<0.001
latitude	−0.1717	0.0435	−3.95	<0.001
town Buxton	0.1914	0.0598	3.20	0.002
town Carlisle	0.3265	0.0885	3.69	<0.001
town Chichester	−0.015	0.136	−0.11	0.914
town Crewe	−0.0628	0.0761	−0.83	0.413
town Durham	0.399	0.106	3.75	<0.001
town Newbury	0.067	0.102	0.66	0.515
town Ripon	0.3421	0.0840	4.07	<0.001
town Royal Leamington Spa	0.0272	0.0874	0.31	0.757
town Stoke-On-Trent	−0.1767	0.0636	−2.78	0.007
town Witney	0	*	*	*

Parameters for factors are differences compared with the reference level:

Factor: Reference level
Town: Bradford

The asterisks (*) in this output indicate places where no value can be calculated. These parameter estimates lead correctly to the sample mean for each town when substituted into

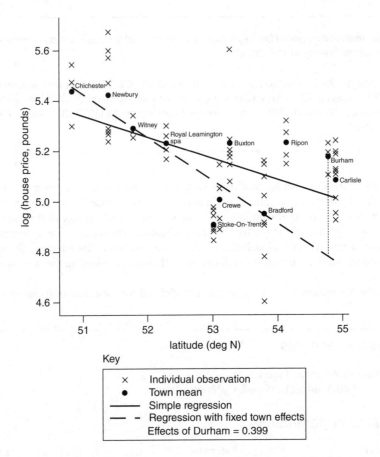

Figure 1.5   Relationship between latitude and house prices in a sample of English towns, comparing the lines of best fit from simple regression analysis and from analysis with town effects treated as fixed.

the formula

$$\text{mean(log(house price))} = 14.18 - 0.1717 \times \text{latitude} + \text{effect of town.}$$

For example, in Durham,

$$\text{mean(log(house price))} = 14.18 - 0.1717 \times 54.7762 + 0.399 = 5.1739.$$

However, the parameter estimates themselves are arbitrary and uninformative, as illustrated in Figure 1.5. The fitted line is arbitrarily specified to pass through the mean values for Bradford and Witney (the first and last towns in the alphabetic sequence), and the effects of the other towns – their vertical distances from the fitted line – are determined accordingly.

The option setting 'PRINT = accumulated' in the FIT statement specifies that the output should include an accumulated anova, which partitions the variation accounted for by the model between its two terms as follows:

Accumulated analysis of variance

Change	d.f.	s.s.	m.s.	v.r.	*F* pr.
+latitude	1	0.70955	0.70955	41.35	<0.001
+town	9	1.23119	0.13680	7.97	<0.001
Residual	53	0.90937	0.01716		
Total	63	2.85011	0.04524		

Despite the arbitrary nature of the parameter estimates, this anova is informative. The term 'latitude' represents the part of the variation in house price that is due to the effect of latitude, and the term 'town' represents the part due to the variation among towns after allowing for latitude, that is, the deviations of the town means from the original line of best fit (not the arbitrary fitted line presented above). Note that the MS for latitude is the same as the corresponding value from Model 1.1:

$$MS_{latitude} = MS_{Regression, \, Model \, 1.1} = 0.70955.$$

That is, the amount of variation explained by latitude is consistent between Models 1.1 and 1.3. Note also that

$$SS_{Total, \, Model \, 1.1} = SS_{Total, \, Model \, 1.3} = 2.850, \text{ allowing for rounding.}$$

Hence Model 1.3 represents a partitioning of the residual term in Model 1.1. Thus

$$DF_{Residual, \, Model \, 1.1} = DF_{town} + DF_{Residual, \, Model \, 1.3}$$
$$62 = 9 + 53$$

and

$$SS_{Residual, \, Model \, 1.1} = SS_{town} + SS_{Residual, \, Model \, 1.3}$$
$$2.141 = 1.23119 + 0.90937, \text{ allowing for rounding.}$$

Part of the variation, formerly unexplained, is now attributed to the effects of towns

1.5   Construction of the appropriate *F* test for the significance of the explanatory variable when groups are present

The accumulated anova can be adapted to provide a realistic assessment of the significance of the effect of latitude. In the GenStat output, both of the MSs for model terms in this anova are tested against the residual MS, that is:

- $F_{latitude} = MS_{latitude}/MS_{Residual} = 0.70955/0.01716 = 41.35$ and
- $F_{town} = MS_{town}/MS_{Residual} = 0.13680/0.01716 = 7.97$.

Table 1.3   Comparison between the F statistics from regression analyses based on individual houses and on town means.

	Regression based on town means	Regression with term for towns (Model 1.3)
Formula for F_{latitude}	$\text{MS}_{\text{Regression}}/\text{MS}_{\text{Residual}}$	$\text{MS}_{\text{latitude}}/\text{MS}_{\text{town}}$
Numerical values	$0.11601/0.02234 = 5.19293$	$0.70955/0.13680 = 5.18677$
$\text{DF}_{\text{Numerator}}$	1	1
$\text{DF}_{\text{Denominator}}$	9	9
p	0.0487	0.0488

F_{town} is highly significant ($p < 0.001$), confirming that there is real variation among the towns in addition to that accounted for by latitude. Consequently, F_{latitude} is misleading: it is much larger than the value of 5.19 obtained when the regression is fitted using the town means. In order to obtain the appropriate value from the present analysis, we need to calculate

$$F_{\text{latitude}} = \frac{\text{MS}_{\text{latitude}}}{\text{MS}_{\text{town}}} = \frac{0.70955}{0.13680} = 5.18677.$$

This is almost exactly equivalent to the F statistic for latitude in the analysis based on the town means, as shown in Table 1.3. (The GenStat output does not give these F and p-values to sufficient precision to reveal the slight differences between them.) The reason why the two sets of values do not agree exactly will be explained in Section 1.9.

1.6   The decision to specify a model term as random: A mixed model

The F values presented in Table 1.3 are based on a comparison of the variation explained by the regression line with the variation of *the town means* about the line, whereas the much larger value in the accumulated anova ($F_{\text{latitude}} = 41.35$) is based on comparison with the variation of *individual values of log(house price)* about their respective town means. When we use the variation of the town means as the basis of comparison, we are regarding these means as values of a *random variable*. One way in which we may justify this is to consider the towns in this study as a representative sample from a large population of towns. Formally speaking, we then assume that the values τ_i in Model 1.3 are independent values of a variable T, such that

$$T \sim N(0, \sigma_T^2), \tag{1.4}$$

that is, they are very like the values ε_{ij}, except that their variance, σ_T^2, is different. This random variation influences our estimate of the effect of latitude: if we were to drop one town from the study and replace it with a newly chosen town, this would have an effect on the slope of the regression line – a larger effect than would be produced by simply taking a new sample of houses from within the same town. This is because there is more variation among town means, even after allowing for the effect of latitude, than among individual houses from the same town: this is made clear by the highly significant value of F_{town}, 7.97.

The assumption that the τ_i are values of a random variable may seem radical, but it is necessary if we are to make any general statement about the relationship between latitude and house prices: we saw earlier (Section 1.4) that if the town effects are specified as fixed, the estimate of the effect of latitude is arbitrary. But if our set of towns is a random sample from a population, the estimate gives a prediction of what would be observed in another town from the same population at a given latitude. In the language of mixed modelling, in order to obtain a meaningful estimate of the effect of latitude on house prices, we must specify the effect of each town as a *random effect* and specify town as a *random-effect term* in the regression model. The effect of latitude itself is a non-random or *fixed effect*. Since our model contains effects of both types, it is a *mixed model*. (The assumption that the random variable T has a normal distribution is not a requirement, but other distributions require more advanced modelling methods – see Chapter 10.)

When deciding whether to specify a factor as random, it is helpful to ask whether the levels studied can be regarded as a representative sample from some large population of levels. In the present case, can the towns sampled be considered representative of the population of English towns? If so, we will be able to predict house prices in a town not included in the sample from the town's latitude and the regression line – acknowledging that the precision of our prediction will be limited not only by the residual variation, but also by the random variation among towns. We would be willing to make such a prediction only if the town were in England and were a substantial, distinct town – not, for example, a village or a suburb of a larger conurbation – that is, if it belonged to the population of which our sample is representative.

However, even if no such population can be specified, it may still be valid to specify a factor as random, and to make inferences about its levels collectively. To determine whether this is the case, it is helpful to ask whether if we lost the values of the response variable from one factor level, we would consider it worthwhile to predict them from the value(s) of the fixed-effect term(s) and the fitted model. In the present case, if we lost the house prices from, say, Durham, would we consider it worthwhile to predict them, albeit with reduced precision, from the latitude of Durham and the regression line? If so, the factor levels comprise an *exchangeable set*, and the factor can be specified as random. A set of levels sampled at random from an infinite population are exchangeable by definition: that is, random sampling is a stronger assumption than exchangeability, but exchangeability is sufficient to permit a term to be specified as random in a mixed model. If we decide that the house prices in different towns are so unconnected that such a prediction would be meaningless, we should specify 'town' as a fixed-effect term and drop 'latitude' from the model. Of course, the regression line will have some predictive value because it is based partly on the data from Durham, but that is cheating: would we still have confidence in its predictive value if our set of towns was very large, so that the contribution of Durham became negligible?

Formally, the concept of an exchangeable set means that although we have some information about the collective properties of the members, we have no information about their individual values – like a deck of playing cards of which one can see only the backs. In the present case, our knowledge of the collective properties of the town effects may be reasonably well represented by Distribution 1.4 even if they are not a random sample from an infinite population: that is, we know that they are centred on the regression line (mean$(T) = 0$) with a characteristic variance, σ_T^2, which we shall estimate during the model fitting process. Our ignorance of their individual values means that we have no prior idea of which towns will lie above the regression line and which below. The concept of exchangeability will be explored further in Sections 2.6 and 4.1 in the context of randomized experimental designs, and Section 5.6, in the context of 'shrunk estimates' of random effects. A more rigorous account of this concept is given by Spiegelhalter, Abrams and Myles (2004, Section 3.4). The question of how to determine which model terms should be specified as random will be discussed more fully, in the context of a wide range of models, in Section 6.3.

1.7 Comparison of the tests in a mixed model with a test of lack of fit

The partitioning of the variation around the regression line into two components, one due to the deviations of the town means from the line and the other to the deviations of individual houses from the town means, is similar to the test for lack of fit described by Draper and Smith (1998, Section 2.1, pp. 49–53). Indeed, the calculations performed to obtain the MSs are identical in the two analyses. However, there is an important difference in the ideas that underlie them, and consequently in the F tests specified, as illustrated in Table 1.4. The angled lines in this table indicate the pairs of MSs that are compared by the two F tests. The purpose of the test of lack of fit is to determine whether the variation among groups of observations at the same value of X (towns in the present case) is significant, or whether it can be absorbed into the residual term. In the mixed-model analysis, on the other hand, the reality of the variation among towns is not in doubt. The question is whether the effect of latitude is significant, or whether it can be absorbed into the 'town' term.

Table 1.4 Comparison between the F test for lack of fit and the mixed-model F test in the analysis of the effect of latitude on house prices in England.

Source of variation	DF	MS	Test of lack of fit		Mixed-model test	
			F	p	F	p
latitude	1	0.70955			5.19	0.0488
town	9	0.13680	7.97	<0.001		
Residual	53	0.01716				

Table 1.5 Comparison between the F tests conducted in ordinary multiple regression analysis and in mixed model analysis of the effects of latitude and town on house prices in England.

Source of variation	DF	MS	Analysis with test of lack of fit		Mixed-model analysis	
			F	p	F	p
latitude	1	0.70955	41.35	<0.001	5.19	0.0488
town	9	0.13680	7.97	<0.001	7.97	<0.001
Residual	53	0.01716				

Even if the test of lack of fit leads to the conclusion that the variation among towns is significant, the two analyses are not equivalent. Because the test of lack of fit treats only the residual term as random, it leads to the use of this term as the denominator in all F tests, whereas the mixed model leads to the use of the random-effect term town as the denominator against which to test the significance of the effect of latitude. The full set of tests specified by the two analyses is therefore as shown in Table 1.5. The dotted angled lines in this table indicate the pairs of MSs that are compared by the additional F tests. In the case considered by Draper and Smith, the additional term required (equivalent to 'town' in the present example) is a quadratic one, to allow for curvature in the response to the explanatory variable, and in this situation the test of lack of fit is correct.

1.8   The use of REsidual Maximum Likelihood (REML) to fit the mixed model

So far we have taken an improvised approach to fitting the mixed model. We have

- fitted two regression models, one without a term for the effects of towns and the other including such a term;

- taken the estimate of the slope from the first model and the MSs from the second;

- obtained the F statistic to test the significance of the effect of latitude from the MSs by hand.

However, the mixed model analysis can be performed in a more unified manner using the criterion of *REsidual Maximum Likelihood* (REML, also known as *REstricted Maximum Likelihood*). The formal meaning of this criterion will be explained in Chapter 11: here, we will simply apply it to the present data. The following GenStat statements specify the mixed model to be fitted:

```
VCOMPONENTS [FIXED = latitude; CADJUST = none] RANDOM = town
REML [PRINT = model, components, Wald, effects] logprice
```

The VCOMPONENTS statement specifies the terms in the model: it is equivalent to the FIT statement in an ordinary regression analysis. The option FIXED specifies the fixed-effect term or terms: in this case, 'latitude'. The constant (the intercept β_0) is also included in the model as a fixed-effect term by default: it does not have to be explicitly specified. The option setting 'CADJUST = none' indicates that no adjustment is to be made to the covariate (explanatory variable) 'latitude' before analysis: by default, a covariate is *centred* by subtracting its mean value from each of its values (see Section 7.2). The *parameter* RANDOM specifies the random-effect term(s): in this case, 'town'. The REML statement specifies the response variate whose variation is to be explained by the model: in this case, 'logprice'. It is equivalent to the MODEL statement in an ordinary regression analysis. Note that the VCOMPONENTS and REML statements are given in the opposite order to their equivalents in ordinary regression analysis. The PRINT option indicates what results from the model-fitting process are to be presented in the output.

The output of these statements is as follows:

REML variance components analysis

Response variate: logprice
Fixed model: Constant + latitude
Random model: town
Number of units: 64

Residual term has been added to model.

Sparse algorithm with AI optimization.
Covariates not centred.

Estimated variance components

Random term	Component	s.e.
town	0.01963	0.01081

Residual variance model

Term	Model (order)	Parameter	Estimate	s.e.
Residual	Identity	Sigma2	0.0171	0.00332

Tests for fixed effects

Sequentially adding terms to fixed model

Fixed term	Wald statistic	n.d.f.	F statistic	d.d.f.	F pr.
latitude	5.04	1	5.04	9.4	0.050

Dropping individual terms from full fixed model

Fixed term	Wald statistic	n.d.f.	F statistic	d.d.f.	F pr.
latitude	5.04	1	5.04	9.4	0.050

Message: denominator DF for approximate F tests are calculated using algebraic deriva-tives ignoring fixed/boundary/singular variance parameters.

Table of effects for Constant

9.497   Standard error: 1.9248

Table of effects for latitude

−0.08147   Standard error: 0.036272

The output begins with a specification of the model fitted and notes that there are 64 observations in the data analysed. It notes that the residual term in the model (ε_{ij}), which was not specified in the GenStat statements, has been added by GenStat. GenStat offers two algorithms for fitting mixed models (see Section 2.5): the output notes that the sparse algorithm with average information (AI) optimization has been used by default. The details of this algorithm need not concern us here: more will be said about them later (Section 11.10). Estimates of *variance components* are then given for the two random-effect terms, 'town' and the residual term. The concept of a variance component will be explained later (Sections 3.2–3.4). Here, we will simply note that the estimate of residual variance from the REML model ($\hat{\sigma}^2 = 0.0171$) is about the same as that from the accumulated anova including the town term ($\hat{\sigma}^2 = 0.01716$): the two models are almost equivalent in their ability to explain the variation in house prices. The significance of the effect of latitude is tested by an F statistic that has nearly the same value (5.04) as the value of F_{latitude} (≈ 5.19) obtained earlier (Section 1.5). The p-values corresponding to these test statistics are also very similar ($p = 0.050$ and 0.049, respectively). The slight difference between them arises not only from the difference in the F values, but also in the difference between the value $DF_{\text{Denominator}} = 9$ obtained earlier and the value obtained in the present analysis, $DF_{\text{Denominator}} = 9.4$. Note that whereas in an ordinary regression analysis DF are always whole numbers, this constraint is removed in a mixed model analysis. The value $DF_{\text{Numerator}} = 1$ is the same in both analyses. Another test statistic, the Wald statistic, is also presented. It is usually related to the F statistic by the formulae

$$\text{Wald statistic} = F \times DF_{\text{Numerator of } F} \tag{1.5}$$

and

$$DF_{\text{Wald statistic}} = DF_{\text{Numerator of } F} \tag{1.6}$$

For a slight exception, see Section 9.6; this will be considered in more detail later (Section 1.9). GenStat provides separate tests of the effects of adding 'latitude' to, and dropping it from, the model, but in the case of this simple model with only one fixed-effect term (apart from the constant), there is no difference between these tests. Finally, the output gives the parameter estimates $\hat{\beta}_0$ and $\hat{\beta}_1$ and their SEs.

The estimates and SEs produced by the different methods of analysis are shown in Table 1.6. The three estimates of the effect of latitude agree closely, as do the three estimates

Table 1.6 Comparison of the parameter estimates and their SEs, obtained from different methods of analysis of the effect of latitude on house prices in England.

Term	Method of analysis					
	Regression analysis ignoring towns (Model 1.1)		Regression analysis on town means		Mixed model analysis (Model 1.3)	
	Estimate	SE_{Estimate}	Estimate	SE_{Estimate}	Estimate	SE_{Estimate}
Constant	9.68	1.00	9.44	1.87	9.497	1.9248
latitude	−0.0852	0.0188	−0.0804	0.0353	−0.08147	0.036272

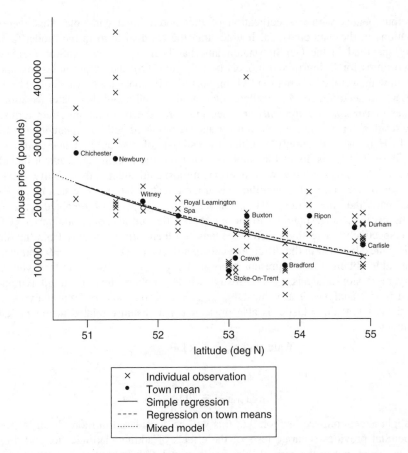

Figure 1.6 Relationship between latitude and house prices in a sample of English towns, comparing the lines of best fit from simple regression analysis, analysis on town means and mixed-model analysis.

of the constant. However, the SE of the estimated effect of latitude from Model 1.1 is much smaller than those from the regression analysis on town means and Model 1.3. This is because the SE for Model 1.1 is based on the assumption that every house is an independent observation. It over-estimates the precision of the line of best fit, just as the F statistic from this model over-estimates the significance of the relationship between house price and latitude.

The three fitted lines can be back-transformed to the original units (pounds), using the formula

$$\text{fitted house price} = 10^{\text{fitted log(house price)}}$$

The resulting curves – together with the town means back-transformed using the same formula – and the original data are displayed in Figure 1.6. The three curves agree closely and all fit the data reasonably well.

1.9   Equivalence of the different analyses when the number of observations per group is constant

The slight discrepancies between the results of the different methods of analysis are due to the unequal numbers of houses observed in different towns. The analysis ignoring towns gives equal weight to each house, whereas the analysis on town means gives equal weight to each town, and hence less weight to each house in towns where a large sample was taken, for example, Carlisle. The analysis of town means can be adjusted to give more weight to such towns by setting an option in the GenStat MODEL statement as follows:

```
MODEL [WEIGHTS = n_houses] meanlogprice
FIT [FPROB = yes; TPROB = yes] meanlatitude
```

The MSs and *F* value then agree precisely with those from the regression based on individual houses with a term for towns (Model 1.3, Table 1.3).

This source of discrepancy is absent when the number of observations in each group is equal. Therefore, in order to compare the different methods of analysis more closely, we will consider a subset of the data from which randomly chosen houses have been removed, so that there are exactly three houses from each town, as shown in Table 1.7.

Table 1.7   Prices of houses in a sample of English towns and their latitudes 'trimmed' to three houses per town.

	A	B	C		A	B	C
1	town!	latitude	price_pounds	18	Durham	54.7762	157000
2	Bradford	53.7947	79950	19	Durham	54.7762	169950
3	Bradford	53.7947	105000	20	Newbury	51.4037	185000
4	Bradford	53.7947	145000	21	Newbury	51.4037	189995
5	Buxton	53.2591	120000	22	Newbury	51.4037	475000
6	Buxton	53.2591	154950	23	Ripon	54.1356	132000
7	Buxton	53.2591	399950	24	Ripon	54.1356	187950
8	Carlisle	54.8923	85000	25	Ripon	54.1356	210000
9	Carlisle	54.8923	128500	26	Royal Leamington Spa	52.2876	147000
10	Carlisle	54.8923	175000	27	Royal Leamington Spa	52.2876	182500
11	Chichester	50.8377	199950	28	Royal Leamington Spa	52.2876	199950
12	Chichester	50.8377	299250	29	Stoke-On-Trent	53.0041	69950
13	Chichester	50.8377	350000	30	Stoke-On-Trent	53.0041	69950
14	Crewe	53.0998	77500	31	Stoke-On-Trent	53.0041	87950
15	Crewe	53.0998	84950	32	Witney	51.7871	179950
16	Crewe	53.0998	112500	33	Witney	51.7871	189950
17	Durham	54.7762	127950	34	Witney	51.7871	220000

The relevant parts of the output produced by the different methods of analysis on this sub-set of the data are as follows. Here is the ordinary regression analysis ignoring the towns (Model 1.1):

Regression analysis

Response variate: logprice
Fitted terms: Constant, latitude

Summary of analysis

Source	d.f.	s.s.	m.s.	v.r.	F pr.
Regression	1	0.283	0.28349	7.84	0.009
Residual	31	1.121	0.03615		
Total	32	1.404	0.04388		

Percentage variance accounted for 17.6.
Standard error of observations is estimated to be 0.190.

Message: the following units have large standardized residuals.

Unit	Response	Residual
6	5.602	2.30

Estimates of parameters

Parameter	Estimate	s.e.	t(31)	t pr.
Constant	9.04	1.37	6.57	<0.001
latitude	−0.0726	0.0259	−2.80	0.009

Ordinary regression analysis on the town means:

Regression analysis

Response variate: meanlogprice
Fitted terms: Constant, meanlatitude

Summary of analysis

Source	d.f.	s.s.	m.s.	v.r.	F pr.
Regression	1	0.0945	0.09450	3.68	0.087
Residual	9	0.2310	0.02567		
Total	10	0.3255	0.03255		

Percentage variance accounted for 21.1.
Standard error of observations is estimated to be 0.160.

Message: the following units have large standardized residuals.

Unit	Response	Residual
10	4.878	−2.04

Estimates of parameters

Parameter	Estimate	s.e.	t(9)	t pr.
Constant	9.04	2.01	4.50	0.001
meanlatitude	−0.0726	0.0378	−1.92	0.087

Ordinary regression analysis with town as a term in the model (Model 1.3, town as a fixed-effect term):

Message: term town cannot be fully included in the model because one parameter is aliased with terms already in the model.

(town Witney) = 26.80 − (latitude)*0.4981 − (town Buxton)*0.2668 + (town Carlisle)* 0.5467 − (town Chichester)*1.473 − (town Crewe)*0.3461 + (town Durham)*0.4889 − (town Newbury)*1.191 + (town Ripon)*0.1698 − (town Royal Leamington Spa)*0.7507 − (town Stoke-On-Trent)*0.3938

Regression analysis

Response variate: logprice
Fitted terms: Constant + latitude + town

Accumulated analysis of variance

Change	d.f.	s.s.	m.s.	v.r.	F pr.
+latitude	1	0.28349	0.28349	14.59	<0.001
+town	9	0.69296	0.07700	3.96	0.004
Residual	22	0.42756	0.01943		
Total	32	1.40402	0.04388		

Mixed-model analysis (Model 1.3, town as a random-effect term):

REML variance components analysis

Response variate: logprice
Fixed model: Constant + latitude
Random model: town
Number of units: 33

Residual term has been added to the model.

Sparse algorithm with AI optimization.
Covariates not centred.

Estimated variance components

Random term	Component	s.e.
town	0.01919	0.01226

Residual variance model

Term	Model (order)	Parameter	Estimate	s.e.
Residual	Identity	Sigma2	0.0194	0.00586

Tests for fixed effects

Sequentially adding terms to fixed model

Fixed term	Wald statistic	n.d.f.	F statistic	d.d.f.	F pr.
latitude	3.68	1	3.68	9.0	0.087

Dropping individual terms from full fixed model

Fixed term	Wald statistic	n.d.f.	F statistic	d.d.f.	F pr.
latitude	3.68	1	3.68	9.0	0.087

Message: denominator DF for approximate F tests are calculated using algebraic derivatives ignoring fixed/boundary/singular variance parameters.

Table of effects for Constant

9.038   Standard error: 2.0068

Table of effects for latitude

−0.07260   Standard error: 0.037835

The different analyses are now equivalent in several ways, namely

- ordinary regression analysis ignoring the towns, ordinary regression analysis on the town means and mixed-model analysis now all give the same fitted line;

- the regression analysis on the town means and the mixed-model analysis give the same value for the SE of the slope, $SE_{latitude} = 0.378$;

- the value of the F statistic from the mixed-model analysis is now identical to the value of F_{town} obtained by hand from Model 1.3 with 'town' as a fixed-effect term, namely, $MS_{latitude}/MS_{town} = 0.28349/0.07700 = 3.68169$. The value $DF_{Denominator} = 9.0$ is now also identical in these two analyses.

The relationship between the F statistic and the Wald statistic deserves further study. Though the values of the two statistics are identical (because $DF_{Numerator} = 1$ – see Equation 1.5), and

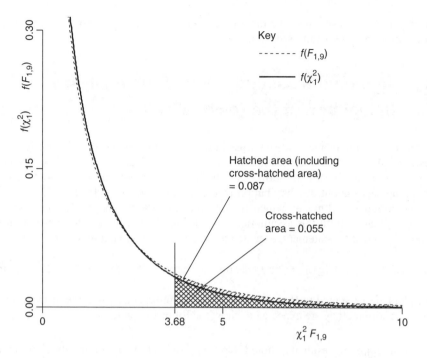

Figure 1.7 Comparison between the distributions of the variables $F_{1,9}$ and χ_1^2.

though both test the null hypothesis that there is no real relationship between latitude and house price, they give rise to somewhat different p-values. If the null hypothesis is true, then the Wald statistic has approximately a χ^2 (chi-square) distribution, in this case with $DF = 1$. Note that the shape of this distribution depends only on the fact that $DF_{latitude} = 1$. However, the shape of the distribution of the F statistic depends also on the value $DF_{town} = 9$, which was used to calculate MS_{town}, the denominator of the F statistic. As specified in Section 1.2, the F statistic is distributed as $F_{DF_{Numerator}, DF_{Denominator}}$: in the present case, as $F_{1,9}$. The difference between the shapes of the χ^2 and F distributions is shown in Figure 1.7: it is because of this difference that the two p-values differ. The p-value is the area under the curve to the right of the observed value of both F and the Wald statistic, 3.68. If the χ^2 distribution is used to perform the test, the area to be considered is the cross-hatched area only, giving $p = 0.055$, whereas if the F distribution is used, it is the whole hatched area, giving $p = 0.087$, the value presented in the GenStat output. The χ^2 approximation assumes that, effectively, the DF in the denominator are infinite. If this were truly the case, the equivalence between the F statistic and the Wald statistic would be exact: that is, when $F_{1,\infty} = 3.68$, $p = 0.055$. In many mixed-model analyses $DF_{Denominator}$ is large, and the use of the limiting value $DF_{Denominator} = \infty$ gives a good approximation to p. However, when there are few DF in the denominator, as in the present case, the approximation is not very close. In this analysis, it is straightforward to obtain the equivalent F statistic with the correct DF, and hence the correct p-value. However, we will see later that there are many mixed-model analyses in which the appropriate DF are not so easily determined. Modern statistical software usually obtains an approximate value by one of several methods (e.g. that of Kenward and Roger, 1997), permitting the calculation of an F statistic, but a simple alternative is to use the Wald statistic but treat the accompanying p-value with scepticism. A case where the determination of $DF_{Denominator}$ is not straightforward will be considered in more detail in Section 4.4.

1.10   Testing the assumptions of the analyses: Inspection of the residual values

In order for the significance tests and SEs in the foregoing analyses to be strictly valid, it is necessary that the ε_{ij} fulfils the assumption stated in Section 1.2: that is, they should be independent values of a variable E that follows Distribution 1.2. GenStat's output has included various warning messages about the distribution of the residual values, and we should examine the residuals from our final model (which are *estimates* of these true residuals) to assess whether they approximately fulfil this assumption. Diagnostic plots of the residuals are produced by the following GenStat statement, executed after the REML statement in Section 1.8:

```
VPLOT [GRAPHICS=high] fittedvalues, normal, halfnormal, histogram
```

These plots are presented in Figure 1.8. If the assumption concerning the distribution of the residuals is correct, they are expected to show the following patterns:

- The histogram of residuals should have approximately a normal distribution (i.e. a bell-shaped, symmetrical distribution).
- In the fitted-value plot, the points should lie in a band of nearly constant width.
- In the normal and half-normal plots, the points should lie nearly on a straight diagonal line from bottom left to top right.

Overall, the plots fit these expectations reasonably well in the present case. However, there are two possible causes for concern, namely

- there are two *outliers*, that is, exceptionally large residual values, one positive and one negative;
- the fitted-value plot shows that the variance of the residuals is larger at some fitted values than at others, that is, the spread of house prices is wider in some towns than others.

To address the first problem, the outliers might be deleted from the data set, but only if there is good reason to think that they are due to an error, or in some other way are very unrepresentative of the population being studied. To address the second, it would be possible to fit a more elaborate model that allowed for variation of the residual variance among towns. This would give a more accurate estimate of the effect of latitude and of its significance, but such refinements lie outside the scope of this chapter. Note that a trend line with a slight positive slope is marked on the fitted-value plot. This may be unexpected, as in an ordinary regression analysis there can be no such trend: the correlation between fitted values and residuals is zero by definition. However, in a mixed model, this constraint is removed. One informal way of understanding this is to note that fitted values based on random effects are 'shrunk' towards the overall mean relative to those based on fixed effects (see Chapter 5). Therefore,

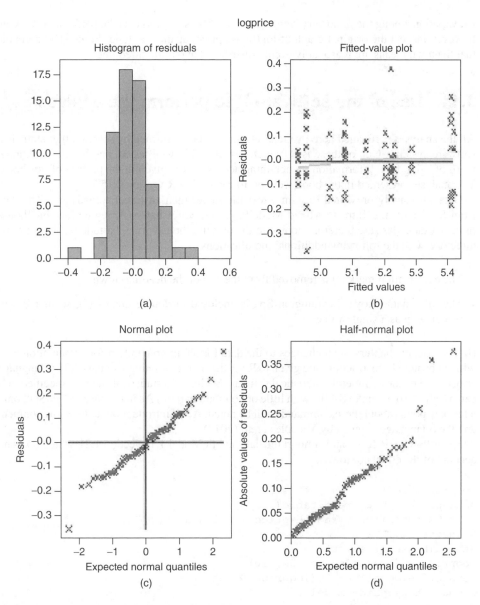

Figure 1.8   (a–d) Diagnostic plots of residuals from the mixed model relating house prices to latitude in a sample of English towns.

when residuals are calculated by subtracting these fitted values from the observed values in a mixed-model analysis, the differences obtained are slightly larger than those obtained using fitted values based on fixed effects. The more extreme the fitted value, the larger this discrepancy, leading to the observed positive association between the fitted values and the residuals. The more strongly shrunk the fitted values, the stronger this positive association tends to be.

In order for the test of F_{latitude} and the Wald test to be valid, it is also necessary that T follows Distribution 1.4, as discussed in Section 1.6. With a sample of only 11 towns and only 9 DF for variation among them, no very rigorous test of this assumption can be performed. We can, however, inspect the original graph of log(house price) against latitude (Figure 1.4) and note that the town means are reasonably evenly distributed about the fitted line.

1.11   Use of the software R to perform the analyses

All the analyses performed here by GenStat can also be performed by several other statistical software systems. One of these is R, which is reasonably well tested, and has the additional merit of being free. Information on obtaining access to R is given in the preface of this book. The analyses presented in the book were performed using R version 2.15.2.

Data are usually presented to R in a text file rather than an Excel spreadsheet. The data from Sheet1 of file 'Intro to Mixed Modelling, 2nd edn\\Chapter 1, regression line\\house price, latitude.xlsx' are therefore copied to the text file 'house price, latitude.txt' in the same directory, with the following additional modifications:

- The exclamation mark (!) is removed from the end of the heading 'town'.

- The town name Royal Leamington Spa is enclosed in double quotes ("), so that R will recognize it as a single value.

(In subsequent chapters, such changes to the data format in preparation for presentation to R will not be noted.) Instructions are given to R via the S programming language. (This language is also used by the commercial statistical software S-Plus, so a statistical analysis specified in R can usually also be run in S-Plus with little or no modification.) As in the case of GenStat, only a few key points about these commands will be noted. A full introduction to the R environment and the S language is given by Venables *et al.* (2012).

The following R commands import the data and fit Model 1.1 (Section 1.2), which takes no account of the effects of towns:

```
rm(list = ls())
houseprice.lat <- read.table(
    "IMM edn 2\\Ch 1\\house price, latitude.txt",
    header = TRUE)
attach(houseprice.lat)
logprice <- log(price_pounds, 10)
houseprice.model1 <- lm(logprice ~ latitude)
summary(houseprice.model1)
```

The preliminary command `rm(list = ls())` deletes any data structures that may be left over from previous work using R: the user should ensure that these are stored elsewhere in case these are needed again. The *function* `read.table()` specifies that data, arranged in a table (i.e. in rows and columns), are to be read. The *arguments* of the function are between the brackets, and the first of these specifies the file in which the data are to be found. The *assignment symbol* (<-) points to the name, 'houseprice.lat', of the *data frame* that is to hold the data within R. The function `attach()` simplifies references to the *lists* (i.e. columns) of data within this data frame in subsequent commands. The function `log()` transforms the house prices to logarithms, and the results are held in the list named 'logprice'. The function `lm()` indicates that a linear model (i.e. an ordinary regression model) is to be fitted, and the argument of this function, '`logprice ~ latitude`', indicates that 'logprice' is the response variable, and that the model to be fitted has the single term 'latitude' (in addition to the constant). The results are stored in the *object* named 'houseprice.model1'. The function `summary()` extracts a summary of the results from this object and displays it. The output of this function is as follows:

```
Call:
lm(formula = logprice ~ latitude)

Residuals:
     Min        1Q    Median        3Q       Max
-0.50115  -0.08019   0.00432   0.10942   0.45372

Coefficients:
             Estimate Std. Error t value Pr(>|t|)
(Intercept)   9.68473    1.00036   9.681 5.14e-14 ***
latitude     -0.08518    0.01879  -4.533 2.71e-05 ***
--
Signif. codes:  0 '***' 0.001 '**' 0.01 '*' 0.05 '.' 0.1 ' ' 1

Residual standard error: 0.1858 on 62 degrees of freedom
Multiple R-squared: 0.249, Adjusted R-squared: 0.2368
F-statistic: 20.55 on 1 and 62 DF,  p-value: 2.71e-05
```

The first item in this output is a statement of the model fitted. This is followed by a summary of the distribution of the residual values: their minimum, quartiles and maximum. The three quartiles are the values that cut off the lowest quarter, half and three-quarters of the distribution, respectively: that is, the median is the second quartile. Next come the estimates of the regression coefficients. These are the values referred to in GenStat's output as parameters. The intercept estimated by R is equivalent to the constant estimated by GenStat. Note that the values produced by the two software systems are the same, as they should be. The statistics associated with these estimates are also the same as those from GenStat, with the addition of asterisks indicating the level of significance: *** for $p < 0.001$, ** for $0.001 < p < 0.01$, and so on. Next comes the residual SE, which is the square root of the value of $MS_{Residual}$ given by GenStat ($0.1858 = \sqrt{0.03453}$), and has the same degrees of freedom, $DF_{Residual} = 62$. The multiple R-squared is a function of the sums of squares given by GenStat ($SS_{Regression}/SS_{Total} = 0.710/2.850 = 0.249$). The F statistic given by R and the associated DF and p-value are the same as those given by GenStat.

The following commands perform the regression analysis on the town means (Section 1.3):

```
rm(list = ls())
houseprice.lat.mean <- read.table(
    "IMM edn 2\\Ch 1\\price, lat, town means.txt",
    header=TRUE)
```

```
attach(houseprice.lat.mean)
houseprice.model2 <- lm(meanlogprice ~ meanlatitude)
summary(houseprice.model2)
```

The format of the output from these commands is the same as that of the output presented above, and again, the results agree numerically with those produced by GenStat.

The following statements fit the ordinary regression model with an additional fixed-effect term for towns (Model 1.3, Section 1.4) and present the accumulated analysis of variance:

```
rm(list = ls())
houseprice.lat <- read.table(
  "IMM edn 2\\Ch 1\\house price, latitude.txt",
  header = TRUE)
attach(houseprice.lat)
logprice <- log(price_pounds, 10)
houseprice.model3 <- lm(logprice ~ latitude + town)
anova(houseprice.model3)
```

Their output is as follows:

```
Analysis of Variance Table

Response: logprice
              Df  Sum Sq Mean Sq F value     Pr(>F)
latitude       1 0.70955 0.70955  41.354 3.710e-08 ***
town           9 1.23119 0.13680   7.973 2.439e-07 ***
Residuals     53 0.90937 0.01716
- - -
Signif. codes:  0 '***' 0.001 '**' 0.01 '*' 0.05 '.' 0.1 ' ' 1
```

Again, the results agree with those produced by GenStat.

The following statements perform the mixed-model analysis of Model 1.3 (Section 1.6) and present the results:

```
library(nlme)
houseprice.model4 <- lme(logprice ~ latitude, random = ~ 1|town)
summary(houseprice.model4)
anova(houseprice.model4)
```

In order to fit this model in R, it is necessary to implement a function, lme() (standing for *linear mixed effects*), that is not in the standard set. To do this, a *package* named 'nlme' must be loaded and this is done by the library() command. The first argument of the function lme() specifies that the response variable is logprice and the fixed-effect model comprises the single term 'latitude'. The second argument specifies that the random-effect model comprises the grouping structure 'town'. A formula specifying random effects to be estimated in conjunction with this grouping structure can be specified to the left of the bar (|). However, in the present case all that is required is a single estimated effect for each group, and this is indicated by the minimal model '1'. The output of the function summary() is as follows:

```
Linear mixed-effects model fit by REML
 Data: NULL
        AIC       BIC    logLik
  -42.12122 -33.61268 25.06061

Random effects:
 Formula: ~1 | town
         (Intercept)  Residual
StdDev:   0.1401131 0.1307764

Fixed effects: logprice ~ latitude
                 Value Std.Error DF  t-value p-value
(Intercept)   9.496793 1.9247991 53  4.933914  0.0000
latitude     -0.081470 0.0362724  9 -2.246063  0.0513
 Correlation:
           (Intr)
latitude -1

Standardized Within-Group Residuals:
       Min           Q1          Med          Q3          Max
-2.7577350  -0.5397447  -0.1141964   0.4664333   2.8930977

Number of Observations: 64
Number of Groups: 11
```

The estimated values of the fixed effects of Intercept and latitude, and their SEs, are the same as those given by GenStat. The output of the function anova() is as follows:

```
             numDF denDF    F-value p-value
(Intercept)      1    53 12712.430  <.0001
latitude         1     9     5.045  0.0513
```

The value of the F statistic for latitude agrees with that given by GenStat, but R gives $DF_{Denominator} = 9$, whereas GenStat gives $DF_{Denominator} = 9.4$. Consequently, the p-value produced by R (=0.0513) is slightly larger than that given by GenStat (=0.050).

The following statements produce a histogram of the residuals, similar to that produced by GenStat (Figure 1.8):

```
resmixedlogprice <- residuals(houseprice.model4)
windows()
hist(resmixedlogprice)
```

The function residuals() extracts the residual values obtained by fitting the model 'houseprice.model4'. The function windows() opens a window for graphical output, and the function hist() produces a histogram of the residual values in this window. The following statements produce other diagnostic plots:

```
windows()
par(pty = "s")
qqnorm(resmixedlogprice)
qqline(resmixedlogprice)
windows()
plot(houseprice.model4)
```

The function `par()` is used here to ensure that the normal plot will be displayed in a square region. The function `qqnorm()` produces this plot (also known as a *quantile–quantile* or *Q–Q* plot, as it is produced by plotting the quantiles of the observed distribution against the quantiles of the corresponding normal distribution), and the function `qqline()` adds a line connecting the first and third quartiles on this plot (Figure 1.9). The function `plot()` plots the residual values against the fitted values from the same model. The line added to the normal plot is the only important difference between the diagnostic plots produced by R and those produced by GenStat.

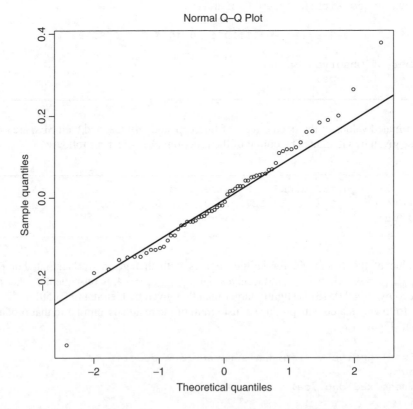

Figure 1.9 Normal plot of residuals from the mixed model relating house prices to latitude in a sample of English towns, produced by R.

1.12   Use of the software SAS to perform the analyses

The analyses performed here can also be performed by the widely used and well-tested statistical software system SAS. Information on obtaining access to SAS is given in the preface of this book: the analyses presented in the book were performed using SAS 9.2. Instructions are given to SAS via the SAS programming language. As in the case of GenStat, only a few key points about these instructions will be noted here: a full introduction to programming in SAS is given in *Getting Started with SAS* (Anonymous, 2007). When SAS has been opened, the preliminary statement

```
OPTIONS HELPBROWSER = SAS;
```

should be given: this streamlines the presentation of the output.

Data can be presented to SAS in an Excel spreadsheet, but the formatting conventions used are slightly different from those for GenStat. The data from Sheet 1 of 'IMM Edn 2\Ch 1\house price, latitude.xlsx' are therefore also presented in the sheet labelled 'for SAS' in the same Excel workbook, with the exclamation mark (!) removed from the end of the heading 'town'. (In subsequent chapters, such changes to the data format in preparation for presentation to SAS will not be noted.) A preliminary statement should be given to indicate the location of the data files. For example, suppose that the folder 'IMM Edn 2' is located in the folder 'C:\Documents and Settings'. The preliminary statement should then be

```
%LET pathname = C:\Documents and Settings;
```

The following SAS statements then import the data and fit Model 1.1 (Section 1.2), which takes no account of the effects of towns:

```
PROC IMPORT OUT = houseprice DBMS = EXCELCS REPLACE
   DATAFILE =
     "&pathname.\IMM edn 2\Ch 1\house price, latitude.xlsx";
   SHEET = "for SAS";
RUN;

DATA houseprice1; SET houseprice;
   logprice = LOG10(price_pounds);
RUN;

ODS RTF;
PROC GLM;
   MODEL logprice = latitude;
RUN;
ODS RTF CLOSE;
```

The PROC IMPORT statement specifies the data set to be imported into SAS. The option setting 'OUT = houseprice' indicates that within the SAS system, this data set is to be given the name 'houseprice'. The option 'DBMS = EXCELCS' indicates the database management system to be used to import the data. The option 'REPLACE' indicates that if a data set named 'houseprice' already exists, it is to be replaced by the imported data set. Every SAS statement ends with a semicolon (;). The SHEET statement indicates the name of the spreadsheet, within the data set, in which the data are held. The RUN statement causes the preceding group of statements to be executed. The DATA step produces a new data set, 'houseprice1', and the SET statement indicates that this is to be formed from the data set 'houseprice'. The function LOG10() transforms the house prices to logarithms, and the results are held in the variable named 'logprice' within the new data set. The statements 'ODS RTF' and 'ODS RTF CLOSE' indicate that the output from the statements in between is to be sent by SAS's Output Delivery System to a Rich Text Format file. PROC GLM fits a *general linear model* (GLM) to the data in the most recently created data set, 'houseprice1'. The MODEL statement indicates that 'logprice' is the response variable, and that the model to be fitted has the single term 'latitude' (in addition to the constant). The output from PROC GLM is as follows:

Number of observations read	64
Number of observations used	64

Source	DF	Sum of squares	Mean square	F value	Pr > F
Model	1	0.70955156	0.70955156	20.55	<0.0001
Error	62	2.14056281	0.03452521		
Corrected total	63	2.85011437			

R-square	Coeff Var	Root MSE	logprice Mean
0.248955	3.607299	0.185810	5.150935

Source	DF	Type I SS	Mean square	F value	Pr > F
latitude	1	0.70955156	0.70955156	20.55	<0.0001

Source	DF	Type III SS	Mean square	F value	Pr > F
Latitude	1	0.70955156	0.70955156	20.55	<0.0001

Parameter	Estimate	Standard error	t value	Pr > |t|
Intercept	9.684730025	1.00035736	9.68	<0.0001
latitude	−0.085176846	0.01878874	−4.53	<0.0001

The number of observations in the data, and the number that were usable (i.e. not missing or otherwise invalid), are first noted. An anova is then presented relating to the whole of the statistical model fitted. The error (residual) MS in the anova agrees with that produced by GenStat. Next come some summary statistics. Among these, R-square and Root MSE correspond to the multiple R-squared and the residual SE produced by R (Section 1.11), respectively. 'logprice mean' is the simple mean of the values of 'logprice', and

$$\text{Coeff Var} = 100 \times \frac{(\text{Root MSE})}{(\text{log price mean})}.$$

Then come two anovas in which the sums of squares are calculated by different methods, Type I and Type III, both of which are equivalent to the accumulated anova produced by GenStat in the case of this simple model. (A case in which there are distinct hypotheses corresponding to the Type I and Type III SSs is discussed in Section 7.4.) The MS and F value presented here for 'latitude' agree with those produced by GenStat. Finally, the parameter estimates and their SEs are presented. These agree with those produced by GenStat.

The following statements perform the regression analysis on the town means (Section 1.3):

```
PROC IMPORT OUT = meanhouseprice
   DATAFILE = "&pathname.\IMM edn 2\Ch 1\price, lat, town means.xlsx"
REPLACE;
   SHEET = "Sheet1";
RUN;

ODS RTF;
PROC GLM;
   MODEL meanlogprice = meanlatitude;
RUN;
ODS RTF CLOSE;
```

The format of the output from `PROC GLM` is the same as that of the output presented above, and again, the results agree numerically with those produced by GenStat.

The following statements fit the ordinary regression model with an additional fixed-effect term for towns (Model 1.3, Section 1.4) and present the accumulated analysis of variance:

```
ODS RTF;
PROC GLM DATA = houseprice1;
   CLASS town;
   MODEL logprice = latitude town;
RUN;
ODS RTF CLOSE;
```

Part of the output produced by `PROC GLM` is as follows:

Source	DF	Type I SS	Mean square	F value	$Pr > F$
latitude	1	0.70955156	0.70955156	41.35	<0.0001
town	9	1.23119443	0.13679938	7.97	<0.0001

Note that in this model, with two terms (*latitude* and *town*), the Type I sums of squares are equivalent to the accumulated anova produced by GenStat.

The following statements perform the mixed-model analysis of Model 1.3 (Section 1.6) and present the results:

```
ODS RTF;
ODS GRAPHICS ON;
PROC MIXED ASYCOV NOBOUND DATA = houseprice1;
   CLASS town;
   MODEL logprice = latitude /CHISQ DDFM = KR HTYPE = 1 SOLUTION
RESIDUAL;
```

```
    RANDOM town;
RUN;
ODS GRAPHICS OFF;
ODS RTF CLOSE;
```

The statements `ODS GRAPHICS ON` and `ODS GRAPHICS OFF` begin and end the output of graphical results by SAS's Output Delivery System. `PROC MIXED` fits a mixed model, the option 'DATA = houseprice1' indicating that the data to be analysed are held in the data set 'houseprice1'. The option setting `ASYCOV` indicates that the asymptotic covariance matrix of covariance parameter estimates is to be presented in the output (see Section 3.7 for an explanation of this matrix), and the option setting `NOBOUND` indicates that no constraints are to be imposed on the estimates of variance components: these estimates will be negative if a negative value gives the best fit. The `CLASS` statement indicates that 'town' is to be treated as a class or categorical variable, equivalent to a factor in GenStat (see Section 1.1). The `MODEL` statement specifies that the response variable is 'logprice' and the fixed-effect model comprises the single term 'latitude'. This model specification is followed by a forward slash (/) that introduces a series of options, as follows:

- `CHISQ` indicates that a chi-square statistic is to be presented for each fixed-effect term.

- 'DDFM = KR' indicates that the value of $DF_{Denominator}$ for each test of a fixed-effect term is to be determined using the Kenward–Roger method (Kenward and Roger, 1997).

- 'HTYPE = 1' indicates the type of null hypothesis to be tested. The alternative settings of this option do not produce informative results in the case of this simple model: for an example in which different types of hypothesis are compared and explained, see Section 7.4.

- `SOLUTION` indicates that the solution to the mixed-model-fitting process is to be presented, that is, the parameter estimates for each term in the fixed-effect model.

- `RESIDUAL` indicates that diagnostic plots of residuals from the model-fitting process are to be produced.

The `RANDOM` statement indicates that the random-effect model consists of the single term 'town'.

The tabular output from `PROC MIXED` is as follows:

Model information	
Data set	WORK.HOUSEPRICE1
Dependent variable	logprice
Covariance structure	Variance components
Estimation method	REML
Residual variance method	Profile
Fixed effects SE method	Kenward-Roger
Degrees of freedom method	Kenward-Roger

Class level information		
Class	**Levels**	**Values**
town	11	Bradford Buxton Carlisle Chichester Crewe Durham Newbury Ripon Royal Leamington Spa Stoke-On-Trent Witney

Dimensions	
Covariance parameters	2
Columns in X	2
Columns in Z	11
Subjects	1
Max Obs per subject	64

Number of observations	
Number of observations read	64
Number of observations used	64
Number of observations not used	0

Iteration history			
Iteration	**Evaluations**	**−2 Res Log Like**	**Criterion**
0	1	−24.00587658	
1	3	−50.12097514	0.00000299
2	1	−50.12121777	0.00000000

Convergence criteria met

Covariance parameter estimates	
Cov Parm	**Estimate**
town	0.01963
Residual	0.01710

Asymptotic covariance matrix of estimates			
Row	**Cov Parm**	**CovP1**	**CovP2**
1	town	0.000115	−1.67E−6
2	Residual	−1.67E−6	0.000011

Fit statistics	
−2 Res Log Likelihood	−50.1
AIC (smaller is better)	−46.1
AICC (smaller is better)	−45.9
BIC (smaller is better)	−45.3

Null model likelihood ratio test		
DF	Chi-square	Pr > ChiSq
1	26.12	<0.0001

Solution for fixed effects					
Effect	Estimate	Standard error	DF	t value	Pr > |t|
Intercept	9.4968	1.9275	9.63	4.93	0.0007
Latitude	−0.08147	0.03632	9.61	−2.24	0.0498

Type 1 tests of fixed effects						
Effect	Num DF	Den DF	Chi-square	F value	Pr > ChiSq	Pr > F
Latitude	1	9.61	5.03	5.03	0.0249	0.0498

The table headed 'Model information' gives details of the model specified. Some of these are self-explanatory, but others require further comment, as follows:

- 'Covariance structure' gives information about the random-effect terms. In the present model, the covariance structure comprises only a variance component for the single random-effect term 'town'. More elaborate covariance structures will be introduced later (Sections 10.9–10.13).

- 'Estimation method' indicates that model fitting is to be performed by the REML method, as indicated in Section 1.8. A fuller account of this method is given in Chapter 11. SAS also permits the specification of the alternative unrestricted maximum likelihood (ML) method, by means of the option setting 'METHOD = ML' in the 'PROC MIXED' statement.

- 'Residual variance method' indicates how the residual variance parameter is dealt with in the model-fitting process, as specified by the parameter NOPROFILE in the PROC MIXED statement

- 'Fixed effects SE method' indicates that the Kenward–Roger method is to be used not only to determine the denominator DF, as indicated in the SAS statements, but also when calculating the SE of the parameter estimates of the fixed-effect model terms.

The table headed 'Class level information' gives the number of distinct values (levels) of each class variable in the model, and the values themselves – in this case the names of the towns. The table headed 'Dimensions' gives technical information about the model specification that will not be considered in detail here. The table headed 'Number of observations' indicates the number of observations in the data, and the number that were usable. The 'Iteration history' indicates the number of iterations of calculation that were required in order to find the best-fitting model, and the results of each iteration. The column headed 'Evaluations' indicates how many times the likelihood was evaluated within each iteration, and the column headed '−2 Res Log Like' gives the final value of $-2\log_e$(residual likelihood) obtained from each iteration. The meaning of this parameter will be explained in outline in Section 11.2. Since the model fitting criterion (whether REML or ML) concerns ML, a larger negative value of the parameter reported indicates a better fit, and the values presented confirm that the fit has improved on each iteration. The column headed 'Criterion' indicates how closely the likelihood has converged to its maximum, a value close to zero indicating close convergence. In the present case, a value of zero is achieved on the final iteration, and hence the output reports that the model-fitting convergence criteria have been met. The table headed 'Covariance parameter estimates' gives the variance component estimates for the random-effect terms, and these agree with those produced by GenStat. The table headed 'Asymptotic covariance matrix of estimates' relates to the precision of the variance component estimates: more will be said about such tables in Section 3.7. The table headed 'Fit statistics' gives four alternative measures of the goodness of fit of the model to the data. More will be said about these in Section 10.11. The table headed 'Null model likelihood ratio test' performs an alternative test of the significance of the effect of 'town' (see Sections 3.12 and 3.17).

The table headed 'Solution for fixed effects' presents the parameter estimate for each fixed-effect term in the model, with its associated statistics. The estimates and their SEs agree with those produced by GenStat. The DF are similar to the $DF_{Denominator}$ in the anova produced by GenStat, but they are not identical because the Kenward–Roger criterion used to calculate them is different from that used by GenStat. The table headed 'Type 1 tests of fixed effects' corresponds to the table of 'Tests for fixed effects', 'Sequentially adding terms to fixed model' in the GenStat output. The chi-square statistic agrees with the Wald statistic produced by GenStat. The value of the F statistic for latitude agrees with that given by GenStat, but SAS gives $DF_{Denominator} = 9.61$, whereas GenStat gives $DF_{Denominator} = 9.4$. Despite this discrepancy in the DF, the p-values given by SAS and GenStat agree to the level of precision given.

PROC MIXED produces diagnostic plots of residuals on several different bases. Those that correspond to the plots produced by GenStat are headed 'Conditional residuals'. These plots closely resemble the histogram and fitted-value plot produced by GenStat, and the normal Q–Q plot produced by R. The non-conditional residuals produced by SAS, known as *marginal residuals*, are based on fitted values that include only the fixed effects: thus these marginal residuals include a contribution due to the departure of each town's mean from the line relating house price to latitude. The consequences of this can be seen in Figure 1.10, in which different groups of marginal residuals, corresponding to the different towns, are seen to have generally positive or generally negative values, whereas every group of conditional residuals is approximately symmetrical around zero.

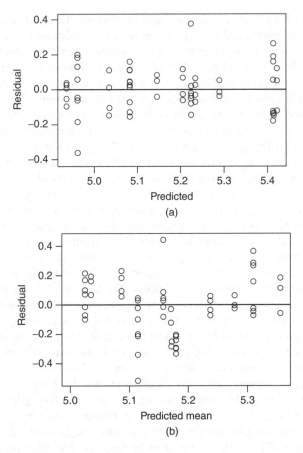

Figure 1.10 Fitted-value plots of residuals from the mixed model relating house prices to latitude in a sample of English towns, produced by SAS. (a) Conditional residuals and (b) marginal residuals.

1.13 Fitting a mixed model using GenStat's Graphical User Interface (GUI)

Most of the mixed models considered in this book can be fitted not only by the execution of statements written in the GenStat language, but also by using GenStat's GUI. We will examine the use of this method for specifying the mixed model fitted above.

When GenStat is first opened, the GUI is mostly occupied by a window headed 'Start Page'. In the present case, this is not needed. When it has been closed, the appearance of the GUI is as illustrated in Figure 1.11. (Some details of the appearance differ from user to user.)

In order to specify the model, the user should proceed as follows. To open the file holding the data, select 'File' in the GenStat main menu across the top of the screen, then, in the 'File' sub-menu, select Open. A window opens headed 'Select Input file … '. In the box in this window labelled 'Files of type:' select 'Other spreadsheet files', then navigate to the directory

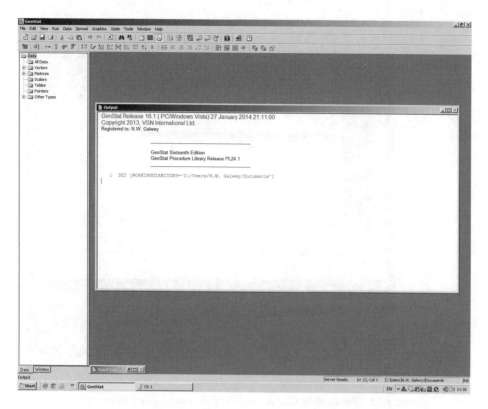

Figure 1.11   The appearance of the GenStat GUI when first opened.

holding the data file required. Select the file – in this case, 'house price, latitude for GUI.xlsx'. The appearance of the GUI is now as shown in Figure 1.12. Click on the button labelled 'Open'. A window opens headed 'Select Excel Worksheet for Import'. Within this window, click on 'S: Sheet1', then click on the button labelled 'Finish'. The specified sheet of the Excel workbook opens, and the appearance of the GUI is as shown in Figure 1.13. Note that the variate 'logprice' has been added to the data prior to import.

In order to specify the mixed model, select 'Stats' in the main menu, then in the 'Stats' sub-menu select 'Mixed Models (REML)', then in the 'Mixed Models (REML)' sub-menu select 'Linear Mixed Models … '. The appearance of the GUI immediately before and immediately after making this last selection is as shown in Figure 1.14 (a, b). Within the window headed 'Linear Mixed Models', in the box labelled 'Y-variate:' enter 'logprice'. In the box labelled 'Fixed Model:' enter 'latitude'. In the box labelled 'Random Model:', enter 'town'. The appearance of the GUI is then as shown in Figure 1.15. In the window headed 'Linear Mixed Models', click on the button labelled 'Run'. The model is fitted, and the results are sent to the output window. When this is brought to the front and appropriately sized, the appearance of the GUI is as shown in Figure 1.16.

Note that the output includes the GenStat commands that have been generated by the GUI. These provide an important *audit trail* – a record of the analysis specified. The results agree with those presented in Section 1.8, but they do not include the estimates of the fixed effects. To

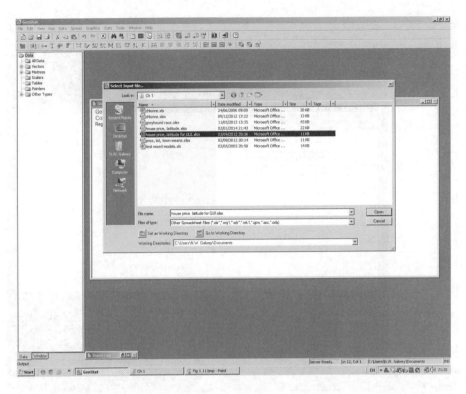

Figure 1.12   The GenStat GUI during selection of the data file to be analysed.

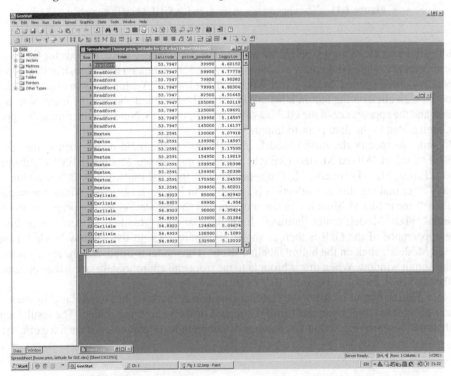

Figure 1.13   The GenStat GUI after opening the data file.

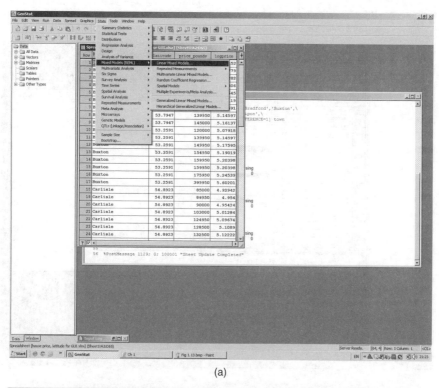

(a)

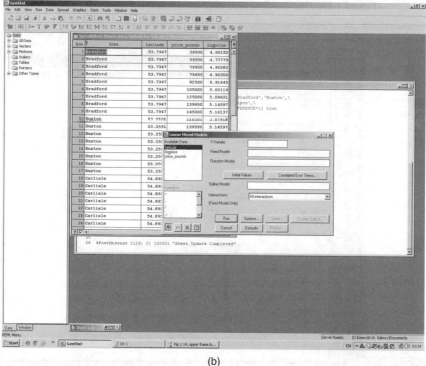

(b)

Figure 1.14   (a,b) The GenStat GUI before and after selection of the menu item 'Linear Mixed Models'.

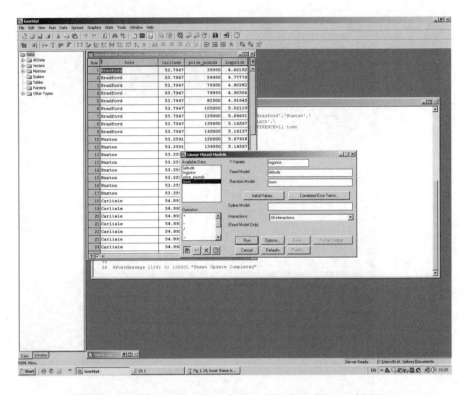

Figure 1.15   The GenStat GUI after specification of the model terms.

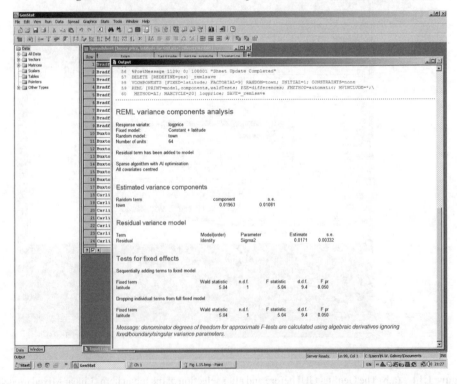

Figure 1.16   The GenStat GUI showing the results of fitting the mixed model.

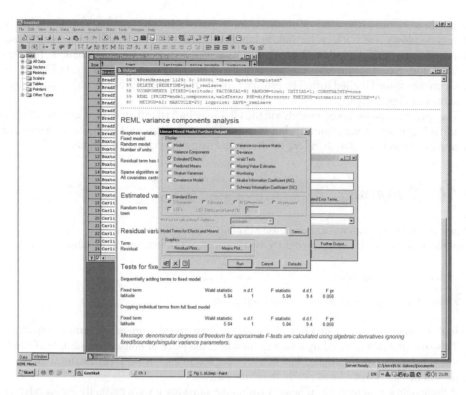

Figure 1.17 The GenStat GUI when the presentation of estimated effects has been specified.

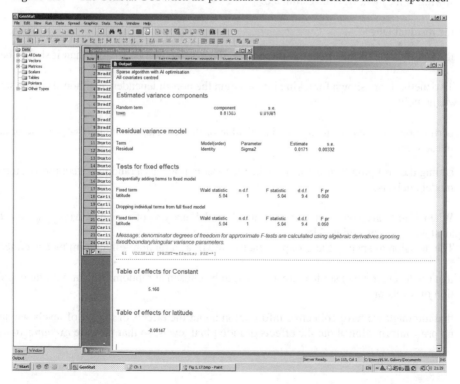

Figure 1.18 The GenStat GUI showing estimates of fixed effects from the mixed model.

obtain these, return to the window headed 'Linear Mixed Models' (currently hidden behind the Output window), and click on the button labelled 'Further Output'. A window opens headed 'Linear Mixed Model Further Output'. In this window, tick the boxes labelled 'Estimated Effects' and 'Standard Errors'. The appearance of the GUI is then as shown in Figure 1.17. Click on the button labelled Run. The estimates of the fixed effects are then added to the output, as shown in Figure 1.18.

The specification of mixed models via the GUI may be easier than writing commands in the GenStat language, at least initially, but it is much more voluminous to illustrate. Subsequent examples in this book will therefore be illustrated using the command language only.

1.14 Summary

When several observations of a response variable Y are taken at each value of an explanatory variable X, we may be able to recognize two sources of random variation, namely

- variation among groups;
- variation among observations within each group.

In this situation, a regression model with more than one random-effect term is required.

This situation is illustrated by a data set in which the price of houses in a sample of English towns is related to their latitude. The houses sampled in each town form a natural group, and houses in the same town lie at nearly the same latitude.

If these data are analysed as if houses within the same town were mutually independent, the following values will be over-estimated:

- the residual DF;
- the significance of the effect of latitude (i.e. the p-value will be smaller than it should be).

Two methods are shown for taking into account the non-independence of houses in the same town, namely

- performing the analysis using the mean value of the response variable ($\log_{10}$(house price)) in each town;
- adding the grouping factor 'town' to the regression model as a random-effect term (a mixed model analysis).

When F tests are constructed in the mixed-model analysis of variance, $MS_{latitude}$ is tested against MS_{town}, not against $MS_{Residual}$.

The decision to specify the grouping factor 'town' as a random-effect term means either

- that the levels (towns) studied are assumed to be chosen at random from an infinite population of levels or
- that although we have collective information about the effects of the set of levels we have no prior information about the effects of individual levels, so that they are *exchangeable*.

The criteria for deciding whether a model term should be specified as random are discussed in Sections 6.3 and 6.7.

A model with one or more fixed-effect terms and one or more random-effect terms in addition to the residual term is called a *mixed model*. In the present case, 'latitude' is the fixed-effect term and 'town' is the random-effect term.

In regression analysis, repeated observations at the same value of the explanatory variable are often used as the basis for a test of goodness of fit. The relationship between this test and the mixed model analysis is explored.

Mixed models can be fitted in a unified manner using the REML criterion.

If the number of observations in each group (each town) is the same, analyses using the town means and mixed-model analysis give exactly equivalent results. If not, the results differ slightly.

A weighted regression analysis of the mean values is equivalent exactly to the mixed model analysis even if the number of observations varies from group to group. More weight is given to those means that are based on larger samples of observations.

The significance of each fixed-effect term in a mixed model can be tested by an F statistic. Alternatively, it can be tested by a Wald statistic, which is a χ^2 statistic, closely related to the F statistic. However, the Wald statistic somewhat over-estimates the significance (i.e. the p-value is smaller than it should be), as it is effectively based on the approximation that $DF_{\text{Denominator of } F \text{ statistic}} = \infty$, so the F statistic should be preferred if available.

The validity of the significance tests used in mixed modelling depends on the assumption that every random-effect term, including the residual term, represents a normally distributed variable. Diagnostic plots of the residuals are used, and other random effects are inspected, to determine whether this assumption is approximately valid.

The use of the software systems GenStat, R and SAS to fit a mixed model is described. Mixed models can be specified either using a command language or via a GUI: for GenStat, both methods are illustrated.

1.15 Exercises

1.1 The spreadsheet in Table 1.8 gives data on the greyhounds that ran in the Kanyana Stake (2) in Western Australia in December 2005. (Data reproduced by kind permission of David Shortte, Western Australian Greyhound Racing Association.)

   (a) Calculate the average speed of each animal in each of its recent races. Plot the speeds against the age of each animal.

   (b) The first value of speed for 'Squeaky Cheeks' (in row 31 of the spreadsheet) is an outlier: it is much lower than the other speeds achieved by this animal. Consider the arguments for and against excluding this value from the analysis of the data.

   For the remainder of this exercise, exclude this outlier from the data.

   (c) Perform a regression analysis with speed as the response variable and age as the explanatory variable, treating each observation as independent. Obtain the equation of the line of best fit and draw the line on your plot of the data.

   (d) Specify a more appropriate regression model for these data, making use of the fact that a group of observations was made on each animal. Fit your model to the data by the ordinary methods of regression analysis. Obtain the accumulated analysis of variance from your analysis.

Table 1.8   Data on greyhounds that ran in the Kanyana Stake (2) in December 2005.

box = box from which the animal started this race; time = time taken to complete recent races (seconds); and distance = distance run in recent races (metres).

	A	B	C	D	E
1	name	box	birthdate	time	distance
2	Leprechaun Kate	1	Jan-04		515
3	Leprechaun Kate	1	Jan-04		525
4	Leprechaun Kate	1	Jan-04	30.89	530
5	Leprechaun Kate	1	Jan-04	31.31	530
6	Leprechaun Kate	1	Jan-04	31.23	530
7	Mystifier	2	Dec-03		430
8	Mystifier	2	Dec-03		450
9	Mystifier	2	Dec-03		457
10	Mystifier	2	Dec-03		457
11	Mystifier	2	Dec-03		430
12	Mystifier	2	Dec-03	24.43	410
13	Mystifier	2	Dec-03	24.35	410
14	Mystifier	2	Dec-03	24.33	410
15	Proudly Agro	3	Feb-03	31.72	530
16	Proudly Agro	3	Feb-03	31.23	530
17	Proudly Agro	3	Feb-03	31.72	530
18	Proudly Agro	3	Feb-03	31.30	530
19	Proudly Agro	3	Feb-03	31.55	530
20	Proudly Agro	3	Feb-03	31.65	530
21	Proudly Agro	3	Feb-03	31.15	530
22	Proudly Agro	3	Feb-03	31.37	530
23	Desperado Lover	4	Jan-02	30.50	509
24	Desperado Lover	4	Jan-02	30.80	509
25	Desperado Lover	4	Jan-02	30.92	509
26	Desperado Lover	4	Jan-02	30.84	509
27	Desperado Lover	4	Jan-02	31.39	509
28	Desperado Lover	4	Jan-02	31.40	530
29	Desperado Lover	4	Jan-02	31.68	530
30	Desperado Lover	4	Jan-02	31.81	530
31	Squeaky Cheeks	5	Nov-03	34.82	530
32	Squeaky Cheeks	5	Nov-03	31.72	530
33	Squeaky Cheeks	5	Nov-03	32.34	530
34	Squeaky Cheeks	5	Nov-03	31.32	530
35	Squeaky Cheeks	5	Nov-03	31.85	530
36	Squeaky Cheeks	5	Nov-03	31.22	530
37	Squeaky Cheeks	5	Nov-03	31.55	530

Table 1.8   (*continued*)

	A	B	C	D	E
38	Squeaky Cheeks	5	Nov-03	31.52	530
39	Beyond the Sea	6	Apr-04	31.37	530
40	Beyond the Sea	6	Apr-04	30.75	530
41	Keith Kaos	7	Feb-03	30.56	509
42	Keith Kaos	7	Feb-03	31.31	530
43	Keith Kaos	7	Feb-03	31.81	530
44	Keith Kaos	7	Feb-03	31.71	530
45	Keith Kaos	7	Feb-03	31.58	530
46	Keith Kaos	7	Feb-03	31.04	530
47	Keith Kaos	7	Feb-03	31.61	530
48	Keith Kaos	7	Feb-03	31.59	530
49	Elza Prince	8	Nov-02		530
50	Elza Prince	8	Nov-02	31.30	530
51	Elza Prince	8	Nov-02	31.55	530
52	Elza Prince	8	Nov-02	31.92	530
53	Elza Prince	8	Nov-02	31.26	530
54	Elza Prince	8	Nov-02	31.73	530
55	Elza Prince	8	Nov-02	31.86	530
56	Elza Prince	8	Nov-02	31.52	530
57	Jarnat Boy	9	Apr-03	31.30	530
58	Jarnat Boy	9	Apr-03	31.90	530
59	Jarnat Boy	9	Apr-03	31.59	530
60	Jarnat Boy	9	Apr-03	31.55	530
61	Jarnat Boy	9	Apr-03	31.28	530
62	Jarnat Boy	9	Apr-03	31.12	530
63	Jarnat Boy	9	Apr-03	32.44	530
64	Jarnat Boy	9	Apr-03	31.64	530
65	Shilo Mist	10	Apr-03	31.71	530
66	Shilo Mist	10	Apr-03	32.30	530
67	Shilo Mist	10	Apr-03	31.49	530
68	Shilo Mist	10	Apr-03	32.17	530
69	Shilo Mist	10	Apr-03	32.14	530
70	Shilo Mist	10	Apr-03	31.74	530
71	Shilo Mist	10	Apr-03	31.98	530
72	Shilo Mist	10	Apr-03	31.82	530

Source: Data reproduced by kind permission of David Shortte, Western Australian Greyhound Racing Association.

(e) Which is the appropriate term against which to test the significance of the effect of age:

   (i) if 'name' is specified as a fixed-effect term?

   (ii) if 'name' is specified as a random-effect term?

   Obtain the F statistic for age using both approaches and obtain the corresponding p-values. Note which test gives the higher level of significance and explain why.

(f) Re-analyse the data by mixed modelling, fitting a model with the same terms but specifying 'name' as a random-effect term. Use the F statistic to test the significance of the effect of age. Also obtain the Wald statistic.

(g) Obtain the equation of the line of best fit from your mixed-model analysis. Draw the line on your plot of the data and compare it with that obtained when every observation was treated as independent.

(h) Obtain a subset of the data comprising only the last two observations on each animal. Repeat your analysis on this subset and confirm that the F statistic for the effect of age obtained by mixed modelling now has the same value as that obtained by hand in Part (e)-ii.

1.2 A chemical product was manufactured at 2-week intervals, and after a period of storage the level of available chlorine, which is known to decline with time, was measured in cartons

Table 1.9   Levels of available chlorine in batches of a chemical product manufactured at 2-week intervals, after a period of storage.

Length of time since production (weeks)	Available chlorine			
8	0.49	0.49		
10	0.48	0.47	0.48	0.47
12	0.46	0.46	0.45	0.43
14	0.45	0.43	0.43	
16	0.44	0.43	0.43	
18	0.46	0.45		
20	0.42	0.42	0.43	
22	0.41	0.41	0.40	
24	0.42	0.40	0.40	
26	0.41	0.40	0.41	
28	0.41	0.40		
30	0.40	0.40	0.38	
32	0.41	0.40		
34	0.40			
36	0.41	0.38		
38	0.40	0.40		
40	0.39			
42	0.39			

Source: Data reproduced by permission of Wiley and Sons, Inc.

of the product (Draper and Smith, 1998, Section 24.3, pp. 518–519). The results obtained are presented in Table 1.9. (Data reproduced by permission of Wiley and Sons, Inc.)

(a) Arrange these data for analysis by GenStat, R or SAS.

(b) Plot the values of available chlorine against the times since production. Obtain the line of best fit relating these two variables, treating each observation as independent, and display it on your plot.

(c) Consider whether observations made at the same time form a natural group. If so, consider whether it is reasonable to specify the variation of the group means around a fitted line as a random variable.

These data will be analysed further in Exercise 7.1.

References

Anonymous (2007) *Getting Started with SAS*, SAS Institute, Cary, NC. Available *via* SAS's Graphical User Interface (GUI).

Draper, N.R. and Smith, H. (1998) *Applied Regression Analysis*, 3rd edn, John Wiley & Sons, Inc., New York 706 pp.

Kenward, M.G. and Roger, J.H. (1997) Small sample inference for fixed effects from restricted maximum likelihood. *Biometrics*, **53**, 983–997.

Payne, R., Murray, D. and Harding, S. (2013a) *An Introduction to the GenStat Command Language*, 16th edn, VSN International, Hemel Hempstead, 137 pp.

Payne, R., Murray, D., Harding, S. *et al.* (2013b) *Introduction to GenStat® for Windows^TM*, 16th edn, VSN International, Hemel Hempstead, 149 pp.

Royall, R.M. (1997) *Statistical Evidence. A Likelihood Paradigm*, Chapman and Hall, London, 191 pp.

Spiegelhalter, D.J., Abrams, K.R. and Myles, J.P. (2004) *Bayesian Approaches to Clinical Trials and Health-Care Evaluation*, John Wiley & Sons, Ltd, Chichester, 391 pp.

Venables, W.N., Smith, D.M. and the R Development Core Team (2012) *An Introduction to R. Notes on R: a Programming Environment for Data Analysis and Graphics. Version 2.15.0.* Available *via* R's Graphical User Interface (GUI).

2

The need for more than one random-effect term in a designed experiment

2.1 The split plot design: A design with more than one random-effect term

In Chapter 1, we examined the idea that more than one random-effect term may be necessary in regression analysis, but this idea arises also in the analysis of designed experiments. In the simplest experimental designs, the fully randomized design leading to the one-way anova and the randomized complete block design leading to the two-way anova, it is sufficient to recognize a single source of random variation, the residual term. In designs with several treatment factors, a single random-effect term is still sufficient. But in designs with more elaborate block structures, it is necessary to recognize the block effects as random, either explicitly by the specification of a mixed model, or implicitly by using an anova protocol specific to the design in question. Such designs are known as *incomplete block designs*, because the blocks do not contain a complete set of treatments. One of the simplest designs of this type is the split plot design.

The theory of the split plot design was worked out in the 1930s in the context of field experiments on crops, and such experiments provide a simple context in which to illustrate it. (The literature from this period is not very accessible, but some account of it is given by Cochran and Cox (1957, Chapter 7, pp. 293–316).) If two *treatment factors*, such as the choice of crop variety and the level of application of nitrogen fertilizer, are to be studied in all combinations, the experimenter may decide to apply each variety.nitrogen combination to several replicate field plots in a randomized block design. Alternatively, he or she may decide to sow each variety on a relatively large area of land, referred to a *main plot*, but to apply each level of nitrogen to a small *sub-plot* within each main plot. This might be done if comparisons between the levels of nitrogen were of more interest than comparisons between the varieties: the former will

Introduction to Mixed Modelling: Beyond Regression and Analysis of Variance, Second Edition. N. W. Galwey.
© 2014 John Wiley & Sons, Ltd. Published 2014 by John Wiley & Sons, Ltd.
Companion website: http://www.wiley.com/go/beyond_regression

be based on comparisons between the sub-plots, which are expected to be more precise than comparisons between the main plots. More often, the use of the split plot design is imposed by a practical constraint – for example, if the seed drill used is too wide to sow more than one variety in each main plot. For a fuller account of split plot designs, see, for example, Mead (1988, Chapter 14, pp. 382–421). For other types of incomplete block design, see Section 7.2, pp. 134–142 and Chapter 15, pp. 422–469 in the same book.

The split plot design is applicable to many other disciplines besides agricultural field experiments – for example, industrial experimentation, pharmacology, physiology, animal nutrition and food science. Here we will illustrate the design with data from an evaluation of four commercial brands of ravioli by nine trained assessors. (Data reproduced by kind permission of Guillermo Hough, DESA-ISETA, Argentina.) The purpose of the study was to identify differences in taste and texture between the brands. Knowledge of such differences is of great commercial importance to food manufacturers, but difficult to obtain: these sensory characteristics must ultimately be assessed by the subjective impressions of a human observer, which vary among individuals, and over occasions in the same individual. However, if the subjective assessment of some aspect of taste or texture (such as saltiness or gumminess) is consistent, for a particular brand, among individuals and over occasions – that is, if the perceived differences between brands are statistically significant – it is safe to conclude that these differences are real. Differences among assessors are of less interest. Different individuals may simply be using different parts of the assessment scale to describe the same sensations: who can say whether food tastes saltier to you than it does to me? However, if there are significant *interactions* between brand and assessor – for example, if the assessor ANA consistently perceives Brand A as saltier than Brand B, whereas GUI consistently ranks these brands in the opposite order – this is of interest to the investigator.

The ravioli were cooked, served into small dishes and presented hot to the assessors. Three replicate evaluations were made, each being completed on a single day: hence each day comprised a block. There may have been uncontrolled and unobserved variation from day to day in the cooking and serving conditions – for example, the temperature of the room may have changed. On each day, the order in which the four brands were presented to the assessors was randomized. However, on any given day, all the assessors received the brands in the same

Day 1	Presentation 1	Brand B	Serving	1	2	3	4	5	6	7	8	9
			Assessor	PER	FAB	HER	MJS	ANA	GUI	ALV	MOI	NOR
	Presentation 2	Brand A	Serving	1	2	3	4	5	6	7	8	9
			Assessor	MJS	ANA	GUI	MOI	FAB	NOR	HER	ALV	PER
	Presentation 3	Brand C	Serving	1	2	3	4	5	6	7	8	9
			Assessor	NOR	FAB	GUI	ALV	MJS	ANA	HER	PER	MOI
	Presentation 4	Brand D	Serving	1	2	3	4	5	6	7	8	9
			Assessor	ALV	HER	MOI	MJS	GUI	PER	ANA	NOR	FAB

Figure 2.1 The arrangement, in a split plot design, of an experiment to compare the perception of aspects of taste and texture of four commercial brands of ravioli by nine trained assessors.
Days 2 and 3 are not shown. They follow the same pattern as Day 1, but the order of brands over presentations, and of assessors over servings within each presentation, is different.

order: for this type of product it is complicated to randomize the order of presentation among assessors. Hence each presentation of a brand comprised a main plot: the brand varied only among presentations, but the whole set of assessors received the brand within each presentation. Each serving, in a single dish, comprised a sub-plot. During each presentation, the servings were shuffled before being taken to the assessors: thus the assessors were informally randomized over the sub-plots within each main plot. (It would have been cumbersome to follow a formal randomization at this stage: it was more important to get the servings to the assessors while they were still hot.) This experimental design is illustrated in Figure 2.1.

Each assessor gave the serving presented to him or her a numerical score for saltiness. The data obtained are displayed in a spreadsheet in Table 2.1. The allocation of assessors to servings within each presentation is an example of a randomization that *might have* occurred, as the *actual* allocation produced by shuffling is unknown.

2.2   The analysis of variance of the split plot design: A random-effect term for the main plots

The algebraic model for the effects in this experimental design (Model 2.1) is

$$y_{ijk} = \mu + \delta_i + \pi_{ij} + \varepsilon_{ijk} + \alpha_{l|ij} + \beta_{m|ijk} + (\alpha\beta)_{lm|ijk} \qquad (2.1)$$

where

$\quad y_{ijk}$ = the evaluation of saltiness of the kth serving in the jth presentation on the ith day,
$\quad \mu$ = the grand mean (overall mean) value of saltiness,
$\quad \delta_i$ = the effect of the ith day, that is, the departure of the mean of the ith day from the grand mean,
$\quad \pi_{ij}$ = the effect of the jth presentation on the ith day, that is, the deviation of the mean of this presentation from the value predicted on the basis of μ, δ_i and $\alpha_{l|ij}$,
$\quad \varepsilon_{ijk}$ = the residual effect, that is, the effect of the kth serving in the ijth day.presentation combination, which is the deviation of y_{ijk} from the value predicted on the basis of μ, δ_i, π_{ij}, $\alpha_{l|ij}$, $\beta_{m|ijk}$ and $(\alpha\beta)_{lm|ijk}$,
$\quad \alpha_{l|ij}$ = the main effect of the lth brand, being the brand served in the ijth day.presentation combination,
$\quad \beta_{m|ijk}$ = the main effect of the mth assessor, being the assessor who evaluated the ijkth day.presentation.serving combination,
$\quad (\alpha\beta)_{lm|ijk}$ = the interaction effect between the lth brand and the mth assessor, being the brand.assessor combination used in the ijkth day.presentation.serving combination.

The *main effect* of a particular brand is its average effect, over all assessors, on the response variable – that is, the average difference between observations on this brand and the grand mean. Similarly, the main effect of a particular assessor is his or her average effect, over all brands. The interaction between brand and assessor is the effect that is specific to that particular brand.assessor combination, but shared by all observations from that combination. It is the departure of the mean for the brand.assessor combination from the value expected on the basis of the two main effects. For a fuller explanation of main effects and interaction effects, see Mead (1988, Section 3.3, pp. 33–36).

The terms on the right hand side of Equation 2.1, μ, δ_i, α_l, and so on, are parameters to be estimated: their estimates are designated μ, $\hat{\delta}_i$, $\hat{\alpha}_l$, and so on. Note that the observation on

Table 2.1 Perceived saltiness of four commercial brands of ravioli by nine trained assessors, investigated in an experiment with a split plot design.

	A	B	C	D	E	F
1	day!	presentation!	serving!	brand!	assessor!	saltiness
2	1	1	1	B	PER	15.59
3	1	1	2	B	FAB	28.95
4	1	1	3	B	HER	8.91
5	1	1	4	B	MJS	6.68
6	1	1	5	B	ANA	0.00
7	1	1	6	B	GUI	24.50
8	1	1	7	B	ALV	33.41
9	1	1	8	B	MOI	11.14
10	1	1	9	B	NOR	24.50
11	1	2	1	A	MJS	40.09
12	1	2	2	A	ANA	11.14
13	1	2	3	A	GUI	22.27
14	1	2	4	A	MOI	13.36
15	1	2	5	A	FAB	44.55
16	1	2	6	A	NOR	22.27
17	1	2	7	A	HER	31.18
18	1	2	8	A	ALV	26.73
19	1	2	9	A	PER	17.82
20	1	3	1	C	NOR	66.82
21	1	3	2	C	FAB	75.73
22	1	3	3	C	GUI	64.59
23	1	3	4	C	ALV	49.00
24	1	3	5	C	MJS	62.36
25	1	3	6	C	ANA	13.36
26	1	3	7	C	HER	53.45
27	1	3	8	C	PER	37.86
28	1	3	9	C	MOI	40.09
29	1	4	1	D	ALV	40.09
30	1	4	2	D	HER	35.64
31	1	4	3	D	MOI	33.41
32	1	4	4	D	MJS	55.68
33	1	4	5	D	GUI	46.77
34	1	4	6	D	PER	6.68
35	1	4	7	D	ANA	0.00
36	1	4	8	D	NOR	55.68

(*continued overleaf*)

Table 2.1   (*continued*)

	A	B	C	D	E	F
37	1	4	9	D	FAB	60.14
38	2	1	1	C	MOI	37.86
39	2	1	2	C	PER	13.36
40	2	1	3	C	HER	22.27
41	2	1	4	C	ALV	35.64
42	2	1	5	C	NOR	33.41
43	2	1	6	C	MJS	57.91
44	2	1	7	C	GUI	62.36
45	2	1	8	C	ANA	11.14
46	2	1	9	C	FAB	89.09
47	2	2	1	D	GUI	31.18
48	2	2	2	D	ALV	31.18
49	2	2	3	D	ANA	8.91
50	2	2	4	D	NOR	17.82
51	2	2	5	D	HER	17.82
52	2	2	6	D	MOI	28.95
53	2	2	7	D	MJS	42.32
54	2	2	8	D	PER	33.41
55	2	2	9	D	FAB	60.14
56	2	3	1	B	NOR	24.50
57	2	3	2	B	ALV	26.73
58	2	3	3	B	GUI	35.64
59	2	3	4	B	HER	26.73
60	2	3	5	B	ANA	0.00
61	2	3	6	B	MOI	15.59
62	2	3	7	B	PER	28.95
63	2	3	8	B	FAB	13.36
64	2	3	9	B	MJS	20.05
65	2	4	1	A	FAB	60.14
66	2	4	2	A	GUI	24.50
67	2	4	3	A	ALV	35.64
68	2	4	4	A	MOI	20.05
69	2	4	5	A	MJS	13.36
70	2	4	6	A	ANA	2.23
71	2	4	7	A	NOR	11.14
72	2	4	8	A	PER	6.68

(*continued overleaf*)

Table 2.1 (*continued*)

	A	B	C	D	E	F
73	2	4	9	A	HER	28.95
74	3	1	1	A	MJS	20.05
75	3	1	2	A	PER	15.59
76	3	1	3	A	GUI	22.27
77	3	1	4	A	MOI	13.36
78	3	1	5	A	FAB	17.82
79	3	1	6	A	ANA	0.00
80	3	1	7	A	HER	24.50
81	3	1	8	A	ALV	28.95
82	3	1	9	A	NOR	13.36
83	3	2	1	C	FAB	60.14
84	3	2	2	C	NOR	62.36
85	3	2	3	C	HER	44.55
86	3	2	4	C	ALV	33.41
87	3	2	5	C	ANA	11.14
88	3	2	6	C	GUI	53.45
89	3	2	7	C	MJS	51.23
90	3	2	8	C	PER	11.14
91	3	2	9	C	MOI	13.36
92	3	3	1	B	HER	17.82
93	3	3	2	B	GUI	31.18
94	3	3	3	B	ALV	37.86
95	3	3	4	B	FAB	13.36
96	3	3	5	B	PER	2.23
97	3	3	6	B	NOR	51.23
98	3	3	7	B	MOI	13.36
99	3	3	8	B	MJS	6.68
100	3	3	9	B	ANA	0.00
101	3	4	1	D	HER	37.86
102	3	4	2	D	PER	26.73
103	3	4	3	D	FAB	46.77
104	3	4	4	D	MOI	17.82
105	3	4	5	D	GUI	35.64
106	3	4	6	D	ALV	44.55
107	3	4	7	D	NOR	17.82
108	3	4	8	D	ANA	0.00
109	3	4	9	D	MJS	55.68

Source: Data reproduced by kind permission of Guillermo Hough, DESA-ISETA, Argentina.

the response variable (y_{ijk}) is identified only by the suffixes i, j and k, not by l and m – that is, it is identified only in terms of the day, presentation and serving to which it relates, not in terms of the experimental treatments (combinations of brand and assessor). The same is true of the residual effect. The effects of day, presentation and serving are grouped together towards the beginning of the model. These terms represent the structure of the *natural variation* in the experiment, independent of the experimental treatments that were imposed: they are sufficient to identify every observation in the experiment. The treatment terms, $\alpha_{l|ij}$, $\beta_{m|ijk}$ and $(\alpha\beta)_{lm|ijk}$, are grouped together at the end of the model. The suffixes of these terms indicate not only the treatment-factor level or combination of levels applied, specified before the bar (|), but also the observation or set of observations to which they were applied, specified after the bar.

As in the models introduced in Chapter 1, it is assumed that the *true* values ε_{ijk} are independent values of a variable E, such that

$$E \sim N(0, \sigma^2). \tag{2.2}$$

The other terms that represent the structure of the natural variation are also assumed to be random-effect terms: that is, it is further assumed that the δ_i are independent values of a random variable Δ, such that

$$\Delta \sim N(0, \sigma_\Delta^2), \tag{2.3}$$

and the π_{ij} are independent values of a random variable Π, such that

$$\Pi \sim N(0, \sigma_\Pi^2). \tag{2.4}$$

The justification for these assumptions will be considered later (Section 2.6). These sources of natural variation are not of intrinsic interest: the aim of the experiment is to compare the treatments applied – the combinations of brand and assessor. The sources of natural variation are *nuisance variables*, which reduce the precision of such comparisons.

Although the concepts underlying Model 2.1 are the same as those underlying the regression models presented in Chapter 1, there is a set of specialized techniques and customs for the analysis of designed, balanced experiments like this. Therefore the model to be fitted is specified not in MODEL and FIT statement like those used previously, but in more specialized statements, as follows:

```
IMPORT 'IMM edn 2\\Ch 2\\ravioli.xlsx'
BLOCKSTRUCTURE day / presentation / serving
TREATMENTSTRUCTURE brand * assessor
ANOVA [FPROB = yes] saltiness
```

The part of the model that relates to the nuisance variables is specified in the BLOCKSTRUC-TURE statement, and that which relates to the variables under investigation is specified in the TREATMENTSTRUCTURE statement. (In experimental design, a group of observations that share a common value of a nuisance variable is referred to as a *block*.) Together, these two statements are analogous to the FIT statement. The composite model terms 'brand*assessor' and 'day/presentation/serving' will be explained below. The ANOVA statement specifies the response variable, in this case saltiness. It is analogous to the MODEL statement.

The output of these statements is as follows. It is quite voluminous, and the first part will be discussed before presenting the rest.

Analysis of variance

Variate: saltiness

Source of variation	d.f.	s.s.	m.s.	v.r.	F pr.
day stratum	2	743.4	371.7	2.67	
day.presentation stratum					
brand	3	9859.9	3286.6	23.59	0.001
Residual	6	835.9	139.3	1.39	
day.presentation.serving stratum					
assessor	8	15474.5	1934.3	19.29	<0.001
brand.assessor	24	6075.3	253.1	2.52	0.002
Residual	64	6417.5	100.3		
Total	107	39406.5			

Message: the following units have large residuals.

day 1 presentation 1		−5.65	s.e.	2.78
day 1 presentation 4 serving 6		−19.80	s.e.	7.71
day 1 presentation 4 serving 8		21.03	s.e.	7.71

Before we consider the numerical values in this anova table, we need to examine the model notation used, which is that of Wilkinson and Rogers (1973). In this notation, the composite model term 'brand*assessor' can be expanded to 'brand + assessor + brand.assessor', and each term in this expanded model appears in the column headed 'Source of variation'. The terms 'brand' and 'assessor' represent the main effects of these factors – that is, the effects represented by α_{llij} and β_{mlijk} in Equation 2.1. The term 'brand.assessor' represents the effects of interaction between these factors, represented by $(\alpha\beta)_{lmlijk}$ in Equation 2.1. The composite model term 'day/presentation/serving' can also be expanded, to 'day + day.presentation + day.presentation.serving', and each term in this expanded model also appears in the column headed 'Source of variation'. The term 'day' represents the main effects of days, but there is no corresponding term for the main effects of presentations. This is because Presentation 1 on Day 1 is not the same presentation as Presentation 1 on Day 2, and it would therefore not be meaningful to obtain the mean for Presentation 1 over days. Because no main effects of presentation are estimated, the term 'day.presentation' does *not* represent day × presentation interaction effects. Instead, it represents the effects of presentations *within* each day. These effects are the variation among presentations which, it is estimated, would have been present even if a single brand had been used throughout the experiment. Similarly, Serving 1 in a particular presentation on a particular day is not the same serving as Serving 1 in any other presentation. Hence no main effects of serving are estimated, and the term 'day.presentation.serving' does not represent day × presentation × serving interaction effects, but the effects of servings within each presentation.

Table 2.2 The strata of a split plot experiment to compare the perception of saltiness of commercial brands of ravioli by trained assessors.

Stratum	Meaning of $MS_{Residual}$
day	Natural variation among days (N.B. there are no other terms in this stratum)
day.presentation	Natural variation among presentations on the same day
day.presentation.serving	Natural variation among servings in the same presentation

Thus the model in the BLOCKSTRUCTURE statement partitions the experiment into three strata, each representing a different source of natural variation. The meaning of $MS_{Residual}$ in each stratum is shown in Table 2.2. Note that the value of $MS_{Residual}$ is greater in higher strata than in lower strata: that is,

$$MS_{Residual, day\ stratum} > MS_{Residual, day.presentation\ stratum} > MS_{Residual, day.presentation.serving\ stratum}$$

$$371.7 > 139.3 > 100.3.$$

This is to be expected: there is more natural variation among days than among presentations on the same day, and more natural variation among presentations than among servings in the same presentation. These sources of variation will be considered in more detail in Section 3.19.

Each of the treatment terms, 'brand', 'assessor' and 'brand.assessor', is tested in the appropriate stratum. Each treatment occurs once in each day, so comparisons between the day means give no information about treatment effects, and no treatment term is tested in the 'day' stratum. Comparisons between brands are based on presentation means, so the main effect of brand is tested in the 'day.presentation' stratum: that is, the F value for brand is given by

$$F_{brand} = \frac{MS_{brand}}{MS_{Residual, day.presentation\ stratum}} = \frac{3286.6}{139.3} = 23.59.$$

Comparisons between assessors, and between brand × assessor interaction effects, are based on individual servings within each presentation, so these effects are tested in the 'day.presentation.serving' stratum: that is,

$$F_{assessor} = \frac{MS_{assessor}}{MS_{Residual, day.presentation.serving\ stratum}} = \frac{1934.3}{100.3} = 19.29,$$

and

$$F_{brand.assessor} = \frac{MS_{brand.assessor}}{MS_{Residual, day.presentation.serving\ stratum}} = \frac{253.1}{100.3} = 2.52.$$

Below the anova table, GenStat gives a warning that two observations have large residual values. We will examine the residual values from this analysis later (Section 2.7).

The rest of the output of these BLOCKSTRUCTURE, TREATMENTSTRUCTURE and ANOVA statements is as follows:

Tables of means

Variate: saltiness
Grand mean 29.28

brand	A	B	C	D			
	21.78	19.22	43.23	32.91			

assessor	ALV	ANA	FAB	GUI	HER	MJS	MOI
	35.27	4.83	47.52	37.86	29.14	36.01	21.53

assessor	NOR	PER
	33.41	18.00

brand	assessor	ALV	ANA	FAB	GUI	HER	MJS
A		30.44	4.46	40.84	23.01	28.21	24.50
B		32.67	0.00	18.56	30.44	17.82	11.14
C		39.35	11.88	74.99	60.13	40.09	57.17
D		38.61	2.97	55.68	37.86	30.44	51.23

brand	assessor	MOI	NOR	PER
A		15.59	15.59	13.36
B		13.36	33.41	15.59
C		30.44	54.20	20.79
D		26.73	30.44	22.27

Standard errors of differences of means

Table	brand	assessor	brand assessor
rep.	27	12	3
s.e.d.	3.213	4.088	8.351
d.f.	6	64	66.70

Except when comparing means with the same level(s) of

brand			8.176
d.f.			64

The main effects of brand are highly significant: that is, the *p*-value associated with F_{brand} is less than 0.001, indicating that we may confidently reject the null hypotheses that the main effects of all brands are zero, that is, that they are all perceived as equally salty. Similarly, the main effects of assessor are highly significant. However, the brand × assessor interaction term is also highly significant, indicating that the two-way table of means for brand.assessor combinations may also be interpreted, and indeed, that the one-way tables should not be interpreted in isolation from this table. The one-way tables show that Brand C was perceived as considerably saltier than the other brands, and that ANA's mean assessment score is strikingly lower than that of the other assessors. The perceived saltiness of Brand C may be important to manufacturers, retailers or consumers. However, ANA's low scores are not necessarily important: as suggested earlier (Section 2.1), they may simply reflect this assessor's use of the assessment scale. The two-way table shows that ANA's scores are also less variable than those of the other assessors – they range only from 0.00 to 11.88, whereas GUI's scores range from 23.01 to 60.13. (Such variation among assessors in the use of the assessment scale can be taken into account by more sophisticated methods of data analysis than the split plot anova presented here, for example, generalized Procrustes analysis – Gower, 1975.) There is also some variation in the ranking of the brands among the assessors: for example, ANA perceived Brand B as the least salty, whereas GUI perceived Brand A as the least salty. Such crossover effects may be important: they may be an obstacle to designing a brand that pleases all consumers. However, all assessors perceived Brand C as the saltiest. The table of standard errors (SEs) of differences between means will be considered in Section 4.3.

2.3   Consequences of failure to recognize the main plots when analysing the split plot design

In order to understand why the natural variation among presentations within the same day must be specified as a term in the model for the analysis of this experiment, it may be helpful to look at what happens if this term is omitted. The model fitted (Model 2.5) is then

$$y_{ij} = \mu + \delta_i + \varepsilon_{ij} + \alpha_{l|ij} + \beta_{m|ij} + (\alpha\beta)_{lm|ij} \tag{2.5}$$

where

y_{ij} = the evaluation of saltiness of the jth serving on the ith day,

μ = the grand mean (overall mean) value of saltiness,

δ_i = the effect of the ith day,

ε_{ij} = the residual effect, that is, the effect of the jth serving on the ith day, which is the deviation of y_{ij} from the value predicted on the basis of μ, δ_i, $\alpha_{l|ij}$, $\beta_{m|ij}$ and $(\alpha\beta)_{lm|ij}$,

$\alpha_{l|ij}$ = the main effect of the lth brand, being the brand served in the jth serving on the ith day,

$\beta_{m|ij}$ = the main effect of the mth assessor, being the assessor who evaluated the jth serving on the ith day,

$(\alpha\beta)_{lm|ij}$ = the interaction effect between the lth brand and the mth assessor, being the brand.assessor combination used in jth serving on the ith day.

The following statements specify and fit this model:

```
BLOCKSTRUCTURE day
TREATMENTSTRUCTURE brand * assessor
ANOVA [FPROB = yes; PRINT = aovtable] saltiness
```

Note that the model in the BLOCKSTRUCTURE statement has been reduced from 'day/ presentation/serving' to 'day': that is, not only the term 'day.presentation' but also 'day. presentation.serving' has been omitted from the model. The latter represents residual variation, and therefore need not be specified explicitly.

The output of the ANOVA statement is as follows:

Analysis of variance

Variate: saltiness

Source of variation	d.f.	s.s.	m.s.	v.r.	F pr.
Day stratum	2	743.4	371.7	3.59	
day.*Units* stratum					
brand	3	9859.9	3286.6	31.72	<0.001
assessor	8	15474.5	1934.3	18.67	<0.001
brand.assessor	24	6075.3	253.1	2.44	0.002
Residual	70	7253.5	103.6		
Total	107	39406.5			

The residual term in this anova is obtained by pooling the terms 'day.presentation' and 'day.presentation.serving' in the anova from Model 2.1 presented in the previous section: that is,

$$DF_{\text{Residual, Model 2.2}} = DF_{\text{Residual, day.presentation stratum, Model 2.1}}$$

$$+ DF_{\text{Residual, day.presentation.serving stratum, Model 2.1}}$$

$$70 = 6 + 64$$

and

$$SS_{\text{Residual, Model 2.2}} = SS_{\text{Residual, day.presentation stratum, Model 2.1}}$$

$$+ SS_{\text{Residual, day.presentation.serving stratum, Model 2.1}}$$

$$7253.5 = 835.9 + 6417.5, \text{ allowing for rounding.}$$

For each model term in the TREATMENTSTRUCTURE statement, an F value is calculated, using MS_{Residual} as the denominator in every case. For example, the F value for brand is given by $MS_{\text{brand}}/MS_{\text{Residual}} = 3286.6/103.6 = 31.72$, and that for assessor by $MS_{\text{assessor}}/MS_{\text{Residual}} = 1934.3/103.6 = 18.67$.

This method of constructing the F tests is clearly seen to be incorrect when the variation among presentations is considered. A single brand is used in each presentation, and any comparison between brands must be made between presentations: hence if only the main effects of brand were to be studied, the analysis could be performed on the presentation means. At this level, the experiment has a randomized block design, each brand being allocated to a random presentation within each day. Hence in this part of the analysis, DF_{Residual} should be

$$\text{No. of presentations} - 1 - DF_{\text{day}} - DF_{\text{brand}} = 12 - 1 - 2 - 3 = 6,$$

not 70. The variation within a presentation is irrelevant to the comparisons between brands: hence it cannot be right to include this variation in the calculation of the MS_{Residual} against which MS_{brand} is tested. We deceive ourselves if we imagine that the 24 servings on each day represent 24 independent observations. The consequences of this mistake are seen when the F statistics for the main effects of brand obtained by the two methods are compared:

- Correct analysis gives $F_{3,6} = 23.59, p = 0.00101$.

- Analysis ignoring presentations gives $F_{3,70} = 31.72, p \approx 0$.

Because the natural variation among servings in the same presentation is less than that among servings in different presentations, pooling these terms under-estimates the $MS_{Residual}$ against which MS_{brand} is tested, and this inflates the F statistic. The excessive number of degrees of freedom in the denominator further exaggerates its significance. This is the same phenomenon as the exaggerated significance obtained when the non-independence of house prices in the same town is ignored, as noted in Sections 1.2 and 1.3.

Whereas any comparison between brands must be made between presentations, any comparison between assessors can be made within a single presentation. Hence it cannot be right to include the variation among presentations in the calculation of the $MS_{Residual}$ against which $MS_{assessor}$ is tested. The analysis ignoring presentations will tend to deflate the value of $F_{assessor}$ and under-estimate its significance. As for the brand.assessor interaction effects, by definition, they sum to zero within each presentation, because all nine assessors are represented in the presentation. Hence variation among presentations is irrelevant to comparisons between these effects also. To summarize: only variation among presentations is relevant to comparisons between brands, whereas only variation within presentations is relevant to comparisons between assessors, and between brand.assessor interaction effects.

2.4 The use of mixed modelling to analyse the split plot design

The foregoing account has emphasized the need to recognize random-effects model terms additional to the residual term when analysing a split plot design. Because the model used (Model 2.1) contains such terms, it is a mixed model, and although a split plot design is usually analysed with specialized tools, such as GenStat's BLOCKSTRUCTURE, TREATMENTSTRUCTURE and ANOVA directives, it can also be fitted by more general tools for mixed-model analysis, such as GenStat's VCOMPONENTS and REML directives. Comparison of the results with the relatively familiar split plot anova (Section 2.2) will help to elucidate the concepts of mixed modelling.

The mixed-model analysis of Model 2.1 can be specified as follows:

```
VCOMPONENTS [FIXED = brand * assessor] \
   RANDOM = day / presentation / serving
REML [PRINT = model, components, Wald, means] saltiness
```

The model formerly specified in the TREATMENTSTRUCTURE statement, 'brand * assessor', is now specified in the option FIXED of the VCOMPONENTS statement, while the model formerly specified in the BLOCKSTRUCTURE statement, 'day/presentation/serving', is now specified in the parameter RANDOM of this statement. The response variable, saltiness, formerly specified in the ANOVA statement, is now specified in the REML statement.

The first part of the output of the REML statement is as follows. (The remainder will be presented in Section 4.5.)

REML variance components analysis

Response variate: saltiness
Fixed model: Constant + brand + assessor + brand.assessor
Random model: day + day.presentation + day.presentation.serving
Number of units: 108

day.presentation.serving used as residual term

Sparse algorithm with AI optimization

Estimated variance components

Random term	Component	s.e.
day	6.5	10.6
day.presentation	4.3	9.2

Residual variance model

Term	Model(order)	Parameter	Estimate	s.e.
day.presentation.serving	Identity	Sigma2	100.3	17.7

Tests for fixed effects

Sequentially adding terms to fixed model

Fixed term	Wald statistic	n.d.f.	F statistic	d.d.f.	F pr.
Brand	70.77	3	23.59	6.0	0.001
assessor	154.32	8	19.29	64.0	<0.001
brand.assessor	60.59	24	2.52	64.0	0.002

Dropping individual terms from full fixed model

Fixed term	Wald statistic	n.d.f.	F statistic	d.d.f.	F pr.
brand.assessor	60.59	24	2.52	64.0	0.002

Message: denominator degrees of freedom for approximate F tests are calculated using algebraic derivatives ignoring fixed/boundary/singular variance parameters.

As in the example in Section 1.8, the output begins with a specification of the model fitted, and notes the number of units (observations) in the data analysed. GenStat detects that there is only one observation of each day.presentation.serving combination, and hence this term should be used as the residual term. Next come estimates of variance components for the terms in the random-effects model: these will be considered later (Section 3.19).

Next come F tests for the terms in the fixed-effects model, when added sequentially to the model, which are identical to those for the treatment terms in the corresponding anova (Section 2.2). Wald statistics are also provided (see Section 1.8). GenStat provides separate tests of the effects of adding and dropping terms to and from the model. However, GenStat detects that it would be inappropriate to drop 'brand' or 'assessor' while retaining 'brand.assessor', and omits the corresponding tests. This is because these main effect terms are *marginal to* their interaction (see Section 7.2).

The means for the treatment terms obtained from the mixed-modelling analysis are the same as those obtained from the corresponding anova.

2.5   A more conservative alternative to the *F* and Wald statistics

An alternative to the F and Wald statistics for testing the significance of a fixed-effect term is provided by the likelihood ratio test statistic proposed by Welham and Thompson (1997). This test tends to under-estimate the significance of the effect (i.e. to over-estimate the p-value), but in situations where the F statistic is not available (e.g. with older software) it provides a useful corrective to the over-estimation of significance by the Wald statistic. Like the F statistic, the likelihood ratio statistic compares the full fixed-effect model with a sub-model (reduced model) in which one or more terms are omitted. For example, the use of this test to determine the significance of the brand.assessor interaction can be specified as follows:

```
VCOMPONENTS [FIXED = brand * assessor] \
    RANDOM = day / presentation / serving
REML [PRINT = deviance; METHOD = Fisher; \
    SUBMODEL = brand + assessor] saltiness
```

Recalling that the fixed-effect model 'brand*assessor' can be expanded to 'brand + assessor + brand.assessor', we see that the SUBMODEL option in the REML statement specifies this model with the brand.assessor term omitted. When the SUBMODEL option is used, it is necessary to fit the model by the *Fisher-scoring algorithm*. This is indicated by the METHOD option: in the analysis reported in Section 1.8, the default *average information* (AI) algorithm was used. These algorithms will be compared in more detail in Section 11.10. The output of these statements is as follows:

Deviance: −2*Log-Likelihood

Submodel: Constant + brand + assessor
Full fixed model: Constant + brand + assessor + brand.assessor

Source	Deviance	d.f.
Submodel	490.54	93
Full model	447.91	69
Change	42.63	24

The change in deviance between the sub-model and the full model is interpreted as a χ^2 statistic, and the difference between the degrees of freedom of the two deviances gives the degrees of freedom for this χ^2 value. In the present case,

$$P(\chi^2_{24} > 42.63) = 0.01096.$$

As expected, this value is larger (less significant, more conservative) than those given by either the Wald statistic ($p = 0.00005$) or the F statistic ($p = 0.00175$).

2.6   Justification for regarding block effects as random

In order to obtain the correct analysis of the split plot design using the mixed modelling directives, we had to specify the block terms as random-effect terms, and we must question whether it is legitimate to regard block effects as random. This may indeed be justified if the blocks of each type (day, presentation within day and serving within presentation) can reasonably be regarded as a representative sample from a much larger population. The days on which the experiment *was* performed are then regarded as a random sample from an effectively infinite population of days on which it *might have been* performed. Likewise, within each day, the presentations that were made are regarded as a random sample of the presentations that might have been made. Finally, the servings within each presentation are regarded as a sample of the servings that might have been served. However, we must then ask whether we can legitimately regard the blocks as having been sampled independently of each other. Can the presentations within a day be considered independent when they were made at successive times, and there may be a trend over time – perhaps towards a decreased perception of saltiness, as palates become jaded? The justification for considering block effects as independent comes from the randomization *of the treatments over blocks*. Because of this, the block effects form an exchangeable set as defined in Section 1.6. Even if there is a known tendency for the last presentation on each day to produce a lower estimate of saltiness, this will not bias the estimate of the difference between any two brands – *provided that* no individual brand is more likely than another to be allocated to this presentation. If the brands are randomized over the presentations, then the presentation effects that *do* occur in combination with Brand A will be a random sample from the presentation effects that *may* occur. The same argument applies to the servings within each presentation: the assessors must be randomized over the servings. The classic account of this role of randomization in experimental design is given by Fisher (1971, Chapter II, Sections 9–10, pp. 17–21; Chapter III, Section 20, pp. 41–44; Chapter IV, Sections 26–28, pp. 62–66).

   This argument based on exchangeability induced by the randomization process provides a strong justification for specifying a block term (or a residual term) as random, regardless of whether the blocks (or experimental units) can be considered as a representative sample from a larger population. Of course, if the block effects comprise a finite set they are not a random sample from a normal distribution, but the significance tests used in the analysis of variance or mixed-model analysis are still approximately valid, provided that the denominator degrees of freedom are fairly large (more than about 30), and provided that the actual distribution of the random effects (in the present case the 108 exchangeable day.presentation.serving effects) is not very unusual (i.e. not very unlike a representative sample from a normal distribution). In the present case, because the assessors were randomly allocated to the servings within each presentation, it is unlikely that a particular assessor was consistently allocated to servings with large day.presentation.serving effects in the same direction. Consequently, on the null hypothesis that there are no real assessor effects, the F statistic for the main effect of assessor will be randomly sampled from something very close to the appropriate F distribution. That is, if all possible allocations of assessors consistent with the experimental design are made, and the F statistic is calculated for each allocation, the resulting distribution of F statistics will be very close to the F distribution. The same argument applies to the test of the brand.assessor interaction, though not to the test of the main effect of brand as it has too few denominator degrees of freedom. Such *randomization tests* are explained in more detail by Nichols and Holmes (2001).

This approximation works because, strictly speaking, it is the estimated means of the levels in each treatment term (i.e. brand means, assessor means and means of brand.assessor combinations), not the individual values, that must be normally distributed in order for the F test to be valid. The assessor and brand.assessor means meet this criterion due to the Central Limit Theorem, which states that the mean of a large number of independent observations from the same distribution is approximately normally distributed, regardless of the distribution of the individual observations. (For a more rigorous account of the Central Limit Theorem, see, for example, Bulmer, 1979, Chapter 7, pp. 115–120.)

The approximation depends critically on the assumption that the random effects within each model term are exchangeable, which is equivalent to the assumption that they are independently sampled from the same distribution. Indeed, the purpose of adding the term 'town' to Model 1.3, or including 'presentation' in Model 2.1, is to recognize that houses in the same town, or servings in the same presentation, are not independent, that is, that houses in different towns, or servings in different presentations, are not exchangeable. As we have seen, failure to do so results in p-values that are very misleading.

It is usual to treat all block terms as random: in particular, in the present case, 'day' is also placed in the random-effect model. However, because each treatment occurs once on each day, no treatment term is tested in the 'day' stratum. Hence the decision to treat 'day' as a random-effect term has no effect on the values of the F statistics or the Wald statistics, or on the values of $SE_{Difference}$. Perhaps because complete blocks (such as days in the present example) have no effect on these statistics, the decision to regard complete block terms as random or fixed is not always considered to be of central importance. It does, however, have an effect on the SEs of treatment means, which will be discussed later (Section 4.6).

2.7   Testing the assumptions of the analyses: Inspection of the residual values

As in the case of the analyses presented in Chapter 1, it is important to check and confirm that the assumptions concerning the distributions of random effects are at least approximately fulfilled. Diagnostic plots of the residual values $\hat{\varepsilon}_{ijk}$ are produced by the statement

```
VPLOT [GRAPHICS=high] fittedvalues, normal, halfnormal, histogram
```

and are presented in Figure 2.2. Although GenStat identified two observations with large residual values, one positive and one negative (Section 2.2), these plots indicate that the distribution of the residuals fits the assumptions of the analysis very well, according to the criteria introduced in Section 1.10.

As there are only 12 day.presentation combinations, with only 6 degrees of freedom, diagnostic plots of the values of $\hat{\tau}_{ij}$ are not as informative of those of $\hat{\varepsilon}_{ijk}$. However, it is worth

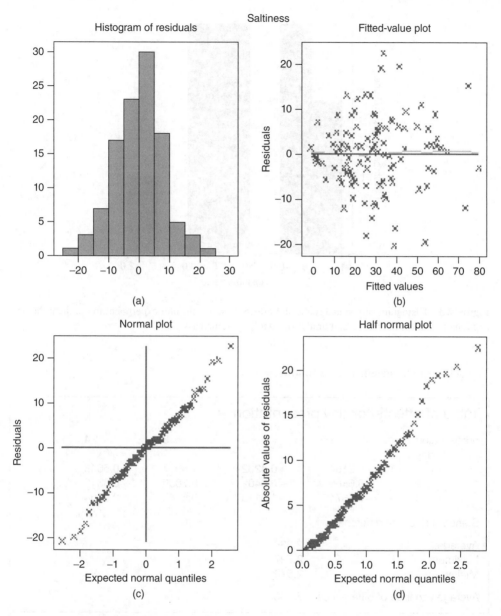

Figure 2.2   (a–d) Diagnostic plots of residuals from a split plot experiment to compare the perception of saltiness of commercial brands of ravioli by trained assessors.

inspecting the values of $\hat{\tau}_{ij}$ and obtaining a histogram of them. The following statement obtains the values:

```
REML [PRINT = effects; PTERMS = day.presentation; \
   METHOD = Fisher] saltiness
```

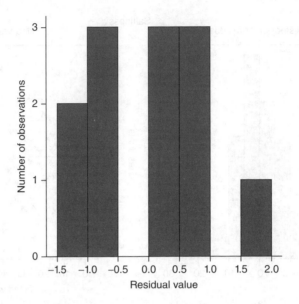

Figure 2.3 Histogram of residual main-plot effects from a split plot experiment to compare the perception of saltiness of commercial brands of ravioli by trained assessors.

The output of this statement is as follows:

Table of effects for day.presentation				
presentation	1	2	3	4
day				
1	−1.2154	0.4262	1.6977	0.5648
2	−0.6865	−0.6402	0.7009	0.3312
3	−0.7574	−1.0113	0.5145	0.0754

Standard errors of differences

Average:	2.680
Maximum:	2.727
Minimum:	2.574

Average variance of differences: 7.186

It is worth noting that unlike fixed effects, and unlike the residuals from a model that contains only fixed effects, these random effects are not required to sum to zero over the levels of other terms in the model. Thus all the effects in Presentation 1 are negative, and all those in Presentations 3 and 4 are positive. This trend reflects the low mean value of the scores given in Presentation 1. It is another example of the positive association between fitted values and random-effect residuals noted in Section 1.10. The histogram of these values (Figure 2.3) does at least show that there are no extreme outliers among the presentation effects, as is expected if the assumptions underlying the analysis are correct.

2.8   Use of R to perform the analyses

The following commands import the data into R, fit Model 2.1 (the correct model for the split plot design) and produce the resulting anova:

```
rm(list = ls())
ravioli <- read.table(
   "IMM edn 2\\Ch 2\\ravioli.txt",
   header=TRUE)
attach(ravioli)
day <- factor(day)
presentation <- factor(presentation)
serving <- factor(serving)
ravioli.model1 <-
   aov(saltiness ~ brand * assessor +
   Error(day / presentation / serving))
summary(ravioli.model1)
model.tables(ravioli.model1, type = "means", se = TRUE)
```

The lists 'day', 'presentation' and 'serving' are not automatically recognized by R as factors, because their values are numerical. The function `factor()` is therefore used to convert them to factors, and each factor is then assigned to the same name as before. The function `aov()` is used to perform an analysis of variance on balanced experimental designs. It is not to be confused with the function `anova()`, which is used to retrieve an accumulated anova from a regression analysis. The model to be fitted is specified as an argument of `aov()` in the same way as in functions `lm()` and `lme()` (Section 1.11), but with an additional feature: within the model specification that comprises the argument of this function, the function `Error()` specifies the block terms. The output of the function `summary()` is as follows:

```
Error: day
          Df Sum Sq Mean Sq F value Pr(>F)
Residuals  2  743.3   371.6

Error: day:presentation
          Df Sum Sq Mean Sq F value  Pr(>F)
brand      3  9859    3286   23.59 0.00101 **
Residuals  6   836     139
—
Signif. codes:  0 '***' 0.001 '**' 0.01 '*' 0.05 '.' 0.1 ' ' 1

Error: day:presentation:serving
               Df Sum Sq Mean Sq F value   Pr(>F)
assessor        8  15475  1934.4  19.291 2.12e-14 ***
brand:assessor 24   6076   253.2   2.525  0.00171 **
Residuals      64   6417   100.3
—
Signif. codes:  0 '***' 0.001 '**' 0.01 '*' 0.05 '.' 0.1 ' ' 1
```

The MS values, the F tests performed and the F and p-values, are the same as those given by GenStat (Section 2.2). The output of the function `model.tables()` is as follows:

```
Tables of means
Grand mean

29.28444

 brand
brand
    A      B      C      D
21.78  19.22  43.23  32.91

 assessor
assessor
  ALV    ANA    FAB    GUI    HER    MJS    MOI    NOR    PER
35.27   4.83  47.52  37.86  29.14  36.01  21.53  33.41  18.00

 brand:assessor
      assessor
brand ALV    ANA    FAB    GUI    HER    MJS    MOI    NOR    PER
    A 30.44   4.46  40.84  23.01  28.21  24.50  15.59  15.59  13.36
    B 32.67   0.00  18.56  30.44  17.82  11.14  13.36  33.41  15.59
    C 39.35  11.88  74.99  60.13  40.09  57.17  30.44  54.20  20.79
    D 38.61   2.97  55.68  37.86  30.44  51.23  26.73  30.44  22.27
Warning message:
In model.tables.aovlist(ravioli.model1, type = "means", se = TRUE) :
  SEs for type 'means' are not yet implemented
```

The mean values are the same as those given by GenStat. However, note that R is not able to give SEs for these means, though a message indicates that this may become possible in future.

The following commands fit Model 2.5, which takes no account of the effects of presentations:

```
ravioli.model2 <- aov(saltiness ~ brand * assessor + Error(day))
summary(ravioli.model2)
```

The values in the resulting anova table are the same as those given by GenStat (Section 2.3), allowing for rounding.

The following commands use mixed modelling to fit Model 2.1 and summarize the results:

```
library(nlme)
ravioli.model3 <- lme(saltiness ~ brand * assessor,
    random = ~ 1|day / presentation)
anova(ravioli.model3)
```

Because there is only one serving per presentation, 'day.presentation.serving' is equivalent to the residual term. However, the function `lme()` is not able to recognize the equivalence, so

'serving' is not mentioned explicitly in the random-effects model. The output of the anova ()
function is as follows:

```
                numDF denDF   F-value p-value
(Intercept)         1    64 249.21723  <.0001
brand               3     6  23.58935  0.0010
assessor            8    64  19.29137  <.0001
brand:assessor     24    64   2.52481  0.0017
```

Note that the term represented as 'brand.assessor' in the GenStat output is represented as
'brand:assessor' in the output of R. Note also that R adds an F test of the null hypothesis
that the constant term in the model (which R fits as an intercept, not as the grand mean) is
zero. However, this null hypothesis is not of interest. The F values and the degrees of freedom
are the same as those obtained by GenStat.

The SE of the difference between any two treatment means can be obtained from the
results of the function lme (), as follows. We first inspect more of the results by giving the
command

summary (ravioli.model3)

Part of the output from this function is as follows:

```
Linear mixed-effects model fit by REML
 Data: NULL
       AIC       BIC     logLik
  658.2372 747.0272 -290.1186

Random effects:
 Formula: ~1 | day
         (Intercept)
StdDev:     2.540342

 Formula: ~1 | presentation %in% day
         (Intercept) Residual
StdDev:     2.082927 10.01351

Fixed effects: saltiness ~ brand * assessor
               Value Std.Error DF   t-value p-value
(Intercept)  30.440000  6.084467 64  5.002903  0.0000
brandB        2.226667  8.351004  6  0.266635  0.7987
brandC        8.910000  8.351004  6  1.066938  0.3270
brandD        8.166667  8.351004  6  0.977926  0.3659
assessorANA -25.983333  8.175994 64 -3.178003  0.0023
assessorFAB  10.396667  8.175994 64  1.271609  0.2081
```

```
assessorGUI           -7.426667  8.175994 64 -0.908350  0.3671
assessorHER           -2.230000  8.175994 64 -0.272750  0.7859
assessorMJS           -5.940000  8.175994 64 -0.726517  0.4702
assessorMOI          -14.850000  8.175994 64 -1.816293  0.0740
assessorNOR          -14.850000  8.175994 64 -1.816293  0.0740
assessorPER          -17.076667  8.175994 64 -2.088635  0.0407
brandB:assessorANA    -6.683333 11.562601 64 -0.578013  0.5653
brandC:assessorANA    -1.486667 11.562601 64 -0.128575  0.8981
brandD:assessorANA    -9.653333 11.562601 64 -0.834876  0.4069
brandB:assessorFAB   -24.506667 11.562601 64 -2.119477  0.0379
brandC:assessorFAB    25.240000 11.562601 64  2.182900  0.0327
brandD:assessorFAB     6.680000 11.562601 64  0.577725  0.5655
  .
  .
  .
brandD:assessorPER     0.743333 11.562601 64  0.064288  0.9489
 Correlation:
                (Intr) brandB brandC brandD assANA assFAB
assGUI assHER assMJS assMOI assNOR
brandB               -0.686
brandC               -0.686  0.500
brandD               -0.686  0.500  0.500
assessorANA          -0.672  0.490  0.490  0.490
assessorFAB          -0.672  0.490  0.490  0.490  0.500
assessorGUI          -0.672  0.490  0.490  0.490  0.500  0.500
assessorHER          -0.672  0.490  0.490  0.490  0.500  0.500
0.500
assessorMJS          -0.672  0.490  0.490  0.490  0.500  0.500
0.500 0.500
assessorMOI          -0.672  0.490  0.490  0.490  0.500  0.500
0.500  0.500   0.500
assessorNOR          -0.672  0.490  0.490  0.490  0.500  0.500
0.500 0.500   0.500  0.500
assessorPER          -0.672  0.490  0.490  0.490  0.500  0.500
0.500 0.500   0.500  0.500   0.500
brandB:assessorANA  0.475 -0.692 -0.346 -0.346 -0.707 -0.354
-0.354 -0.354 -0.354 -0.354 -0.354
brandC:assessorANA  0.475 -0.346 -0.692 -0.346 -0.707 -0.354
-0.354 -0.354 -0.354 -0.354 -0.354
brandD:assessorANA  0.475 -0.346 -0.346 -0.692 -0.707 -0.354
-0.354 -0.354 -0.354 -0.354 -0.354
  .
  .
  .
<remaining rows and columns of the 36 × 36 correlation matrix
among the fixed effects>
```

Any treatment mean, or the difference between any pair of means, can be expressed as a *linear combination* of the fixed effects in this output. For example, the difference between the means for 'Brand C, Assessor FAB' and 'Brand B, Assessor ANA' can be expressed as

$0 \times$ (Intercept)$+$

$(-1) \times$ effect(brandB) $+ 1 \times$ effect(brandC) $+ 0 \times$ effect(brandD) $+$

$(-1) \times$ effect(assessorANA) $+ 1 \times$ effect(assessorFAB) $+ 0 \times$ effect(assessorGUI)

$+ \cdots + 0 \times$ effect(assessorPER) $+$

$(-1) \times$ effect(brandB : assessorANA) $+ 0 \times$ effect(brandC : assessorANA) $+$

$0 \times$ effect(brandD : assessorANA) $+$

$0 \times$ effect(brandB : assessorFAB) $+ 1 \times$ effect(brandC : assessorFAB)

$+ 0 \times$ brandD : assessorFAB $+ \cdots + 0 \times$ effect(brandD : assessorPER)

The coefficients in this expression can be placed in a list by the following command:

```
cont.coef.C.FAB.v.B.ANA <-
c(0, -1, 1, 0, -1, 1, 0, 0, 0, 0, 0, 0, -1, 0, 0, 0, 1, 0,
rep(0, times = 18))
```

Note that all the coefficients not presented explicitly in the expression are zeros. The following commands display each coefficient alongside the corresponding treatment effect:

```
trt.eff <- ravioli.model3$coefficients$fixed
print(data.frame(
  cont.coef.C.FAB.v.B.ANA, round(trt.eff, 3)))
```

The output of this command is as follows:

```
                    cont.coef.C.FAB.v.B.ANA round.trt.eff..3.
(Intercept)                               0            30.440
brandB                                   -1             2.227
brandC                                    1             8.910
brandD                                    0             8.167
assessorANA                              -1           -25.983
assessorFAB                               1            10.397
assessorGUI                               0            -7.427
assessorHER                               0            -2.230
assessorMJS                               0            -5.940
assessorMOI                               0           -14.850
assessorNOR                               0           -14.850
assessorPER                               0           -17.077
brandB:assessorANA                       -1            -6.683
brandC:assessorANA                        0            -1.487
brandD:assessorANA                        0            -9.653
```

brandB:assessorFAB	0	-24.507
brandC:assessorFAB	1	25.240
brandD:assessorFAB	0	6.680
.		
.		
.		
brandD:assessorPER	0	0.743

The following command evaluates the expression using matrix algebra:

```
eff.C.FAB.v.B.ANA <- t(cont.coef.C.FAB.v.B.ANA) %*% trt.eff
```

The following command applies the coefficients to the covariance matrix among the fixed effects (closely related to the correlation matrix in the output of the function summary()), to obtain the SE of the expression:

```
se.C.FAB.v.B.ANA <- sqrt(t(cont.coef.C.FAB.v.B.ANA) %*%
    summary(ravioli.model3)$varFix %*% cont.coef.C.FAB.v.B.ANA)
```

The following command displays the results of these calculations:

```
print(data.frame(eff.C.FAB.v.B.ANA, se.C.FAB.v.B.ANA))
```

The output of this command is as follows:

```
   eff.C.FAB.v.B.ANA se.C.FAB.v.B.ANA
1           74.98667         8.351004
```

The difference between the treatment means is correct, and its SE agrees with the value given by GenStat.

This method for obtaining $SE_{Difference}$ is not elegant, but it does give the user great control over the combination of effects to be considered. It is used to obtain SEs of other means and differences in Exercise 4.3.

The following statements produce diagnostic plots of the residuals equivalent to those produced by GenStat, with the exception of the half-normal plot:

```
resmixedsaltiness <- residuals(ravioli.model3)
windows()
hist(resmixedsaltiness)
windows()
par(pty = "s")
qqnorm(resmixedsaltiness)
qqline(resmixedsaltiness)
windows()
plot(ravioli.model3)
```

Details of the functions used to produce diagnostic plots are given in Section 1.11.

2.9   Use of SAS to perform the analyses

The following SAS statements import the data, fit Model 2.1 (the correct model for the split plot design) and produce the resulting anova:

```
PROC IMPORT OUT = ravioli DBMS = EXCELCS REPLACE
   DATAFILE = "&pathname.\IMM edn 2\Ch 2\ravioli.xlsx";
   SHEET = "for SAS";
RUN;


ODS RTF;
PROC ANOVA;
   CLASS day brand assessor;
   MODEL saltiness = day brand day*brand assessor brand*assessor;
   TEST H = brand E = day*brand;
   MEANS brand assessor brand*assessor;
RUN;
ODS RTF CLOSE;
```

PROC ANOVA specifies that an anova of a designed experiment is to be performed. The CLASS statement indicates which variables are to be treated as class variables when the model is specified. The MODEL statement indicates that the response variable is 'saltiness', and specifies the model terms that correspond to the experimental design. The notation in which these are given has important differences from that used by GenStat. The term 'brand*assessor' indicates combinations of values of 'brand' and 'assessor', and therefore represents the brand × assessor interaction *provided that the main effects of 'brand' and 'assessor' have been specified earlier in the model*, as they have in this case. Similarly, 'day*brand' represents the day × brand interaction, and it is not immediately apparent why this term is present in the model: it was not there in the GenStat specification. The reason is that each combination of values of 'day' and 'brand' identifies an individual presentation, so the term 'day*brand' will obtain the variation due to presentations after eliminating the main effects of 'day' and 'brand'. In GenStat this was done by the term 'day.presentation', but SAS's PROC ANOVA would not recognize that the variation due to the main effect of 'brand' should be eliminated from this term. PROC ANOVA must be told explicitly that the main effect of 'brand' is to be tested against 'day*brand', and this is done by the TEST statement, in which H stands for hypothesis and E for error. The MEANS statement indicates that means for each level of 'brand' and 'assessor', and each 'brand*assessor' combination, are to be presented.

Part of the output produced by PROC ANOVA is as follows:

Source	DF	Sum of squares	Mean square	F value	Pr > F
Model	43	32989.00066	767.18606	7.65	<0.0001
Error	64	6417.51660	100.27370		
Corrected total	107	39406.51726			

R-square	Coeff Var	Root MSE	saltiness Mean
0.837146	34.19442	10.01368	29.28453

MSE, mean-square error.

Source	DF	Anova SS	Mean square	F value	Pr > F
day	2	743.38566	371.69283	3.71	0.0300
brand	3	9859.86887	3286.62296	32.78	<0.0001
day*brand	6	835.94831	139.32472	1.39	0.2325
assessor	8	15474.52808	1934.31601	19.29	<0.0001
brand*assessor	24	6075.26976	253.13624	2.52	0.0017

Tests of hypotheses using the anova MS for day*brand as an error term					
Source	DF	Anova SS	Mean square	F value	Pr > F
brand	3	9859.868868	3286.622956	23.59	0.0010

		Saltiness	
Level of brand	N	Mean	Std Dev
A	27	21.7777444	13.0363929
B	27	19.2201344	12.9892818
C	27	43.2258589	22.0162006
D	27	32.9143770	17.6925827

SS, sum of square.

The first anova table shows that the residual mean square from the overall model fitted (namely $MS_{Residual, day.presentation.serving stratum}$) is 100.27370, which agrees with the value produced by GenStat. The next table gives some summary statistics, namely

- $R^2 = SS_{Model}/SS_{Corrected\ Total} = 32989.00066/39406.51726 = 0.837146$;

- the coefficient of variation, $100 \times (Root\ MSE)/(saltiness\ Mean) = 100 \times 10.01368/29.28453 = 34.1944$;

- $Root\ MSE = \sqrt{(residual\ mean\ square)} = \sqrt{100.27370} = 10.01368$;

- the mean value of 'saltiness'.

The next anova table gives the mean squares and d.f. for each term in the model. These agree with those produced by GenStat ('day*brand' corresponding to the residual term in the day.presentation stratum), but the F statistics are produced by testing all terms against the residual mean square, which is appropriate for 'assessor' and 'brand*assessor' but not for 'brand'. The correct test of 'brand', against 'day*brand', specified in the TEST statement, is presented in a separate table, in which the value obtained agrees with that produced by GenStat. The table of means for each brand follows, and tables of means for each assessor and each brand*assessor combination are also produced, but are not shown here. The means agree with those produced by GenStat: the values in the accompanying column headed 'Std Dev' in such a table will be explained later (Section 4.8).

PROC ANOVA has been rendered largely obsolete by PROC MIXED, but is still useful to demonstrate the traditional analysis of variance of the split plot design. The following statements use PROC MIXED, and hence mixed modelling, to fit Model 2.1:

```
ODS RTF;
ODS GRAPHICS ON;
PROC MIXED ASYCOV NOBOUND;
   CLASS day presentation serving brand assessor;
   MODEL saltiness = brand assessor brand*assessor /CHISQ
      DDFM = KR HTYPE = 1 RESIDUAL;
   RANDOM day presentation(day);
   LSMEANS brand assessor brand*assessor;
RUN;
ODS GRAPHICS OFF;
ODS RTF CLOSE;
```

In PROC MIXED, the MODEL statement indicates the response variable 'saltiness' and the fixed-effect model terms, and the RANDOM statement indicates the random-effect terms. The term 'presentation(day)' represents the effect of 'day' within each level of 'presentation' and thus corresponds to the GenStat term 'day.presentation' where the main effect of 'day' is also included in the model but that of 'presentation' is not. The LSMEANS statement indicates that means are to be presented as before.

Part of the output from PROC MIXED is as follows:

Convergence criteria met.

Covariance parameter estimates	
Cov Parm	Estimate
day	6.4547
presentation(day)	4.3390
Residual	100.27

Asymptotic covariance matrix of estimates				
Row	Cov Parm	CovP1	CovP2	CovP3
1	day	111.59	−19.9706	
2	presentation(day)	−19.9706	83.7614	−34.9125
3	Residual		−34.9125	314.21

Type 1 tests of fixed effects						
Effect	Num DF	Den DF	Chi-square	F value	Pr > ChiSq	Pr > F
brand	3	6	70.77	23.59	<0.0001	0.0010
assessor	8	64	154.32	19.29	<0.0001	<0.0001
brand*assessor	24	64	60.59	2.52	<0.0001	0.0017

Least squares means							
Effect	brand	assessor	Estimate	Standard error	DF	t value	Pr > |t|
brand	A		21.7777	2.7040	6.35	8.05	0.0001
brand	B		19.2201	2.7040	6.35	7.11	0.0003
brand	C		43.2259	2.7040	6.35	15.99	<0.0001
brand	D		32.9144	2.7040	6.35	12.17	<0.0001
assessor		ALV	35.2652	3.2969	17.4	10.70	<0.0001
assessor		ANA	4.8263	3.2969	17.4	1.46	0.1610
assessor		FAB	47.5151	3.2969	17.4	14.41	<0.0001
assessor		GUI	37.8632	3.2969	17.4	11.48	<0.0001
assessor		HER	29.1400	3.2969	17.4	8.84	<0.0001
assessor		MJS	36.0080	3.2969	17.4	10.92	<0.0001
assessor		MOI	21.5299	3.2969	17.4	6.53	<0.0001
assessor		NOR	33.4092	3.2969	17.4	10.13	<0.0001
assessor		PER	18.0039	3.2969	17.4	5.46	<0.0001
brand*assessor	A	ALV	30.4391	6.0846	61.7	5.00	<0.0001
brand*assessor	A	ANA	4.4555	6.0846	61.7	0.73	0.4668
brand*assessor	A	FAB	40.8352	6.0846	61.7	6.71	<0.0001
⋮							⋮
brand*assessor	D	MOI	26.7279	6.0846	61.7	4.39	<0.0001
brand*assessor	D	NOR	30.4400	6.0846	61.7	5.00	<0.0001
brand*assessor	D	PER	22.2736	6.0846	61.7	3.66	0.0005

The table headed 'Covariance parameter estimates' gives the estimates of the variance components corresponding to each term in the random-effect model, and the table headed 'Asymptotic covariance matrix of estimates' gives the covariances between these estimates. The values obtained agree with those from GenStat, and such tables are explained more fully in Section 3.7. The table headed 'Type 1 tests of fixed effects' corresponds to the tests for fixed effects sequentially adding terms to the fixed model produced by GenStat: the $DF_{Denominator}$ values are exactly those indicated by the correct anova of the split plot design (performed by PROC ANOVA above, but laid out more clearly in the GenStat output in Section 2.2), and the F and p-values agree with those produced by GenStat. The chi-square values agree with the Wald statistics produced by GenStat. The table headed 'Least square means' (of which the first and last few rows are shown) gives the mean for each level of 'brand' and 'assessor', and each brand.assessor combination. The values obtained agree with those produced by GenStat. The SEs presented are for the means themselves, rather than for the differences between means as in the GenStat output. More will be said about the choice of SEs to accompany means in Section 4.6. The t values and accompanying degrees of freedom and p-values relate to the null hypothesis that the true value of the corresponding mean is zero, but this is not usually a hypothesis of interest.

PROC MIXED produces diagnostic plots of residuals on several different bases. Those that correspond to the plots produced by GenStat are headed 'Conditional residuals for saltiness'.

2.10 Summary

The need for mixed models occurs not only in regression analysis, but also in the analysis of designed experiments.

In a fully randomized design (leading to one-way anova) or a randomized complete block design (leading to two-way anova), it is sufficient to recognize the residuals as a random-effect term.

However, in designs with more elaborate block structures, it is necessary to recognize the block effects as random. One of the simplest designs of this type is the split plot design.

The application of mixed modelling to the analysis of designed experiments is illustrated by a split plot experiment to assess the sensory characteristics of ravioli. The treatment factors are brand and assessor. The block factors are day (complete block), presentation (main plot) and serving (sub-plot). The brand varies only among main plots (not within each main plot), but the assessor varies among sub-plots within each main plot.

If the main plots are not recognized when the experiment is analysed, the results will be incorrect in the following ways:

- The value of $MS_{Residual}$ against which MS_{brand} is tested will be underestimated.

- The value of $DF_{Residual}$ for the test statistic F_{brand} will be exaggerated.

- The significance of 'brand' will be overestimated (i.e. the p-value will be smaller than it should be).

Mixed modelling can be used as an alternative to the analysis of variance to analyse the split plot design. The relationship between these two methods of specifying the analysis is shown in Table 2.3.

The Wald statistic over-estimates the significance of a fixed-effect term (see also Sections 1.9 and 1.14). A more conservative alternative to the F and Wald statistics for testing the significance of a fixed-effect term is provided by the likelihood ratio statistic proposed by Welham and Thompson (1997). It is useful in situations where the F statistic is not available.

An argument justifying the decision to regard all block terms as random is presented. This argument is based on the fact that the treatments are randomized over the blocks, so that the block effects form an exchangeable set as defined in Section 1.6.

Table 2.3 The relationship between the specification of the analysis of a split plot design by anova and by mixed modelling.

Terms		Method of specification of terms	
In the general case	In the example presented	In anova	In mixed model
A*B[a]	brand*assessor	Treatment structure	Fixed-effect model
block/mainplot/subplot	day/presentation/serving	Block structure	Random-effect model

[a] A and B are the treatment factors.

2.11   Exercises

2.1   A field experiment to investigate the effect of four levels of nitrogen fertilizer on the yield of three varieties of oats was laid out in a split plot design (Yates, 1937). The experiment comprised six blocks of land, each divided into three main plots. Each main plot was sown with a single variety of oats. Each variety was sown on one main plot within each block: within the block, the allocation of varieties to main plots was randomized. Each main plot was divided into four sub-plots. A different level of nitrogen fertilizer was applied to each of the four sub-plots within each main plot: within the main plot, the allocation of fertilizer levels to sub-plots was randomized. The results obtained are presented in Table 2.4.

(a) Make a sketch of the field layout of Block 1. N.B. It is not possible to determine the orientation of the sub-plots relative to the main plots from the information provided: make a sensible assumption about this.

(b) Analyse the experiment by the methods of analysis of variance. Determine whether each of the following treatment terms is significant according to the F test:

   (i) variety

   (ii) nitrogen level

   (iii) variety × nitrogen level interaction.

(c) Analyse the experiment by mixed modelling. Confirm that the Wald statistic for each treatment term has the expected relationship to the corresponding F statistic. Explain the difference in the p-values between each F statistic and the corresponding Wald statistic.

2.2   An experiment was conducted to compare the effects of three types of oil on the amount of wear suffered by piston rings in an engine (Bennett and Franklin, 1954, Section 8.62, pp. 542–543). Five piston rings in the engine were weighed at the beginning and end of each 12-hour test run, and the weight loss of each ring was calculated. The engine was charged with a new oil at the beginning of each test run. The experiment comprised five replications, so there were 15 test runs in total. The results obtained are presented in Table 2.5. (Data reproduced by the kind permission of Wiley and Sons, Inc.) The oil type and the rings are to be regarded as treatment factors.

(a) The experiment has a split plot design. Identify the block, main-plot and sub-plot factors. Identify the treatment factor that varies only among main plots, and the treatment factor that varies among sub-plots within each main plot.

(b) What steps should have been taken during the planning of this experiment to ensure that treatment effects are not confounded with any other trends?

(c) Arrange the data for analysis by GenStat, R or SAS.

(d) Analyse the experiment by the methods of analysis of variance. Identify the three terms that belong to the treatment structure, and determine whether each is significant according to the F test.

(e) Analyse the experiment by mixed modelling. Confirm that the Wald statistic for each term in the treatment structure has the expected relationship to the corresponding F statistic.

Table 2.4   Results of an experiment to determine the effect of nitrogen fertilizer on the yield of oat varieties, laid out in a split plot design.

	A	B	C	B	D	E
1	block	mainplot	subplot	variety	nitrogen (cwt)	yield
2	1	1	1	Marvellous	0.6	156
3	1	1	2	Marvellous	0.4	118
4	1	1	3	Marvellous	0.2	140
5	1	1	4	Marvellous	0	105
6	1	2	1	Victory	0	111
7	1	2	2	Victory	0.2	130
8	1	2	3	Victory	0.6	174
9	1	2	4	Victory	0.4	157
10	1	3	1	Golden rain	0	117
11	1	3	2	Golden rain	0.2	114
12	1	3	3	Golden rain	0.4	161
13	1	3	4	Golden rain	0.6	141
14	2	1	1	Marvellous	0.4	104
15	2	1	2	Marvellous	0	70
16	2	1	3	Marvellous	0.2	89
17	2	1	4	Marvellous	0.6	117
18	2	2	1	Victory	0.6	122
19	2	2	2	Victory	0	74
20	2	2	3	Victory	0.2	89
21	2	2	4	Victory	0.4	81
22	2	3	1	Golden rain	0.2	103
23	2	3	2	Golden rain	0	64
24	2	3	3	Golden rain	0.4	132
25	2	3	4	Golden rain	0.6	133
26	3	1	1	Golden rain	0.2	108
27	3	1	2	Golden rain	0.4	126
28	3	1	3	Golden rain	0.6	149
29	3	1	4	Golden rain	0	70
30	3	2	1	Marvellous	0.6	144
31	3	2	2	Marvellous	0.2	124
32	3	2	3	Marvellous	0.4	121
33	3	2	4	Marvellous	0	96
34	3	3	1	Victory	0	61
35	3	3	2	Victory	0.6	100
36	3	3	3	Victory	0.2	91

(continued overleaf)

Table 2.4 (*continued*)

	A	B	C	B	D	E
37	3	3	4	Victory	0.4	97
38	4	1	1	Marvellous	0.4	109
39	4	1	2	Marvellous	0.6	99
40	4	1	3	Marvellous	0	63
41	4	1	4	Marvellous	0.2	70
42	4	2	1	Golden rain	0	80
43	4	2	2	Golden rain	0.4	94
44	4	2	3	Golden rain	0.6	126
45	4	2	4	Golden rain	0.2	82
46	4	3	1	Victory	0.2	90
47	4	3	2	Victory	0.4	100
48	4	3	3	Victory	0.6	116
49	4	3	4	Victory	0	62
50	5	1	1	Golden rain	0.6	96
51	5	1	2	Golden rain	0	60
52	5	1	3	Golden rain	0.4	89
53	5	1	4	Golden rain	0.2	102
54	5	2	1	Victory	0.4	112
55	5	2	2	Victory	0.6	86
56	5	2	3	Victory	0	68
57	5	2	4	Victory	0.2	64
58	5	3	1	Marvellous	0.4	132
59	5	3	2	Marvellous	0.6	124
60	5	3	3	Marvellous	0.2	129
61	5	3	4	Marvellous	0	89
62	6	1	1	Victory	0.4	118
63	6	1	2	Victory	0	53
64	6	1	3	Victory	0.6	113
65	6	1	4	Victory	0.2	74
66	6	2	1	Golden rain	0.6	104
67	6	2	2	Golden rain	0.4	86
68	6	2	3	Golden rain	0	89
69	6	2	4	Golden rain	0.2	82
70	6	3	1	Marvellous	0	97
71	6	3	2	Marvellous	0.2	99
72	6	3	3	Marvellous	0.4	119
73	6	3	4	Marvellous	0.6	121

Table 2.5 Weight loss of piston rings during test runs of an engine in an experiment to compare the effects of three types of oil.

The values presented are logarithms of loss of weight in grams × 100.

Replicate	1			2			3			4			5		
Oil Ring	A	B	C	A	B	C	A	B	C	A	B	C	A	B	C
1	1.782	1.568	1.507	1.642	1.539	1.562	1.682	1.616	1.630	1.654	1.680	1.740	1.496	1.626	1.558
2	1.306	1.223	1.240	1.346	1.064	1.334	1.322	1.369	1.428	1.532	1.452	1.408	1.354	1.466	1.478
3	1.149	1.029	1.068	1.090	0.778	1.136	1.176	1.053	1.202	1.233	1.193	1.228	1.038	1.167	1.330
4	1.025	0.919	0.982	1.012	0.690	1.021	0.930	0.935	1.057	0.992	0.973	1.093	0.924	0.974	0.996
Oil ring	1.110	1.093	1.094	1.000	0.733	0.987	0.892	0.845	1.029	0.940	0.786	1.060	0.863	0.881	0.968

Source: Data reproduced by kind permission of Wiley and Sons, Inc.

    (f) The mean log-transformed weight loss is smallest during test runs with Oil B. Is it safe to conclude that this type of oil causes less wear than the other two?

    N.B. The sums of squares obtained in the analysis of these data do not agree exactly with those presented by Bennett and Franklin.

References

Bennett, C.A. and Franklin, N.L. (1954) *Statistical Analysis in Chemistry and the Chemical Industry*, John Wiley & Sons, Inc., New York, 724 pp.

Bulmer, M.G. (1979) *Principles of Statistics*, 2nd edn, Dover Publications, New York, 252 pp.

Cochran, W.G. and Cox, G.M. (1957) *Experimental Designs*, 2nd edn, John Wiley & Sons, Inc., New York, 611 pp.

Fisher, R.A. (1971) in *The Design of Experiments*, Oliver and Boyd, London, 245 pp. Re-issued in *Statistical Methods, Experimental Design and Scientific Inference* (ed. J.H. Bennett (1990)), Oxford University Press, Oxford.

Gower, J.C. (1975) Generalized Procrustes analysis. *Psychometrika*, **40**, 33–51.

Mead, R. (1988) *The Design of Experiments: Statistical Principles for Practical Application*, Cambridge University Press, Cambridge, 620 pp.

Nichols, T.E. and Holmes, A.P. (2001) Nonparametric permutation tests for functional neuroimaging: a primer with examples. *Human Brain Mapping*, **15**, 1–25.

Welham, S.J. and Thompson, R. (1997) Likelihood ratio tests for fixed model terms using residual maximum likelihood. *Journal of the Royal Statistical Society, Series B*, **59**, 701–714.

Wilkinson, G.N. and Rogers, C.E. (1973) Symbolic description of factorial models for analysis of variance. *Applied Statistics*, **22**, 392–399.

Yates, F. (1937) The Design and Analysis of Factorial Experiments. Technical Communication No. 35 of the Commonwealth Bureau of Soils, Commonwealth Agricultural Bureaux, Farnham Royal.

3

Estimation of the variances of random-effect terms

3.1 The need to estimate variance components

In Chapters 1 and 2, we recognized random-effect terms and took them into account when fitting models to data, in order to interpret the fixed-effect terms more reliably. In Chapter 1, the emphasis was on the effect of latitude (a fixed-effect term) on house prices: the additional variation among towns (a random-effect term) was treated as a nuisance variable that reduced the precision with which the effect of latitude was estimated, and that must be taken into account in order to assess the statistical significance of this effect realistically. In Chapter 2, the emphasis was on the effects of brand and assessor (fixed-effect terms) on the perceived saltiness of ravioli, and the additional variation among days, presentations and servings (random-effect terms) was treated as a set of nuisance variables. However, a random-effect term may be of interest in its own right. Sometimes the means of individual levels of a random-effect factor are of interest, but even if this is not the case, it may still be useful to estimate the variance of the effects among the population of levels from which the sample studied was drawn. This chapter examines the methods for estimating the variance due to each random-effect term, and the interpretation of the results.

3.2 A hierarchical random-effects model for a three-stage assay process

In an assay process, several sources of random variation commonly occur, and the relative magnitude of these has an important bearing on the cost and accuracy of the process. We will examine this problem in a data set obtained from delivery batches of a chemical paste (Davies

Introduction to Mixed Modelling: Beyond Regression and Analysis of Variance, Second Edition. N. W. Galwey.
© 2014 John Wiley & Sons, Ltd. Published 2014 by John Wiley & Sons, Ltd.
Companion website: http://www.wiley.com/go/beyond_regression

and Goldsmith, 1984, Section 6.5, pp. 135–141). Three casks from each batch were sampled, and two analytical tests were performed on the contents of each cask to determine the 'strength' (%) of the paste. The data are displayed in the spreadsheet in Table 3.1. (Data reproduced by the kind permission of Pearson Education Ltd.)

A natural model for these data (Model 3.1) is

$$y_{ijk} = \mu + \delta_i + \gamma_{ij} + \varepsilon_{ijk} \tag{3.1}$$

Table 3.1 'Strength' (%) of a chemical paste, assessed by two tests on each of three casks in each of 10 delivery batches.

	A	B	C	D		A	B	C	D
1	delivery!	cask!	test!	strength	32	6	1	1	63.4
2	1	1	1	62.8	33	6	1	2	64.9
3	1	1	2	62.6	34	6	2	1	59.3
4	1	2	1	60.1	35	6	2	2	58.1
5	1	2	2	62.3	36	6	3	1	60.5
6	1	3	1	62.7	37	6	3	2	60.0
7	1	3	2	63.1	38	7	1	1	62.5
8	2	1	1	60.0	39	7	1	2	62.6
9	2	1	2	61.4	40	7	2	1	61.0
10	2	2	1	57.5	41	7	2	2	58.7
11	2	2	2	56.9	42	7	3	1	56.9
12	2	3	1	61.1	43	7	3	2	57.7
13	2	3	2	58.9	44	8	1	1	59.2
14	3	1	1	58.7	45	8	1	2	59.4
15	3	1	2	57.5	46	8	2	1	65.2
16	3	2	1	63.9	47	8	2	2	66.0
17	3	2	2	63.1	48	8	3	1	64.8
18	3	3	1	65.4	49	8	3	2	64.1
19	3	3	2	63.7	50	9	1	1	54.8
20	4	1	1	57.1	51	9	1	2	54.8
21	4	1	2	56.4	52	9	2	1	64.0
22	4	2	1	56.9	53	9	2	2	64.0
23	4	2	2	58.6	54	9	3	1	57.7
24	4	3	1	64.7	55	9	3	2	56.8
25	4	3	2	64.5	56	10	1	1	58.3
26	5	1	1	55.1	57	10	1	2	59.3
27	5	1	2	55.1	58	10	2	1	59.2
28	5	2	1	54.7	59	10	2	2	59.2
29	5	2	2	54.2	60	10	3	1	58.9
30	5	3	1	58.8	61	10	3	2	56.6
31	5	3	2	57.5					

Source: Numerical data on the compressive strength of Portland cement, presented. Davies and Goldsmith (1984). Data reproduced by kind permission of Pearson Education Ltd.

where

y_{ijk} = the value of strength from the kth test on the jth cask from the ith delivery,

μ = the grand mean (overall mean) value of strength,

δ_i = the effect of the ith delivery,

γ_{ij} = the effect of the jth cask from the ith delivery,

ε_{ijk} = the effect of the kth test in the ijth delivery.cask combination.

It is natural to treat delivery, cask within delivery and test within cask all as random-effect terms – that is, to make the following assumptions:

- that the δ_i are independent values of a random variable Δ, such that

$$\Delta \sim N(0, \sigma_\Delta^2), \tag{3.2}$$

- that the γ_{ij} are independent values of a random variable Γ, such that

$$\Gamma \sim N(0, \sigma_\Gamma^2), \tag{3.3}$$

and

- that the ε_{ijk} are independent values of a random variable E, such that

$$E \sim N(0, \sigma_E^2). \tag{3.4}$$

The following GenStat statements read the data and analyse them according to Model 3.1:

```
IMPORT 'IMM edn 2\\Ch 3\\paste strength.xlsx'
BLOCKSTRUCTURE delivery / cask / test
ANOVA [FPROB = yes] strength
```

Note that because all the terms in this model are random-effect terms, they are all specified in the BLOCKSTRUCTURE statement: there is no TREATMENTSTRUCTURE statement. The output of the ANOVA statement is as follows:

Analysis of variance

Variate: strength

Source of variation	d.f.	s.s.	m.s.	v.r.	F pr.
delivery stratum	9	247.4027	27.4892	1.57	
delivery.cask stratum	20	350.9067	17.5453	25.88	
delivery.cask.test stratum	30	20.3400	0.6780		
Total	59	618.6493			

Message: the following units have large residuals.

delivery 5	−4.2	s.e. 2.0
delivery 4 cask 3	4.9	s.e. 2.4
delivery 9 cask 2	5.3	s.e. 2.4

> ## Tables of means
>
> Variate: strength
> Grand mean: 60.1

As in the analysis of the split-plot design (Model 2.1, Section 2.2), each random-effect term defines a stratum in the anova table. Like the random-effect part of the model for the split-plot design, Model 3.1 is *hierarchical*: its terms are *nested*, one within another. This hierarchical structure is illustrated diagrammatically in Figure 3.1. (For clarity, the detail of the lower strata is shown only for the first five deliveries.)

Each of the variables Δ, Γ and E contributes to the total variation in paste strength, and the variances σ_Δ^2, σ_Γ^2 and σ_E^2 are referred to as *components* of the total variance. Each term in the model is tested against the term nested within it, that is, the term below it in the hierarchy, and hence in the anova table. Thus

- $F_{\text{delivery}} = MS_{\text{delivery}}/MS_{\text{delivery.cask}} = 27.4892/17.5453 = 1.57$ and

- $F_{\text{delivery.cask}} = MS_{\text{delivery.cask}}/MS_{\text{delivery.cask.test}} = 17.5453/0.6780 = 25.88$.

As all the model terms are included in the GenStat BLOCKSTRUCTURE statement, they are all assumed by GenStat to represent nuisance variables. Hence the p-values corresponding to these F statistics (see Section 1.2) are not given, and no tables of means are presented (except the grand mean). However, the p-values can be looked up by many standard software systems, including GenStat. They are presented, together with the information required to obtain them, in Table 3.2. There is a suggestion that the paste strength varies between deliveries, but this

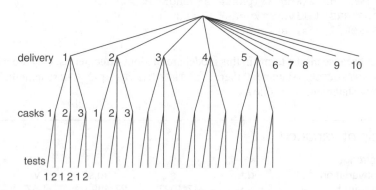

Figure 3.1   The hierarchical structure of the experiment to assess the effects of delivery batches, casks and tests on the strength of a chemical paste.

Table 3.2 p-Values from the anova of the strength of a chemical paste in a hierarchical design.

Term	$DF_{\text{Numerator}}$	$DF_{\text{Denominator}}$	F	p
delivery	9	20	1.57	0.19
delivery.cask	20	30	25.88	<0.0001

Table 3.3 Expected mean squares of the terms in the anova of the strength of a chemical paste in a hierarchical design.

Source of variation	MS	Expected MS
delivery	$MS_{delivery}$	$n_\Gamma \cdot n_E \sigma_\Delta^2 + n_E \sigma_\Gamma^2 + \sigma_E^2$
delivery.cask	$MS_{delivery.cask}$	$n_E \sigma_\Gamma^2 + \sigma_E^2$
delivery.cask.test	$MS_{delivery.cask.test}$	σ_E^2

falls short of significance. However, there is highly significant variation among casks within a delivery, relative to the variation between tests on the same cask.

3.3 The relationship between variance components and stratum mean squares

It is natural to enquire not only into the significance of the different sources of variation, but also into their magnitude. This may be of importance for various purposes – for example, to decide whether an allowance for variation between deliveries should be written into the specification of a manufacturing process or the terms of a contract, or to decide how many casks should be sampled, and how many tests conducted on each cask, in future deliveries. Broadly speaking, the 'delivery' stratum in the preceding anova represents the variation among deliveries, the 'delivery.cask' stratum represents variation among casks within each delivery and the 'delivery.cask.test' stratum represents variation among tests within the same cask. However, $MS_{delivery}$ cannot be used directly as an estimate of σ_Δ^2, nor $MS_{delivery.cask}$ as an estimate of σ_Γ^2: the relationship between mean squares (MSs) and variance components is somewhat more elaborate. In order to explore it, we need to introduce the concept of the *expected MS* – the mean value of the MS that would be obtained over a long run of similar experiments, sampled from the same underlying distributions.
    Let

$$n_\Gamma = \text{the number of casks within each delivery} = 3 \qquad (3.5)$$

$$n_E = \text{the number of tests within each cask} = 2. \qquad (3.6)$$

The relationships between the true values of the variance components and the expected MSs in the present experiment are then as shown in Table 3.3.
    The simple formulae in this table are applicable only in the case of a balanced experimental design, where n_Γ is constant over deliveries, and n_E is constant over casks within each delivery and over deliveries. We will not here derive them formally, but we will note some intuitively reasonable properties that they have. These will be reviewed first in relation to the comparison between $MS_{delivery.cask}$ and $MS_{delivery.cask.test}$, as follows:

- The variation among tests in the same cask ($MS_{delivery.cask.test}$) is not influenced by the variation among casks (σ_Γ^2) or among deliveries (σ_Δ^2).

- However, the variation among casks in the same delivery ($MS_{delivery.cask}$) is influenced by the variation among tests within each cask (σ_E^2). This means that even if there is no real variation among casks within each delivery (i.e. if $\sigma_\Gamma^2 = 0$), there will still be some variation among the observed cask means, due to the random effects of tests.

- If n_E is large, then the influence of σ_E^2 on $\text{MS}_{\text{delivery.cask}}$ is small relative to that of σ_Γ^2. In the limiting case, if an infinite number of tests were performed on each cask, then their mean would be the true mean value for the cask, and $\text{MS}_{\text{delivery.cask}}$, which is closely related to the variance among the cask means, would be determined only by σ_Γ^2.

- These expected MSs justify the F test of $\text{MS}_{\text{delivery.cask}}$ against $\text{MS}_{\text{delivery.cask.test}}$. If the hypothesis

$$H_0 : \sigma_\Gamma^2 = 0 \tag{3.7}$$

is true, then

$$\text{expected MS}_{\text{delivery.cask}} = \text{expected MS}_{\text{delivery.cask.test}}$$

and

$$\text{expected value of } F = \frac{(\text{expected MS}_{\text{delivery.cask}})}{(\text{expected MS}_{\text{delivery.cask.test}})} = 1.$$

The F statistic then follows the F distribution with the appropriate degrees of freedom.

- The greater the value of n_E, the greater the power of the F test. For a given value of σ_Γ^2, a greater value of n_E gives a greater expected value of F, and a greater probability that H_0 will be rejected.

Similar considerations apply to the comparison between $\text{MS}_{\text{delivery}}$ and $\text{MS}_{\text{delivery.cask}}$, as follows:

- The variation among casks in the same delivery ($\text{MS}_{\text{delivery.cask}}$) is not influenced by the variation among deliveries (σ_Δ^2).

- However, the variation among deliveries ($\text{MS}_{\text{delivery}}$) is influenced by the variation among casks within each delivery (σ_Γ^2) and among tests within each cask (σ_E^2). This means that even if there is no real variation among deliveries (i.e. if $\sigma_\Delta^2 = 0$), there will still be some variation among the observed delivery means, due to the random effects of casks and/or tests.

- If n_Γ is large, then the influence of σ_Γ^2 on $\text{MS}_{\text{delivery}}$ is small relative to that of σ_Δ^2. In the limiting case, if an infinite number of casks were sampled in each delivery, then their mean would be the true mean value for the delivery, and $\text{MS}_{\text{delivery}}$, which is closely related to the variance among the delivery means, would be determined only by σ_Δ^2.

- These expected MSs justify the F test of $\text{MS}_{\text{delivery}}$ against $\text{MS}_{\text{delivery.cask}}$. If the hypothesis

$$H_0 : \sigma_\Delta^2 = 0 \tag{3.8}$$

is true, then

$$\text{expected MS}_{\text{delivery}} = \text{expected MS}_{\text{delivery.cask}}$$

and

$$\text{expected value of } F = \frac{(\text{expected MS}_{\text{delivery}})}{(\text{expected MS}_{\text{delivery.cask}})} = 1.$$

The F statistic then follows the F distribution with the appropriate degrees of freedom.

- The greater the value of n_Γ, the greater the power of the F test. For a given value of σ^2_Δ, a greater value of n_Γ gives a greater expected value of F, and a greater probability that H_0 will be rejected.

3.4   Estimation of the variance components in the hierarchical random-effects model

The formulae for the expected MSs, together with the observed MSs from the anova table, can be used to obtain estimates, $\hat{\sigma}^2_\Delta$, $\hat{\sigma}^2_\Gamma$ and $\hat{\sigma}^2_E$, of the true variances σ^2_Δ, σ^2_Γ and σ^2_E respectively. Thus

$$\hat{\sigma}^2_E = \text{MS}_{\text{delivery.cask.test}} = 0.6780. \tag{3.9}$$

Similarly,

$$n_E \hat{\sigma}^2_\Gamma + \hat{\sigma}^2_E = \text{MS}_{\text{delivery.cask}} = 17.5453. \tag{3.10}$$

Rearranging Equation 3.10, and substituting the numerical value of n_E and of the estimate $\hat{\sigma}^2_E$ from Equations 3.6 and 3.9, respectively, we obtain

$$\hat{\sigma}^2_\Gamma = \frac{\text{MS}_{\text{delivery.cask}} - \text{MS}_{\text{delivery.cask.test}}}{n_E} = \frac{17.5453 - 0.6780}{2} = 8.43365. \tag{3.11}$$

Similarly,

$$n_\Gamma n_E \hat{\sigma}^2_\Delta + n_E \hat{\sigma}^2_\Gamma + \hat{\sigma}^2_E = \text{MS}_{\text{delivery}} = 27.4892. \tag{3.12}$$

Rearranging Equation 3.12, and substituting the numerical values of n_Γ and n_E and of the estimates $\hat{\sigma}^2_E$ and $\hat{\sigma}^2_\Gamma$ from Equations 3.5, 3.6, 3.9 and 3.11, respectively, we obtain

$$\hat{\sigma}^2_\Delta = \frac{\text{MS}_{\text{delivery}} - 2 \times \hat{\sigma}^2_\Gamma - \hat{\sigma}^2_E}{n_\Gamma n_E} = \frac{27.4892 - 2 \times 8.43365 - 0.6780}{3 \times 2} = 1.657317. \tag{3.13}$$

These variance component estimates can be produced directly by the anova directives BLOCK-STRUCTURE and ANOVA. The statement required is as follows:

```
ANOVA [PRINT = stratumvariances] strength
```

The output of this statement is given below.

Estimated stratum variances

Variate: strength

Stratum	variance	effective d.f.	variance component
delivery	27.49	9.000	1.66
delivery.cask	17.55	20.000	8.43
delivery.cask.test	0.68	30.000	0.68

The variance component estimates can also be produced by the mixed modelling directives. The statements required are as follows:

```
VCOMPONENTS RANDOM = delivery / cask / test
REML [PRINT = model, components] strength
```

Note that as the model contains no fixed-effect terms, the option FIXED in the VCOMPONENTS statement is not set. Similarly, as the F test and Wald test are applicable only to fixed-effect terms, the setting 'Wald' is omitted from the PRINT option in the REML statement. The output of these statements is given below.

REML variance components analysis

Response variate:	strength
Fixed model:	Constant
Random model:	delivery + delivery.cask + delivery.cask.test
Number of units:	60

delivery.cask.test used as residual term.
Sparse algorithm with average information (AI) optimization.

Estimated variance components

Random term	Component	s.e.
delivery	1.6573	2.3494
delivery.cask	8.4337	2.7755

Residual variance model

Term	Model(order)	Parameter	Estimate	s.e.
delivery.cask.test	Identity	Sigma2	0.678	0.1751

The estimates of the variance components for delivery and delivery.cask agree with the values derived from the MSs in the anova. GenStat identifies delivery.cask.test as the residual term (because there is only one observation for each combination of these factors – see Section 2.4), and the estimated variance component for this term is presented as the *residual variance*. It agrees with $MS_{delivery.cask.test}$ in the anova. A standard error (SE) is presented with each estimated variance component. However, whereas estimated means are normally distributed around the true value (provided that the assumptions underlying the analysis are true – see Section 1.10), estimated variances are not: a variance cannot be less than zero, and the distribution of an estimated variance must therefore be asymmetrical. Consequently, the SE of a variance is not as informative as that of a mean. Nevertheless, these values do give a rough indication of the precision with which the variance components have been estimated. For example, we can say that $\hat{\sigma}_{\Delta}^2 = 1.6573 \pm 2.3494$. We note that in this case the SE of the estimate is larger than the estimate itself, which is consistent with the finding that the F value for the effect of deliveries is non-significant. We can also say that $\hat{\sigma}_{\Gamma}^2 = 8.4337 \pm 2.7755$, and can be fairly confident that the true variance among casks within each delivery lies somewhere in the range indicated. Likewise, $\hat{\sigma}_{E}^2 = 0.678 \pm 0.1751$.

3.5   Design of an optimum strategy for future sampling

Knowledge of the magnitude of each variance component can be used to design an optimum strategy for future sampling. This will be of importance if, for example, it is desired to estimate the mean paste strength of future deliveries to a certain degree of precision, without incurring unnecessary cost. The investigator can choose the number of casks to sample from each delivery (n_{Γ}), and the number of tests to perform on each cask (n_E). Should he or she perform more tests on fewer casks, or fewer tests on more casks, in order to deploy a given amount of effort to best effect?

In order to answer this question, the cost of sampling is expressed as a function of three components, which, it is assumed, can be treated as constants. The first is the fixed costs, which are the same regardless of the number of casks sampled or the number of tests performed on each cask, and which do not concern us further. The remaining two components are

- c_{Γ}, the average cost of obtaining samples from each cask, after the fixed costs have been met, regardless of the number of tests performed on the cask and

- c_E, the average cost of each test on a cask, after the cost of obtaining samples from it has been met.

It is assumed that $c_{\Gamma} \geq 0$ and $c_E \geq 0$. The total cost of sampling is then given by

$$k = c_{\Gamma} n_{\Gamma} + c_E n_{\Gamma} n_E = n_{\Gamma}(c_{\Gamma} + c_E n_E). \tag{3.14}$$

The true mean paste strength in the ith delivery is

$$\mu_i = \mu + \delta_i,$$

and an estimate of this is given by

$$\hat{\mu}_i = \frac{\displaystyle\sum_{j=1}^{n_{\Gamma}} \sum_{k=1}^{n_E} y_{ijk}}{n_{\Gamma} n_E}. \tag{3.15}$$

Note that this is the ordinary mean of the observed values, not the 'shrunk estimate' introduced in Section 5.2. The SE of this estimate is given by

$$
SE_{\hat{\mu}_i} = \sqrt{\frac{\sigma_\Gamma^2}{n_\Gamma} + \frac{\sigma_E^2}{n_\Gamma n_E}} = \sqrt{\frac{1}{n_\Gamma}\left(\sigma_\Gamma^2 + \frac{\sigma_E^2}{n_E}\right)}.
\tag{3.16}
$$

More will be said about the calculation of the SE of a parameter estimate, and the meaning of this statistic, in Chapter 4. For the moment, it is sufficient to note that $SE_{\hat{\mu}_i}$ indicates the precision with which $\hat{\mu}_i$ estimates μ_i. $SE_{\hat{\mu}_i}$ is small, and the precision of $\hat{\mu}_i$ is high, when

- the amount of variation among individual observations, determined by the variance components σ_Γ^2 and σ_E^2, is small and
- the sample size, indicated by n_Γ and n_E, is large.

From Equation 3.14,

$$
n_\Gamma = \frac{k}{c_\Gamma + c_E n_E}.
\tag{3.17}
$$

That is, if we stipulate that the total cost k is to be kept constant, the investigator is not able to vary n_Γ and n_E independently: any choice of n_E enforces a corresponding value of n_Γ. Substituting the expression for n_Γ from Equation 3.17 into Equation 3.16, we obtain

$$
SE_{\hat{\mu}_i} = \sqrt{\frac{c_\Gamma + c_E n_E}{k}\left(\sigma_\Gamma^2 + \frac{\sigma_E^2}{n_E}\right)},
\tag{3.18}
$$

and n_E is the only term in this expression that can be manipulated by the investigator. The problem then is to find the value of n_E that minimizes the value of $SE_{\hat{\mu}_i}$, and it can be shown that this is given by

$$
n_E = \sqrt{\frac{c_\Gamma \sigma_E^2}{c_E \sigma_\Gamma^2}}
\tag{3.19}
$$

(Snedecor and Cochran, 1989, Section 21.10, pp. 447–450). That is, the number of tests made on each cask should be large, and the number of casks sampled should be correspondingly small, if

- the cost of sampling each cask (c_Γ) is high;
- the cost of each test on a cask already sampled (c_E) is low;
- the variation among tests within each cask (σ_E^2) is large;
- the variation among casks within each delivery (σ_Γ^2) is small.

For example, suppose the cost of sampling each cask is small, say $0.3 \times$ the cost of each test within a cask already sampled. This cost might represent a few moments of the sampler's time, to open the cask. Then

$$
\frac{c_\Gamma}{c_E} = 0.3.
$$

Substituting this value and the estimates $\hat{\sigma}_E^2$ and $\hat{\sigma}_\Gamma^2$ obtained in Equations 3.9 and 3.11, respectively, into Equation 3.19, we obtain

$$n_E = \sqrt{\frac{0.3 \times 0.678}{8.4337}} = 0.0241.$$

That is, there is no justification for performing more than one test per cask. On the other hand, if the cost of sampling each cask is large (e.g. if the whole contents of the cask is rendered unusable by sampling), say

$$\frac{c_\Gamma}{c_E} = 30,$$

then

$$n_E = \sqrt{\frac{30 \times 0.678}{8.4337}} = 2.41,$$

and two or three tests should be performed on each cask.

This approach can be extended to assay processes with more than two stages (Snedecor and Cochran, 1967, Section 17.12, pp. 533–534). For example, suppose that the mean strength of the paste in a future series of deliveries is to be estimated. The investigator must now decide not only how many tests to perform on each cask, but also how many casks to sample within each delivery. There is now an additional component to the cost of sampling, namely

- c_Δ, the cost of obtaining samples from each delivery.

It is assumed that $c_\Delta \geq 0$. The total cost of sampling is then given by

$$k = c_\Delta n_\Delta + c_\Gamma n_\Delta n_\Gamma + c_E n_\Delta n_\Gamma n_E = n_\Delta(c_\Delta + (c_\Gamma + c_E n_E)n_\Gamma), \qquad (3.20)$$

where

n_Δ = number of deliveries sampled,

which is an equation of the same form as Equation 3.14. Let

$$\hat{\mu} = \frac{\displaystyle\sum_{i=1}^{n_\Delta}\sum_{j=1}^{n_\Gamma}\sum_{k=1}^{n_E} y_{ijk}}{n_\Delta n_\Gamma n_E} = \text{estimated grand mean strength of over deliveries, casks and tests.}$$

$$(3.21)$$

Then

$$\mathrm{SE}_{\hat{\mu}} = \sqrt{\frac{\sigma_\Delta^2}{n_\Delta} + \frac{\sigma_\Gamma^2}{n_\Delta n_\Gamma} + \frac{\sigma_E^2}{n_\Delta n_\Gamma n_E}} = \sqrt{\frac{1}{n_\Delta}\left(\sigma_\Delta^2 + \frac{\left(\sigma_\Gamma^2 + \frac{\sigma_E^2}{n_E}\right)}{n_\Gamma}\right)}, \qquad (3.22)$$

which is an equation of the same form as Equation 3.16, and the value of $\mathrm{SE}_{\hat{\mu}}$ is minimized when

$$n_\Gamma = \sqrt{\frac{c_\Delta\left(\sigma_\Gamma^2 + \frac{\sigma_E^2}{n_E}\right)}{(c_\Gamma + c_E n_E)\sigma_\Delta^2}}, \qquad (3.23)$$

which is an equation of the same form as Equation 3.19. The argument concerning the number of tests per cask is unchanged, and when the value of n_E has been decided as described above, it can be substituted from Equation 3.17 into Equation 3.23. This equation shows that the number of casks to be sampled in each delivery should be large, and the number of deliveries to be sampled correspondingly small, if

- the cost of sampling each delivery (c_Δ) is high;
- the variation among deliveries (σ_Δ^2) is low.

For example, suppose that the cost of sampling each cask is $0.3 \times$ the cost of an additional test within a cask already sampled, as discussed above. In practice, the number of tests performed within each cask should then be set to the lowest value physically possible,

$$n_E = 1.$$

Suppose further that c_Δ, the cost of sampling each delivery, is $3.5 \times$ the cost of each test. This might reflect the cost of sending a sampler to attend the delivery. Substituting these values, and the other values obtained earlier, into Equation 3.23 gives

$$n_\Gamma = \sqrt{\frac{3.5 \times \left(17.5453 + \frac{0.678}{1}\right)}{(0.3 + 1 \times 1) \times 27.4892}} = 1.7848,$$

indicating that two casks should be sampled from each delivery.

3.6   Use of R to analyse the hierarchical three-stage assay process

The following commands import the data, fit Model 3.1 and produce the resulting anova:

```
rm(list = ls())
pastestrength <- read.table(
    "IMM edn 2\\Ch 3\\paste strength.txt",
    header=TRUE)
attach(pastestrength)
delivery <- factor(delivery)
cask <- factor(cask)
test <- factor(test)
pastestrength.model1aov <- aov(strength ~
    Error(delivery / cask / test))
summary(pastestrength.model1aov)
```

The output from these commands is as follows:

```
Error: delivery
          Df Sum Sq Mean Sq F value Pr(>F)
Residuals  9  247.4   27.49

Error: delivery:cask
          Df Sum Sq Mean Sq F value Pr(>F)
Residuals 20  350.9   17.55

Error: delivery:cask:test
          Df Sum Sq Mean Sq F value Pr(>F)
Residuals 30  20.34   0.678
```

The degrees of freedom, sum of squares and MS for each term are the same as those produced by GenStat, and the F tests comparing the MS for each term with that for the term below can be conducted by hand.

The following commands perform a mixed-model analysis on the same model:

```
library(nlme)
pastestrength.model1lme <- lme(strength ~ 1,
    random = ~ 1|delivery / cask)
summary(pastestrength.model1lme)
```

The term 'delivery.cask.test' is equivalent to the residual term, and therefore 'test' is not mentioned in the model (see Section 2.8). The output of these statements is as follows:

```
Linear mixed-effects model fit by REML
 Data: NULL
       AIC      BIC    logLik
  254.9907 263.3009 -123.4954

Random effects:
 Formula: ~1 | delivery
          (Intercept)
StdDev:     1.287347

 Formula: ~1 | cask %in% delivery
          (Intercept)   Residual
StdDev:      2.90406 0.8234126
```

```
Fixed effects: strength ~ 1
              Value Std.Error DF  t-value p-value
(Intercept) 60.05333 0.6768643 30 88.72285       0

Standardized Within-Group Residuals:
        Min              Q1             Med              Q3             Max
-1.479769086 -0.515579619  0.009496961  0.471991368  1.389687052

Number of Observations: 60
Number of Groups:
        delivery cask %in% delivery
            10                    30
```

The standard deviation (StdDev or SD) given by R for the model component '~1 | delivery' is the square root of the variance component for the term 'delivery': thus

$$1.287347^2 = 1.6573 = \hat{\sigma}_\Delta^2.$$

Similarly, the SD for model component '~1 | cask %in% delivery' is the square root of the variance component for the term 'delivery.cask': thus

$$2.90406^2 = 8.4337 = \hat{\sigma}_\Gamma^2.$$

The residual SD is the square root of the variance component for test within cask: thus

$$0.8234126^2 = 0.678 = \hat{\sigma}_E^2.$$

These values agree with those given by GenStat (Section 3.4), allowing for rounding error.

3.7   Use of SAS to analyse the hierarchical three-stage assay process

The following SAS statements import the data, fit Model 3.1 and produce the resulting anova:

```
PROC IMPORT OUT = pastestrength DBMS = EXCELCS REPLACE
    DATAFILE = "&pathname.\IMM edn 2\Ch 3\paste strength.xlsx";
    SHEET = "for SAS";
RUN;

ODS RTF;
PROC ANOVA;
    CLASS delivery cask test;
    MODEL strength = delivery cask(delivery);
    TEST H = delivery E = cask(delivery);
RUN;
ODS RTF CLOSE;
```

The model term 'cask(delivery)' represents the effect of 'cask' within each level of 'delivery'. The effect of test within each cask within each delivery is given by the residual term, and this term is not specified explicitly in the model. The test of 'delivery' against 'cask(delivery)' is not performed by default, and is specified explicitly in the TEST statement.

The output from PROC ANOVA is as follows:

Source	DF	Sum of squares	Mean square	F value	Pr > F
Model	29	598.3093333	20.6313563	30.43	<0.0001
Error	30	20.3400000	0.6780000		
Corrected total	59	618.6493333			

R-square	Coeff Var	Root MSE	strength Mean
0.967122	1.371127	0.823408	60.05333

Source	DF	Anova SS	Mean square	F value	Pr > F
delivery	9	247.4026667	27.4891852	40.54	<0.0001
cask(delivery)	20	350.9066667	17.5453333	25.88	<0.0001

Tests of hypotheses using the Anova MS for cask(delivery) as an error term					
Source	DF	Anova SS	Mean square	F value	Pr > F
delivery	9	247.4026667	27.4891852	1.57	0.1926

The degrees of freedom, sum of squares and MS for the 'Error' term in the first anova agree with the values produced by GenStat for 'delivery.cask.test'. Those in the second anova table agree with the values produced by GenStat for 'delivery' and 'delivery.cask', as does the F test comparing $MS_{delivery.cask}$ with $MS_{delivery.cask.test}$ in this anova table ($F = 25.88$). However, the F test for 'delivery' in this anova table is obtained by comparing $MS_{delivery}$ with $MS_{delivery.cask.test}$, and is inappropriate unless delivery.cask is regarded as a fixed-effect term. Even then, if the effect of this term is found to be significant relative to the effect of delivery.cask.test, the interpretation of the main effects of delivery presents problems of the kind encountered in the experiment to evaluate the saltiness of brands of ravioli (Section 2.2). The correct F test for this term, obtained by comparing $MS_{delivery}$ with $MS_{delivery.cask}$, is given in the second anova table ($F = 1.57$).

The following commands perform a mixed-model analysis on the same model:

```
ODS RTF;
PROC MIXED ASYCOV NOBOUND;
   CLASS delivery cask test;
   MODEL strength = /CHISQ DDFM = KR HTYPE = 1;
   RANDOM delivery cask(delivery);
RUN;
ODS RTF CLOSE;
```

Again the term 'cask(delivery)' represents the effect of 'cask' within each level of 'delivery', and the effect of test within each cask within each delivery is given by the residual term.

Part of the output from PROC MIXED is as follows:

Convergence criteria met.

Covariance parameter estimates	
Cov Parm	**Estimate**
delivery	1.6573
cask(delivery)	8.4337
Residual	0.6780

Asymptotic covariance matrix of estimates				
Row	**Cov Parm**	**CovP1**	**CovP2**	**CovP3**
1	delivery	5.5196	−2.5653	
2	cask(delivery)	−2.5653	7.7036	−0.01532
3	Residual		−0.01532	0.03065

The variance component estimates are given in the table headed 'Covariance parameter estimates', and agree with those given by GenStat. These estimates are not mutually independent: an increase in the value of one could be compensated for by a decrease in the value of another, and the table headed 'Asymptotic covariance matrix of estimates' gives estimates of these covariances between the variance component estimates. The values on the leading diagonal of this table correspond to the SEs of the variance component estimates produced by GenStat. Thus

- SE(var. comp. for 'delivery') $= \sqrt{\text{Cov Parm(Row 1, CovP1)}} = \sqrt{5.5196} = 2.3494$

- SE(var. comp. for 'delivery.cask') $= \sqrt{\text{Cov Parm(Row 2, CovP2)}} = \sqrt{7.7036} = 2.7755$

- SE(var. comp. for 'delivery.cask.test') $= \sqrt{\text{Cov Parm(Row 3, CovP3)}} = \sqrt{0.03065} = 0.1751$.

and the values produced by SAS approximately agree with those produced by GenStat.

3.8 Genetic variation: A crop field trial with an unbalanced design

There are many other situations, besides the optimization of assay procedures, in which it is useful to estimate the magnitude of the different variance components that contribute to a response variable. In particular, variance components are routinely estimated in studies related to plant and animal breeding. The amount of genetic variation among the individuals of a species of crop plant or domesticated animal can be compared to the amount of variation due to non-genetic causes in a ratio called the *heritability*. Estimates of these two variance components can also be used to give a prediction of the potential for genetic improvement.

We will explore this issue in data from a field trial of barley breeding lines (reproduced by the kind permission of Reg Lance, Department of Agriculture, Western Australia). The lines studied were derived from a cross between two parent varieties, 'Chebec' and 'Harrington'. They were 'doubled-haploid' lines, which means they were obtained by a laboratory technique that ensures that

- all plants within the same breeding line are genetically identical and

- each plant, when self-fertilized, will 'breed true', that is, will produce progeny genetically identical to itself.

These features improve the precision with which genetic variation among the lines can be estimated. The trial considered here, conducted in Western Australia in 1995, was arranged in two randomized blocks. Within each block, each line occupied a single rectangular field plot. All lines were present in Block 1, but due to limited seed stocks, some were absent from Block 2. The grain yield (g/m^2) was measured in each field plot. The data obtained are displayed in the spreadsheet in Table 3.4.

A natural model for these data (Model 3.24) is

$$y_{ij} = \mu + \delta_i + \varepsilon_{ij} + \phi_{k|ij} \tag{3.24}$$

where

y_{ij} = the grain yield of the *j*th plot in the *i*th block,
μ = the grand mean value of grain yield,
δ_i = the effect of the *i*th block,
ε_{ij} = the effect of the *j*th plot in the *i*th block,
$\phi_{k|ij}$ = the effect of the *k*th breeding line, being the line sown in the *ij*th block.plot combination.

It is also natural to treat the block and residual terms as random-effect terms: that is, to assume that the δ_i are independent values of a random variable Δ, such that

$$\Delta \sim N(0, \sigma_\Delta^2), \tag{3.25}$$

and the ε_{ij} are independent values of a random variable E, such that

$$E \sim N(0, \sigma_E^2). \tag{3.26}$$

The following GenStat statements might be used to read these data and to attempt to analyse them according to Model 3.24:

```
IMPORT 'IMM edn 2\\Ch 3\\barley progeny.xlsx'; SHEET = 'Sheet1'
BLOCKSTRUCTURE block
TREATMENTSTRUCTURE line
ANOVA [FPROBABILITY = yes; PRINT = aovtable, information] \
    yield_g_m2
```

Table 3.4　Yields of doubled-haploid breeding lines of barley, assessed in a field trial with an unbalanced randomized block design.

	A	B	C		A	B	C
1	block!	line!	yield_g_m2	37	1	12	436.59
2	1	2	483.33	38	1	24	671.22
3	1	39	145.84	39	1	51	692.55
4	1	41	321.84	40	1	6	849.66
5	1	4	719.14	41	1	14	910.76
6	1	79	317.63	42	1	80	487.86
7	1	76	344.48	43	1	46	724.01
8	1	78	260.02	44	1	42	793.43
9	1	35	374.28	45	1	47	192.43
10	1	25	428.61	46	1	37	895.30
11	1	67	407.25	47	1	27	731.87
12	1	17	551.84	48	1	54	809.41
13	1	30	353.29	49	1	32	669.16
14	1	73	355.30	50	1	70	996.19
15	1	64	647.92	51	1	9	774.84
16	1	72	165.76	52	1	8	636.45
17	1	44	517.52	53	1	77	357.94
18	1	18	366.24	54	1	22	340.65
19	1	82	251.30	55	1	11	644.83
20	1	29	606.37	56	1	58	521.67
21	1	23	605.75	57	1	19	622.72
22	1	38	641.42	58	1	49	830.57
23	1	31	166.75	59	1	28	679.92
24	1	40	410.87	60	1	50	721.13
25	1	57	181.97	61	1	74	489.31
26	1	63	562.90	62	1	7	907.38
27	1	83	280.44	63	1	68	325.96
28	1	36	800.35	64	1	59	553.46
29	1	66	687.92	65	1	16	210.71
30	1	10	764.88	66	1	33	770.23
31	1	13	541.15	67	1	53	559.14
32	1	62	730.48	68	1	75	617.33
33	1	81	315.63	69	1	20	632.46
34	1	43	678.46	70	1	61	611.52
35	1	26	580.22	71	1	1	717.78
36	1	45	519.88	72	1	71	595.86

Table 3.4 (*continued*)

	A	B	C		A	B	C
73	1	55	555.29	109	2	31	226.12
74	1	69	467.24	110	2	71	569.61
75	1	60	572.90	111	2	65	713.43
76	1	21	514.62	112	2	7	820.08
77	1	65	818.74	113	2	58	435.34
78	1	34	673.43	114	2	69	378.89
79	1	5	798.99	115	2	34	639.11
80	1	48	786.06	116	2	11	516.84
81	1	56	522.61	117	2	44	873.18
82	1	3	873.04	118	2	49	823.25
83	1	15	600.06	119	2	54	859.36
84	1	52	603.04	120	2	72	258.59
85	2	64	681.64	121	2	9	587.07
86	2	27	762.42	122	2	23	817.51
87	2	29	932.33	123	2	48	645.10
88	2	68	385.47	124	2	59	634.58
89	2	16	240.00	125	2	78	260.26
90	2	33	846.85	126	2	61	472.44
91	2	67	702.58	127	2	45	575.76
92	2	51	746.11	128	2	18	265.37
93	2	66	846.05	129	2	19	423.76
94	2	50	885.67	130	2	35	554.69
95	2	36	1054.70	131	2	6	755.05
96	2	80	478.12	132	2	40	568.31
97	2	17	959.25	133	2	8	299.92
98	2	26	639.39	134	2	1	591.19
99	2	63	755.90	135	2	10	756.63
100	2	74	551.41	136	2	42	552.53
101	2	56	435.62	137	2	46	627.25
102	2	81	303.72	138	2	37	552.70
103	2	43	836.82	139	2	60	284.72
104	2	79	439.17	140	2	38	540.68
105	2	14	934.72	141	2	13	475.10
106	2	32	836.95	142	2	24	463.22
107	2	70	904.90	143	2	39	212.66
108	2	41	538.00				

Source: Reproduced by kind permission of Reg Lance, Department of Agriculture, Western Australia.

However, the ANOVA statement produces the following output:

Fault 11, code AN 1, statement 1 on line 7

Command: ANOVA [FPROBABILITY = yes; PRINT = aovtable, information] yield_g_m2
Design unbalanced – cannot be analysed by ANOVA.
Model term line (non-orthogonal to term block) is unbalanced, in the block.*Units* stratum.
Note, though, that the terms are nearly orthogonal (average efficiency factor = 1.0000). So it
may be worth checking their factor values if you were expecting the design to be balanced.

Because some breeding lines are present in both blocks but others only in Block 1, the design
is unbalanced and analysis of variance cannot proceed.

3.9   Production of a balanced experimental design by 'padding' with missing values

Such an unbalanced design can be analysed by mixed modelling (see Section 3.10). However,
a less rigorous method of overcoming the problem of imbalance – or at least disguising it – is
to 'pad' the spreadsheet with rows representing the missing breeding lines in Block 2. A missing value (represented by an asterisk (*)) is inserted for 'yield_g_m2' in each row added. To
facilitate this, the spreadsheet is de-randomized and sorted by the identifying numbers of the
breeding lines. The padded, de-randomized spreadsheet is as shown in Table 3.5.

This spreadsheet is stored as 'Sheet2' in the same workbook as the previous one. The same
statements are used to analyse the padded data, with the substitution of 'Sheet2' for 'Sheet1'
in the IMPORT statement. The output of the ANOVA statement is now as follows:

Analysis of variance

Variate: yield_g_m2

Source of variation	d.f.	m.v.	s.s.	m.s.	v.r.	F pr.
block stratum	1		7015	7015	0.52	
block.*Units* stratum						
line	82		6351920	77462	5.78	<0.001
Residual	58	(24)	777747	13409		
Total	141	(24)	6179665			

Message: the following units have large residuals.

block 1 *units* 8	174.77	s.e. 68.45
block 1 *units* 17	−197.20	s.e. 68.45
block 1 *units* 37	177.80	s.e. 68.45
block 1 *units* 44	−171.33	s.e. 68.45
block 2 *units* 8	−174.77	s.e. 68.45
block 2 *units* 17	197.20	s.e. 68.45
block 2 *units* 37	−177.80	s.e. 68.45
block 2 *units* 44	171.33	s.e. 68.45

Table 3.5 Spreadsheet holding yields of doubled-haploid breeding lines of barley, 'padded' to produce a balanced randomized block design.

	A	B	C		A	B	C		A	B	C
1	block!	line!	yield_g_m2	38	1	37	895.30	75	1	74	489.31
2	1	1	717.78	39	1	38	641.42	76	1	75	617.33
3	1	2	483.33	40	1	39	145.84	77	1	76	344.48
4	1	3	873.04	41	1	40	410.87	78	1	77	357.94
5	1	4	719.14	42	1	41	321.84	79	1	78	260.02
6	1	5	798.99	43	1	42	793.43	80	1	79	317.63
7	1	6	849.66	44	1	43	678.46	81	1	80	487.86
8	1	7	907.38	45	1	44	517.52	82	1	81	315.63
9	1	8	636.45	46	1	45	519.88	83	1	82	251.30
10	1	9	774.84	47	1	46	724.01	84	1	83	280.44
11	1	10	764.88	48	1	47	192.43	85	2	1	591.19
12	1	11	644.83	49	1	48	786.06	86	2	2	*
13	1	12	436.59	50	1	49	830.57	87	2	3	*
14	1	13	541.15	51	1	50	721.13	88	2	4	*
15	1	14	910.76	52	1	51	692.55	89	2	5	*
16	1	15	600.06	53	1	52	603.04	90	2	6	755.05
17	1	16	210.71	54	1	53	559.14	91	2	7	820.08
18	1	17	551.84	55	1	54	809.41	92	2	8	299.92
19	1	18	366.24	56	1	55	555.29	93	2	9	587.07
20	1	19	622.72	57	1	56	522.61	94	2	10	756.63
21	1	20	632.46	58	1	57	181.97	95	2	11	516.84
22	1	21	514.62	59	1	58	521.67	96	2	12	*
23	1	22	340.65	60	1	59	553.46	97	2	13	475.10
24	1	23	605.75	61	1	60	572.90	98	2	14	934.72
25	1	24	671.22	62	1	61	611.52	99	2	15	*
26	1	25	428.61	63	1	62	730.48	100	2	16	240.00
27	1	26	580.22	64	1	63	562.90	101	2	17	959.25
28	1	27	731.87	65	1	64	647.92	102	2	18	265.37
29	1	28	679.92	66	1	65	818.74	103	2	19	423.76
30	1	29	606.37	67	1	66	687.92	104	2	20	*
31	1	30	353.29	68	1	67	407.25	105	2	21	*
32	1	31	166.75	69	1	68	325.96	106	2	22	*
33	1	32	669.16	70	1	69	467.24	107	2	23	817.51
34	1	33	770.23	71	1	70	996.19	108	2	24	463.22
35	1	34	673.43	72	1	71	595.86	109	2	25	*
36	1	35	374.28	73	1	72	165.76	110	2	26	639.39
37	1	36	800.35	74	1	73	355.30	111	2	27	762.42

Table 3.5   (*continued*)

	A	B	C		A	B	C		A	B	C
112	2	28	*	131	2	47	*	150	2	66	846.05
113	2	29	932.33	132	2	48	645.10	151	2	67	702.58
114	2	30	*	133	2	49	823.25	152	2	68	385.47
115	2	31	226.12	134	2	50	885.67	153	2	69	378.89
116	2	32	836.95	135	2	51	746.11	154	2	70	904.90
117	2	33	846.85	136	2	52	*	155	2	71	569.61
118	2	34	639.11	137	2	53	*	156	2	72	258.59
119	2	35	554.69	138	2	54	859.36	157	2	73	*
120	2	36	1054.70	139	2	55	*	158	2	74	551.41
121	2	37	552.70	140	2	56	435.62	159	2	75	*
122	2	38	540.68	141	2	57	*	160	2	76	*
123	2	39	212.66	142	2	58	435.34	161	2	77	*
124	2	40	568.31	143	2	59	634.58	162	2	78	260.26
125	2	41	538.00	144	2	60	284.72	163	2	79	439.17
126	2	42	552.53	145	2	61	472.44	164	2	80	478.12
127	2	43	836.82	146	2	62	*	165	2	81	303.72
128	2	44	873.18	147	2	63	755.90	166	2	82	*
129	2	45	575.76	148	2	64	681.64	167	2	83	*
130	2	46	627.25	149	2	65	713.43				

Source: Reproduced by kind permission of Reg Lance, Department of Agriculture, Western Australia.

The option setting 'PRINT = aovtable, information' in the ANOVA statement specifies that only the anova table and summary information such as details of any large residuals are to be presented. In particular, the tables of treatment means are not to be included in the output: they are voluminous and are not needed here.

This anova, like that of the split-plot design presented in Section 2.2, contains a stratum for each random-effect term. However, a slight difference in the way these strata have been specified in the present case should be noted. The term 'block' was explicitly mentioned in the BLOCKSTRUCTURE statement (Section 3.8), but the term 'block.*Units*' was not. It relates to the variation among individual values (referred to as *units*) within each block. There was no implicitly defined term like this in the anova of the split-plot design, because in the specification of that analysis, the units (servings) were explicitly mentioned in the BLOCKSTRUCTURE statement. In the present case, a factor named 'plot' might be defined to number the plots within each block. The model in the BLOCKSTRUCTURE statement could then be modified to 'block/plot', and the 'block.*Units*' stratum would be re-labelled as 'block.plot'. The anova would be numerically unchanged.

The column holding the degrees of freedom in the anova table contains information on the number of missing values (abbreviated to m.v.). The data set comprises 83 breeding lines in two blocks, so if there were no missing values, the total degrees of freedom would be given by

$$DF_{total} = 83 \times 2 - 1 = 165.$$

However, there are 24 missing values, as noted in brackets in the 'Total' row of the anova, so

$$DF_{total} = 165 - 24 = 141.$$

How should the missing degrees of freedom be distributed between the terms 'line' and 'Residual'? There is no breeding line for which all observations are missing. Therefore DF_{line} is unaffected by the missing values, and the residual degrees of freedom are reduced from

$$DF_{Residual} = \text{No. of units} - 1 - DF_{line} - DF_{block} = 166 - 1 - 82 - 1 = 82$$

to

$$DF_{Residual} = \text{No. of units} - 1 - DF_{line} - DF_{block} - \text{No. of missing values}$$
$$= 166 - 1 - 82 - 1 - 24 = 58.$$

Estimates of these missing values are obtained in order to proceed with the analysis, and the resulting anova is an approximation. This is revealed by the fact that the sums of squares for the various terms do not add up to the total sum of squares:

$$7015 + 6351920 + 777747 = 7136682 \neq 6179665.$$

The appropriate null hypothesis here is that there is no 'real', consistent variation in grain yield among the breeding lines: that is, all the observed variation is due to the same sources as that among observations on the same line. On this hypothesis the expected value of the F statistic (variance ratio, v.r. – see Section 1.2) for the term 'line' is 1. The observed value, 5.78, is considerably larger than this, and the associated p-value is small (<0.001), indicating that real variation is present – provided that the assumptions that underlie the analysis of variance are correct. Some observations are noted as having large residual values, which may cast doubt on these assumptions: however, we will postpone a fuller exploration of the residuals until we have arrived at an analysis that can be applied to the original, unbalanced data set. We will, however, take a closer look at this F statistic, as we will shortly be comparing it with other significance tests of the same null hypothesis (Sections 3.11–3.13). Values of F that provide significant evidence against the null hypothesis lie in the hatched region of Figure 3.2 – the upper tail of the F distribution. Values of F well below 1 are also extreme and improbable, lying in the lower tail of the distribution. However, they do not provide evidence against the null hypothesis: any variation among the breeding lines will tend to *increase* the value of F.

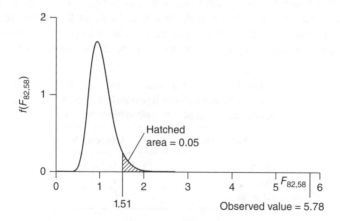

Figure 3.2 Distribution of the variable $F_{82,58}$, showing the critical value and critical region for significance at the 5% level.

3.10 Specification of a treatment term as a random-effect term: The use of mixed-model analysis to analyse an unbalanced data set

The cross Chebec × Harrington could produce many other progeny lines besides those studied here, and the lines in this field trial may reasonably be regarded as a random sample from this population of potential lines. Thus it is reasonable to specify 'line' as a random-effect term – that is, to assume that in Model 3.24, the ϕ_k are independent values of a random variable Φ, such that

$$\Phi \sim N(0, \sigma_G^2).\tag{3.27}$$

The variance component σ_G^2 represents the genetic variation among the lines and is given this symbol, in preference to σ_Φ^2, to connect it to the genetics literature. It can be estimated by considering the expected MS for each term in the analysis of variance. These are as shown in Table 3.6. Thus (following the method presented in Section 3.4),

$$MS_{block} = 83\hat{\sigma}_\Delta^2 + \hat{\sigma}_E^2 = 7015,$$

$$MS_{line} = 2\hat{\sigma}_G^2 + \hat{\sigma}_E^2 = 77462$$

and

$$MS_{Residual} = \hat{\sigma}_E^2 = 13409.$$

Rearranging these expressions, and substituting for $\hat{\sigma}_E^2$, we obtain

$$\hat{\sigma}_\Delta^2 = \frac{MS_{block} - MS_{Residual}}{83} = \frac{7015 - 13409}{83} = -77.0361$$

$$\hat{\sigma}_G^2 = \frac{MS_{line} - MS_{Residual}}{2} = \frac{77462 - 13409}{2} = 32026.5.$$

The estimate of the variance component due to blocks is negative and small compared with the other components. We may decide that the best estimate of this component is zero, that is, that there is no real difference between the mean grain yield values in the two blocks: that is, we may place a *constraint* upon the variance component estimate. The interpretation of

Table 3.6 Expected mean squares of the terms in the anova of yields of breeding lines of barley in a balanced randomized block design.

Source of variation	Expected MS
block	$83\sigma_\Delta^2 + \sigma_E^2$
line	$2\sigma_G^2 + \sigma_E^2$
Residual	σ_E^2

negative variance component estimates is discussed further in Section 3.12. The estimate of the variance due to breeding lines is about double the residual variance.

So far, the random-effect terms that we have encountered have been treated as block terms: when data have been analysed using anova directives, they have been included in the model in the BLOCKSTRUCTURE statement. However, it is not appropriate to treat 'line' as a block term. The variable Φ is not a nuisance variable: we are interested in knowing whether the variation among breeding lines is real (i.e. we are interested in testing its significance) and in estimating its variance component σ_G^2. In due course (Section 5.1) we shall want to estimate the mean yield of each breeding line. The proper place for 'line' is among the treatment terms. GenStat's anova directives require that random-effect terms be treated as block terms, but the mixed modelling directives are not subject to this constraint. Moreover, unlike the anova directives, they do not require a balanced experimental design: hence they can be used on the original, 'unpadded' data in the spreadsheet stored as 'Sheet1'. Thus the following statements will perform the mixed-model analysis that corresponds to the anova in Section 3.9, except that the unpadded data are used and 'line' is specified as a random-effect term:

```
IMPORT 'IMM edn 2\\Ch 3\\barley progeny.xlsx'; SHEET = 'Sheet1'
VCOMPONENTS RANDOM = block + line
REML [PRINT = model, components, deviance, means; \
   PTERMS = 'constant'] yield_g_m2
```

The option setting 'PTERMS = 'constant'' specifies that the estimated value of the constant (the overall mean) is to be printed, but not the means for levels of 'block' or 'line'.

The output of the REML statement above is as follows:

REML variance components analysis

Response variate:	yield_g_m2
Fixed model:	Constant
Random model:	block + line
Number of units:	142

Residual term has been added to model.
Sparse algorithm with AI optimization.

Estimated variance components

Random term	Component	s.e.
block	15	325
line	30645	6242

Residual variance model

Term	Model(order)	Parameter	Estimate	s.e.
Residual	Identity	Sigma2	13222	2431

Deviance: −2*Log-Likelihood

Deviance   d.f.
1613.24   138

Note: deviance omits constants which depend on fixed model fitted.

Table of predicted means for Constant

572.6   Standard error: 21.84

The estimates of the three variance components, $\hat{\sigma}_\Delta^2 = 15$, $\hat{\sigma}_G^2 = 30645$ and $\hat{\sigma}_E^2 = 13222$, are similar to the corresponding values obtained from the anova on the balanced, 'padded' experimental design, but not identical. As the values from the mixed-model analysis are based on the true, unbalanced experimental design, they are the more reliable.

3.11   Comparison of a variance component estimate with its standard error

As before (Section 3.4), each variance component in the output of the REML statement is accompanied by an SE: for example, the variation among breeding lines is estimated to be $\hat{\sigma}_G^2 = 30645 \pm 6242$. This SE provides an indication of the precision with which the variance component is estimated. It may also provide a tentative indication of whether any variation is really accounted for by the term under consideration: that is, whether the null hypothesis

$$H_0 : \sigma_G^2 = 0 \qquad (3.28)$$

can be rejected. If H_0 is true, the ratio

$$z = \frac{\hat{\sigma}_G^2}{\text{SE}(\hat{\sigma}_G^2)} \qquad (3.29)$$

is expected to have a value of about ± 1, on average. In the present case,

$$z = \frac{30646}{6251} = 4.90.$$

This is considerably larger than the expected value, suggesting that H_0 is false.

If $\hat{\sigma}_G^2$ were an observation of a normally distributed variable, this ratio could be made the basis for a formal significance test. z would then be approximately an observation of a *standard normal variable*: that is, of a variable Z such that

$$Z \sim N(0, 1). \qquad (3.30)$$

(For an account of the normal distribution, see Section 1.2.) Figure 3.3 shows that a value of Z greater than 1.645 provides significant evidence, at the 5% level, against H_0. The observed value z is well above this value, and, if the assumptions underlying this test were fulfilled, it would provide highly significant evidence of variation among the breeding lines. However, $\hat{\sigma}_\Phi^2$

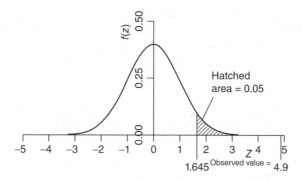

Figure 3.3   The standard normal distribution, showing the critical value and critical region for significance at the 5% level.

is *not* an observation of a normally distributed variable. As noted in Section 3.4, estimates of a variance component do not follow a normal distribution at all reliably, and their SEs should not be used as the basis of a formal significance test. It is nevertheless worth comparing each variance component estimate with its SE informally. In the case of the variation among blocks, the ratio is

$$z = \frac{15}{325} = 0.0462.$$

As this is well below 1, we might consider dropping the term 'block' from the model fitted and proceed to conduct a formal significance test to decide whether this was justified (see Sections 3.12 and 3.13). However, many statisticians would argue that if a term represents part of the design of the experiment – as in the present case – it should be retained, regardless of its significance or the magnitude of its variance component.

   Although an SE is presented for the residual variance component, $\hat{\sigma}_E^2$, it is not appropriate to test the significance of this component, even informally. It is known with certainty that there is some variation due to this source.

3.12   An alternative significance test for variance components

There is an alternative method for testing the significance of variance components, based on the *deviance* presented in the output of the REML statement in Section 3.10. Deviance is a measure of the goodness of fit of the model to the data: the better the fit, the smaller the deviance. By comparing the deviance in the output in Section 3.10 with that obtained when the term under consideration is omitted from the random-effects model, the contribution made by this term to the fit of the model can be assessed. The model fitted above is referred to as the *full model*. It is compared with the *reduced model*, from which the term 'line' is omitted, by the following statements:

```
VRACCUMULATE [PRINT = *]
VCOMPONENTS RANDOM = block
REML [PRINT = *] yield_g_m2
VRACCUMULATE [deviance, dfrandom, change]
```

The first VRACCUMULATE statement initiates the accumulation of summary information on the results from random-effect models in successive REML statements, and the option setting 'PRINT = *' prevents the presentation of any results at this stage. The VCOMPONENTS and REML statements fit the reduced model. The output of the second VRACCUMULATE statement is as follows:

Accumulated summary of REML random models

	Deviance	Random d.f.	Change in deviance	Change in random d.f.	Change chi-prob
block + line	1872.38	3	*	*	*
− line	1912.06	2	39.68	1	0.000

Note: omits constant, −log(det(X′X)), that depends only on the fixed model.

The deviances of the full and reduced models are presented, and a note indicates that a constant has been omitted in the calculation of these deviances. This constant cancels out when the difference between the deviances (the change in deviance) is calculated, *provided that* the deviances were obtained from models with the same fixed-effect terms. This condition is met in the present case. Because the full and reduced models differ by a single random-effect term, the degrees of freedom of these two deviances differ by 1. This is also a requirement, if two models are to be compared by the method described here.

The interpretation of this difference between the deviances requires some explanation. It can be treated approximately as a χ^2 statistic with 1 degree of freedom. A large value of this statistic will be obtained from a data set in which the estimate of the variance component due to lines, $\hat{\sigma}_G^2$, is large and positive. Such a data set will give an F statistic that lies in the upper tail of the distribution presented in Figure 3.2 (Section 3.9), providing evidence against the null hypothesis in Equation 3.28. However, it is also possible to obtain a data set in which $\hat{\sigma}_G^2$ is large *and negative*, and such data will also give a large *positive* difference between the deviances. Negative values of $\hat{\sigma}_G^2$ are obtained from data sets that give an F statistic in the lower tail of the distribution, and as noted earlier (Section 3.9), such an occurrence does *not* provide evidence against the null hypothesis, unless we are willing to consider the possibility that the true value of the variance component is negative. An approximate adjustment for this can be made by halving the probability value associated with the χ^2 statistic. Formal arguments concerning the validity of this adjustment are given by Stram and Lee (1994) and by Crainiceanu and Ruppert (2004).

The deviance for each model is given by $-2 \times \log_e(\text{likelihood})$ (for an explanation of the concept of likelihood, see Section 11.2): hence the difference between the deviances is proportional to the logarithm of the ratio between the likelihoods for the full and reduced models, and this significance test is referred to as the *likelihood ratio test*. In the basic form in which it is presented here, it is valid only when the fixed-effect model is unchanged, and the models to be compared differ by the inclusion or omission of a single random-effect term. Possible extension to the inclusion or omission of several random-effect terms is discussed by Stram and Lee (1994), but is not straightforward. One possibility is to interpret the difference in deviance as a chi-square statistic with degrees of freedom equal to the number of terms omitted. Stram and Lee state that this is common, but 'asymptotically conservative': that is, when applied to large samples it tends to produce an excessively large p-value – hence the halving of the p-value in the single-term case. An analogous correction in the more general case is to interpret the difference in deviance as a mixture of chi-square distributions with different degrees of freedom, but determination of the appropriate proportions in the mixture is difficult. Methods for more general comparisons between mixed models are presented in Section 10.11.

In the present case, we obtain

$$P(\chi_1^2 > 39.68) = 2.99 \times 10^{-10}$$

and hence, approximately,

$$P(\text{deviance}_{\text{reduced model}} - \text{deviance}_{\text{full model}} > 39.68) = \frac{1}{2} \times 2.99 \times 10^{-10} = 1.50 \times 10^{-10}.$$

As before, the variation among breeding lines is found to be highly significant.

The significance of the variation among blocks can also be tested in this way. The following statements obtain the deviances from the full model and from the reduced model excluding the term 'block':

```
VCOMPONENTS RANDOM = block + line
REML [PRINT = *] yield_g_m2
VRACCUMULATE [METHOD = restart; PRINT = *]
VCOMPONENTS RANDOM = line
REML [PRINT = *] yield_g_m2
VRACCUMULATE [deviance, dfrandom, change]
```

The option setting 'METHOD = restart' in the first VRACCUMULATE statement indicates that results are to be accumulated only from the immediately preceding REML statement onward. The output of the second VRACCUMULATE statement is as follows:

Accumulated summary of REML random models

	Deviance	Random d.f.	Change in deviance	Change in random d.f.	Change chi-prob
block + line	1872.38	3	*	*	*
− block	1872.38	2	0.00	1	0.963

Note: omits constant, −log(det(X′X)), that depends only on the fixed model.

There is no difference between the deviances, to the degree of precision presented in the GenStat output. However, the deviance from the reduced model can be saved, and printed with more precision, by the following statements:

```
VKEEP [DEVIANCE = devreduced]
PRINT devreduced; DECIMALS = 6
```

The output of the PRINT statement is as follows:

```
devreduced
1613.243236
```

When the precise deviance from the full model is obtained by the same method, we obtain the result

$$\text{deviance}_{\text{reduced model}} - \text{deviance}_{\text{full model}} = 1613.243236 - 1613.241029 = 0.002207,$$

and

$$P(\chi_1^2 > 0.002207) = 0.9625.$$

Hence

$$P((\text{deviance}_{\text{reduced model}} - \text{deviance}_{\text{full model}}) > 0.002207) = \frac{1}{2} \times 0.9625 = 0.4813.$$

As before, the difference between the blocks is found to be non-significant.

3.13   Comparison among significance tests for variance components

The two types of significance test for random effects introduced above (Sections 3.9 and 3.12), and the informal comparison of a variance component estimate with its SE (Sections 3.11), are summarized and compared in Table 3.7. The p-values for all these tests depend on distributional assumptions that are rarely if ever precisely true, and are better approximations in some cases than others. They are likely to be particularly unreliable for very small values in the upper tail of the distribution, where the relative values given by the numerical approximations used are not very precise. However, these p-values do illustrate some important general comparisons among the tests, as follows:

- Of the three tests compared, the F test from the anova makes the fullest use of the available information and is the most accurate *in the circumstances in which it is valid*. This is reflected by the fact that it gives the highest level of significance (i.e. the smallest value of p) for the effect of 'line'. However, in the present case, the data had to be 'padded' in order to conduct this test, which means that it is not strictly valid, and there are many other experimental designs and types of data set for which the F test is unavailable.

- The ratio z does not provide a reliable significance test, due to the non-normality of the distribution of the estimated variance component. It is nevertheless a useful tool. It does not require the fitting of a reduced model – an important consideration when there are several

Table 3.7   Comparison of different significance tests in the statistical analysis of yields of breeding lines of barley in a randomized block design.

Term	\multicolumn{4}{c}{Basis of test}			
	\multicolumn{2}{c}{F from anova on 'padded' data}	$z = \hat{\sigma}_{\text{term}}^2/\text{SE}(\hat{\sigma}_{\text{term}}^2)$	\multicolumn{2}{c}{$\chi^2 = \text{deviance}_{\text{reduced model}} - \text{deviance}_{\text{full model}}$}
	Test statistic	p		Test statistic	p
block $F_{1,58} = 0.52$		0.4737	0.0462	0.002207	0.4813
line $F_{82,58} = 5.78$		3.43×10^{-11}	4.90	39.68	1.50×10^{-10}

random-effect terms, any of which might be dropped from the model, giving a large number of models to be explored and compared. If a variance component estimate is much smaller than its SE, as in the case of the 'block' term in the present analysis, its significance can be tested formally, and the possibility of dropping it from the model can be considered.

- The likelihood ratio test is 'asymptotic': that is, it does not depend on the degrees of freedom of the model strata concerned, which means that it does not fully take account of the consequences of limited sample size. This makes it less reliable than the F test. Nevertheless, it is a good substitute for the F test when this is not available.

3.14   Inspection of the residual values

The validity of the F test presented in Section 3.9, and of the alternative significance test introduced in Section 3.12, depends on the assumption that the random variable E has a normal distribution (Distribution 3.26). This assumption can be explored by obtaining diagnostic plots of residuals, using the following statement immediately after the fitting of the full model:

```
VPLOT [GRAPHICS=high] fittedvalues, normal, halfnormal, histogram
```

The output of this statement is as shown in Figure 3.4. These plots show slight indications of departure from the assumptions of the analysis. The histogram of residuals is slightly skewed, and its tails are slightly compressed relative to those of a normal distribution: that is, there are *fewer* extreme residual values than would be expected by chance. There is a definite trend in the fitted-value plot, larger fitted values being more commonly associated with positive residuals. This reflects the fact that the residual variance is substantial relative to the other variance components – it is of the same order of magnitude as the 'line' variance component – and hence the shrinkage of the estimated values (which will be discussed in Chapter 5) is substantial (see Section 1.10). The normal and half-normal plots conform well to the assumptions, except for a slight flattening at both ends of the former and the upper end of the latter, again indicating a slight deficit of extreme values.

3.15   Heritability: The prediction of genetic advance under selection

It is often of interest to compare the magnitude of the different components of variance that contribute to the total variance of a response variable. For example, in the genetic improvement of plants and animals, the genetic (i.e. heritable) and environmental (i.e. non-heritable) sources of variation are routinely compared by means of a quantity called the *heritability*. The method for calculating heritability depends on the genetic structure of the population considered – inbred, random-mating, back-crossed, and so on. Here we will demonstrate a method of calculation presented by Allard (1960, Chapter 9, pp. 94–98) which is appropriate to the barley field trial under consideration. The estimated heritability of a trait is defined as

$$h^2 = \frac{\widehat{\sigma}_G^2}{\widehat{\sigma}_P^2} \tag{3.31}$$

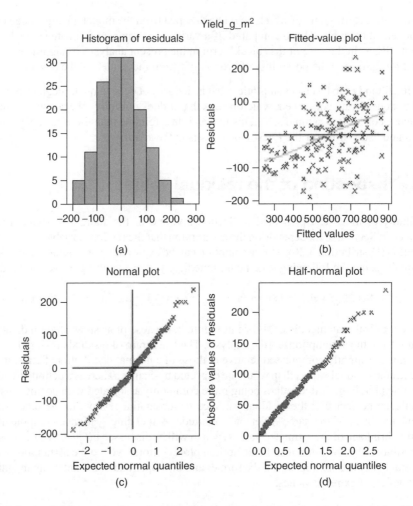

Figure 3.4 (a–d) Diagnostic plots of residuals from the mixed-model analysis of yields of breeding lines of barley in an unbalanced randomized block design.

where

$\hat{\sigma}_G^2$ = genetic component of variance, that is, the part of the variation in the organism's *phenotype* (its observable traits) that is due to its *genotype*, that is, to genetic effects and

$\hat{\sigma}_P^2$ = phenotypic variance, that is, the variance due to the combined effects of genotype and environment.

Note that h^2, not h, stands for heritability. The variance component estimates given by the mixed-model analysis can be used to obtain the phenotypic variance, as follows:

$$\hat{\sigma}_P^2 = \hat{\sigma}_G^2 + \frac{\hat{\sigma}_E^2}{r} \qquad (3.32)$$

where
   r = the number of replications per line,

   whence

$$h^2 = \frac{\hat{\sigma}_G^2}{\hat{\sigma}_G^2 + \frac{\hat{\sigma}_E^2}{r}} \tag{3.33}$$

The presence of r in the formula for heritability perhaps requires explanation, as this coefficient is a feature of the experimental design, not of the sources of variation under investigation. The heritability is a measure of the proportion of variance *among the estimated breeding-line means* that is genetic in origin. The more replicate observations are made on each line (i.e. the higher the value of r), the more reliable are these means, and the higher the heritability. Conversely, the variance component for the block term, $\hat{\sigma}_\Delta^2 = 15$, is absent from the formula for heritability, although this is one of the sources of variation under investigation. This is because the block effects contribute equally to each estimated breeding-line mean and therefore do not contribute to the variation among them. In the present case, $r=1$ for 25 of the lines and $r=2$ for 59 of them. This is not an ideal situation, but we can deal with it fairly satisfactorily by using the average value of

$$r = \frac{25 \times 1 + 59 \times 2}{25 + 59} = 1.70.$$

Substitution of numerical values into Equations 3.32 and 3.33 then gives

$$\hat{\sigma}_P^2 = 30645 + \frac{13222}{1.70} = 38422$$

$$h^2 = \frac{30645}{30645 + \frac{13222}{1.70}} = 0.798 \text{ or } 79.8\%.$$

   The heritability can be used to calculate the *expected genetic advance under selection* in a plant or animal breeding programme. This is given by the formula

$$G_s = i\sigma_P h^2 \tag{3.34}$$

where
   i = an index of the intensity of selection.

The index i is defined in relation to the standard normal distribution (Distribution 3.30, Section 3.11). It is determined by the fraction (k) of the population that is to be selected. For example, suppose that the highest-yielding 5% of breeding lines are to be selected from the present field trial. The hatched area of Figure 3.5 is the corresponding fraction of a standard normal distribution – that is, an area of 0.05 at the upper end of the distribution. The mean of the specified part of the distribution can be obtained by numerical evaluation of the function

$$i = \frac{\int_{Z_{1-k}}^{\infty} Zf(Z)\mathrm{d}Z}{k} \tag{3.35}$$

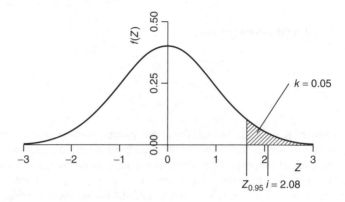

Figure 3.5 The standard normal distribution, showing the mean value of a selected part of the distribution.

where

Z_{1-k} = the $(1 - k)$th quantile of the distribution of Z, that is, the value that cuts off a fraction k at the upper end of the distribution.

This integral has been tabulated, (for example by Becker, 1975), but for values of k in the range 0.1–0.001, a reasonable approximation is given by

$$i = 1.13 + 0.73\log_{10}\left(\frac{1}{k}\right). \tag{3.36}$$

Substitution of

$$k = 0.05$$

into Equation 3.36 gives

$$i = 2.08,$$

and substituting this value into Equation 3.34 we obtain the following expression for the expected genetic advance under selection:

$$G_s = 2.08 \times \sqrt{38757\,\mathrm{g}^2/\mathrm{m}^4} \times 0.798 = 326.8\,\mathrm{g/m}^2.$$

The estimated grand mean grain yield of all the breeding lines, given by the estimate of the constant in the output from the REML statement presented in Section 3.10, is 572.6 g/m². Hence the expected mean grain yield that would be obtained if the selected fraction of the breeding lines were sown in a new field trial is

$$572.6\,\mathrm{g/m}^2 + 326.8\,\mathrm{g/m}^2 = 899.4\,\mathrm{g/m}^2.$$

This is an advance of

$$\frac{326.8}{572.6} \times 100 = 57.1\%.$$

The mean grain yield of the highest-yielding 5% of the lines in the present field trial (i.e. the highest-yielding four lines) is 918.5 g/m². The expected mean in a future trial falls short of this value because the presence of environmental variation interferes with the crop breeder's attempt to select on the genetic variation among the breeding lines. We will return to this 'shrinkage' of predicted values, relative to our naïve expectation, in Chapter 5. In fact, the shrinkage due to residual variation is greater than this simple comparison of the expected future means and the present means suggests. If the heritability of grain yield were 100%, the expected genetic advance would be

$$G_s = 2.08 \times \sqrt{38757\,\mathrm{g^2/m^4}} \times 1 = 409.5\,\mathrm{g/m^2},$$

giving an expected mean grain yield of the selected lines in a future trial of

$$572.6\,\mathrm{g/m^2} + 409.5\,\mathrm{g/m^2} = 982.1\,\mathrm{g/m^2}.$$

If the distribution of the breeding-line means were exactly normal, this would be the same as the mean value of the selected lines in the present trial. The distribution of the breeding-line means in this field trial is superimposed on the normal distribution with the same mean and variance in Figure 3.6. The upper tail of this distribution is somewhat shorter than that of the normal distribution: the normal distribution leads us to expect one or two breeding lines with a mean yield above $1000\,\mathrm{g/m^2}$, whereas none are observed. The absence of such lines reduces the mean yield of the breeding lines in this tail.

Just as the formula for heritability does not include the variance due to blocks, it does not take into account the difference in growing conditions between the present location and season

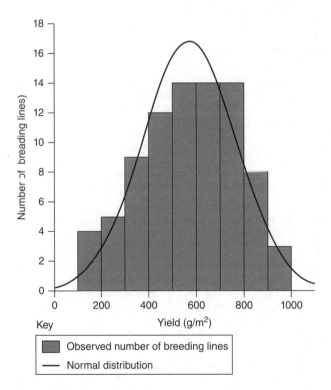

Figure 3.6 Distribution of the mean yields of breeding lines of barley, and the normal distribution with the same mean and variance.

and the environment in which a future trial might be conducted. Thus the method presented here for obtaining the expected mean for a future trial is somewhat misleading. All lines might yield more or less in the new environment, but this would not interfere with the validity of the breeder's selection. It is more precise to consider G_s as an indication of the *difference* expected between the mean of the selected breeding lines and that which would be obtained from the full set of lines studied, if the full set were grown again. Even this interpretation takes no account of genotype×environment interaction. It may be that the response of the selected lines to the new environment will be different from that of the unselected lines, so that the gap between the two groups will be wider or narrower than in the present field trial. Thus any calculation of the heritability and the expected genetic advance under selection in a substantial crop breeding programme has to take account of components of variance due to genotype×year, genotype×location and genotype×year×location interaction effects, as well as the effects considered in this simple example.

3.16   Use of R to analyse the unbalanced field trial

The following commands import the 'unpadded' data into R, fit Model 3.24 and produce the resulting anova:

```
rm(list = ls())
barleyprogeny.unbalanced <- read.table(
    "IMM edn 2\\Ch 3\\barley progeny.txt",
    header=TRUE)
attach(barleyprogeny.unbalanced)
fline <- factor(line)
fblock <- factor(block)
barleyprogeny.model1aov <- aov(yield_g_m2 ~ fline +
    Error(fblock))
summary(barleyprogeny.model1aov)
```

The output of these commands is as follows:

```
Error: fblock
        Df Sum Sq Mean Sq
fline   1   58080   58080


Error: Within
            Df   Sum Sq Mean Sq  F value    Pr(>F)
fline       82 5343839   65169     4.86  1.33e-09 ***
Residuals   58  777747   13409
—
Signif. codes:  0 '***' 0.001 '**' 0.01 '*' 0.05 '.' 0.1 ' ' 1
```

Whereas the GenStat analysis of variance directives recognizes the design as unbalanced and refuses to produce an anova table (Section 3.8), R partitions the variation among lines into a component that occurs between the blocks and a component that occurs within each block. However, the interpretation of this anova is not straightforward: no F test is available for the between-blocks component of the variation due to 'line', and that for the within-blocks component does not represent all the variation due to this term.

The 'padded', balanced version of the same data set in a form suitable for analysis by R is held in the file 'barley progeny padded.txt'. The missing values are represented by the code 'NA'. When the padded data are analysed using the function aov(), the same results are obtained as from the unpadded data. This shows that GenStat's anova directives and R's function aov() deal with the missing values in fundamentally different ways: GenStat's directives estimate them, whereas R's function excludes the units concerned from the design.

The function lme(), *as implemented in the package 'nlme'*, is not able to accept both of the terms 'fline' and 'fblock' in its random-effect argument, so it cannot be used to perform the mixed-model analysis on the original, unbalanced data set. In order to fit a random-effect model with both these terms, another package, named 'lme4', available from the R website, must be installed.[1] The following commands then perform the mixed-model analysis on the unbalanced data set:

```
rm(list = ls())
barleyprogeny.unbalanced <- read.table(
    "IMM edn 2\\Ch 3\\barley progeny.txt",
    header=TRUE)
attach(barleyprogeny.unbalanced)
fline <- factor(line)
fblock <- factor(block)

library(lme4)
barleyprogeny.model1lmer <-
    lmer(yield_g_m2 ~ 1 + (1|fblock) + (1|fline),
    data = barleyprogeny.unbalanced)
summary(barleyprogeny.model1lmer)
```

The function library() loads the package 'lme4'. The function lmer() is similar to lme(), but can fit a wider range of models. The syntax for model specification is similar to that for lm() and lme() presented in Section 1.11, but the fixed-effect and random-effect models are specified in a single argument of the function. The fixed-effect model comes first. In the present case, the only fixed-effect term is the intercept, which is represented by the constant '1'. The term '(1|fblock)' specifies that the factor 'fblock' is a random-effect term: we will see later (Section 7.6) how this notation can be extended to accommodate random factor × fixed variate interactions. The factor 'fline' is specified as a random-effect term in the same way. The next argument of the function, 'data', names the data frame holding the data to be analysed.

---

[1] One current method for doing this is as follows. With your computer connected to the internet, in the main menu of the R GUI, select 'Packages'. In the 'Packages' sub-menu, select 'Set CRAN mirror ... '. A window opens headed 'CRAN mirror', containing a list of locations world-wide. Select a location near you, then click on 'OK'. Next, in the main menu select 'Packages' again, then select 'Install package(s) ... '. A window opens headed 'Packages', containing a list of packages. Select 'lme4', then click on 'OK'. Further windows may open asking questions: if so, answer 'Yes'.

The output of the function `summary()` is as follows:

```
Linear mixed model fit by REML
Formula: yield_g_m2 ~ 1 + (1 | fblock) + (1 | fline)
   Data: barleyprogeny.unbalanced
 AIC  BIC logLik deviance REMLdev
 1880 1892 -936.2    1880    1872
Random effects:
 Groups    Name             Variance    Std.Dev.
 fline     (Intercept) 3.0667e+04 175.119636
 fblock    (Intercept) 8.2083e-06   0.002865
 Residual              1.3226e+04 115.003263
Number of obs: 142, groups: fline, 83; fblock, 2

Fixed effects:
            Estimate Std. Error t value
(Intercept)   572.47      21.67   26.42
```

The estimates of the variance components, $\hat{\sigma}_{\Delta}^2 = 0.0000082083$, $\hat{\sigma}_G^2 = 30667$ and $\hat{\sigma}_E^2 = 13236$, are similar to those given by GenStat (Section 3.10), but not identical. The square root of each variance estimate – the SD – is also presented.

The following commands fit the reduced model omitting the term 'fline'

```
barleyprogeny.model2lmer <-
    lmer(yield_g_m2 ~ 1 + (1|fblock),
    data = barleyprogeny.unbalanced)
summary(barleyprogeny.model2lmer)
```

The output of these commands is as follows:

```
Linear mixed model fit by REML
Formula: yield_g_m2 ~ 1 + (1 | fblock)
   Data: barleyprogeny.unbalanced
 AIC  BIC logLik deviance REMLdev
 1918 1927 -956.1    1920    1912
Random effects:
 Groups    Name             Variance    Std.Dev.
 fblock    (Intercept) 6.4217e-05 8.0135e-03
 Residual              4.3827e+04 2.0935e+02
Number of obs: 142, groups: fblock, 2

Fixed effects:
            Estimate Std. Error t value
(Intercept)   581.57      17.57    33.1
```

Values of deviance calculated on two different bases are presented in this output, namely the maximum likelihood deviance (deviance) and the *residual maximum likelihood* deviance (REMLdev). The latter is the one we require in order to compare the full and reduced models. Thus we obtain

$$\text{deviance}_{\text{reduced model}} - \text{deviance}_{\text{full model}} = 1912 - 1872 = 40.$$

Note that although neither of the individual deviances agrees with that obtained by GenStat, the difference between them agrees to the degree of accuracy given. This reflects the fact that a deviance includes an arbitrary constant dependent on the fixed-effects model and emphasizes the importance of ensuring that the fixed-effect models are the same, and specified in the same way, when deviances are compared.

The following commands fit the alternative reduced model, omitting the term 'fblock':

```
barleyprogeny.model3lmer <-
   lmer(yield_g_m2 ~ 1 + (1|fline),
   data = barleyprogeny.unbalanced)
summary(barleyprogeny.model3lmer)
```

The output of these commands is as follows:

```
Linear mixed model fit by REML
Formula: yield_g_m2 ~ 1 + (1 | fline)
   Data: barleyprogeny.unbalanced
  AIC  BIC logLik deviance REMLdev
 1878 1887 -936.2    1880    1872
Random effects:
 Groups   Name        Variance Std.Dev.
 fline    (Intercept) 30667    175.12
 Residual             13226    115.00
Number of obs: 142, groups: fline, 83

Fixed effects:
            Estimate Std. Error t value
(Intercept)   572.47      21.67   26.42
```

From the values of 'REMLdev' from this alternative reduced model and the full model, we obtain

$$\text{deviance}_{\text{reduced model}} - \text{deviance}_{\text{full model}} = 1872 - 1872 = 0,$$

which agrees to the degree of accuracy given with the value provided by GenStat.

3.17 Use of SAS to analyse the unbalanced field trial

The following statements import the 'unpadded' data into SAS, fit Model 3.24 and produce the resulting anova:

```
PROC IMPORT OUT = barley DBMS = EXCELCS REPLACE
   DATAFILE = "&pathname.\IMM edn 2\Ch 3\barley progeny.xlsx";
   SHEET = "sheet1 SAS";
RUN;

ODS RTF;
PROC ANOVA;
   CLASS block line;
   MODEL yield_g_m2 = block line;
RUN;
ODS RTF CLOSE;
```

Part of the output of PROC ANOVA is as follows:

Source	DF	Sum of squares	Mean square	F value	Pr > F
Model	83	5455070.452	65723.740	5.26	<0.0001
Error	58	724594.860	12493.015		
Corrected total	141	6179665.312			

R-square	Coeff Var	Root MSE	yield_g_m2 mean
0.882745	19.21892	111.7722	581.5735

Source	DF	Anova SS	Mean square	F value	Pr > F
block	1	58079.912	58079.912	4.65	0.0352
line	82	5396990.540	65816.958	5.27	<0.0001

Whereas the GenStat anova directives recognize the design as unbalanced and refuse to produce an anova table (Section 3.8), SAS calculates the MS due to the difference between blocks regardless of the fact that the blocks contain different sets of breeding lines, and calculates the MS due to variation among the lines only on the basis of the variation within each block. Hence a different result would be obtained if the terms 'block' and 'line' were included in the model in the opposite order, and the anova does not have an unambiguous interpretation. The 'padded', balanced version of the same data set in a form suitable for analysis by SAS is held in the sheet 'sheet2 SAS' in the same Excel workbook as the unpadded data. The missing values are represented by an asterisk (*). When the padded data are analysed using PROC ANOVA, the same results are obtained as from the unpadded data. This shows that GenStat's analysis of variance directives and SAS's PROC ANOVA deal with the missing values in fundamentally different ways: GenStat's directives estimate them, whereas the SAS procedure excludes the units concerned from the design.

The following statements perform a mixed-model analysis of the unpadded data:

```
ODS RTF;
PROC MIXED ASYCOV NOBOUND DATA = barley;
   CLASS block line;
   MODEL yield_g_m2 = /DDFM = KR HTYPE = 1 SOLUTION;
   RANDOM block line;
RUN;
ODS RTF CLOSE;
```

The SOLUTION option in the MODEL statement specifies that the estimate of the intercept (which is the only parameter in the fixed-effect model) should be printed.

Part of the output from PROC MIXED is as follows:

Convergence criteria met.

Covariance parameter estimates	
Cov Parm	Estimate
block	14.6071
line	30645
Residual	13222

Asymptotic covariance matrix of estimates				
Row	Cov Parm	CovP1	CovP2	CovP3
1	block	105590	−148626	−20617
2	line	−148626	39032099	−3497283
3	Residual	−20617	−3497283	5863391

Fit statistics	
−2 Res Log Likelihood	1872.4
AIC (smaller is better)	1878.4
AICC (smaller is better)	1878.6
BIC (smaller is better)	1874.5

Solution for fixed effects					
Effect	Estimate	Standard error	DF	t value	$Pr > |t|$
Intercept	572.65	22.4507	13.8	25.51	<0.0001

The variance component estimates, given in the table headed 'Covariance parameter estimates', agree with those produced by GenStat, as does the intercept.

The following statements fit the reduced model omitting the term 'line':

```
ODS RTF;
PROC MIXED ASYCOV NOBOUND DATA = barley;
    CLASS block line;
    MODEL yield_g_m2 = /DDFM = KR HTYPE = 1 SOLUTION;
    RANDOM block;
RUN;
ODS RTF CLOSE;
```

Part of the output from `PROC MIXED` is now as follows:

Fit statistics	
−2 Res Log Likelihood	1912.1
AIC (smaller is better)	1916.1
AICC (smaller is better)	1916.1
BIC (smaller is better)	1913.4

The value labelled '−2 Res Log Likelihood' is the deviance obtained by fitting this model. Thus we obtain:

$$\text{deviance}_{\text{reduced model}} - \text{deviance}_{\text{full model}} = 1912.1 - 1872.4 = 39.7.$$

Note that although neither of the individual deviances agrees with that obtained by GenStat, the difference between them does. This reflects the fact that a deviance includes an arbitrary constant dependent on the fixed-effects model and emphasizes the importance of ensuring that the fixed-effect models are the same, and specified in the same way, when deviances are compared.

The following statements fit the alternative reduced model, omitting the term 'block':

```
ODS RTF;
PROC MIXED ASYCOV NOBOUND;
    CLASS block line;
    MODEL yield_g_m2 = /DDFM = KR HTYPE = 1 SOLUTION;
    RANDOM line;
RUN;
ODS RTF CLOSE;
```

Part of the output of `PROC MIXED` is now as follows:

Fit statistics	
−2 Res Log Likelihood	1872.4
AIC (smaller is better)	1876.4
AICC (smaller is better)	1876.5
BIC (smaller is better)	1881.2

Thus we obtain

$$\text{deviance}_{\text{reduced model}} - \text{deviance}_{\text{full model}} = 1872.4 - 1872.4 = 0.0$$

which agrees with the value given by GenStat, to the degree of accuracy given.

3.18 Estimation of variance components in the regression analysis on grouped data

We now return to the model of the relationship between latitude and house prices in England, introduced in Chapter 1 (Model 1.3), to interpret the variance components for the random-effect terms. The output from fitting this model (Section 1.8) gives the following estimates:

- Variance component for the effect of towns: $\hat{\sigma}_T^2 = 0.01963$, $SE_{\hat{\sigma}_T^2} = 0.01081$.

- Residual variance: $\hat{\sigma}^2 = 0.0171$, $SE_{\hat{\sigma}^2} = 0.00332$.

We can take the square root of each of these variance component estimates in order to convert them to SDs in the units of the original model (Equation 1.3), log (house price in pounds), and then take the antilogarithm to obtain the multiple of house price represented by the SD, namely

- SD for the effect of towns: $\hat{\sigma}_T = \sqrt{0.01963} = 0.14011$

$$10^{\hat{\sigma}_T} = 10^{0.14011} = 1.381$$

- residual SD: $\hat{\sigma} = \sqrt{0.0171} = 0.13076$

$$10^{\hat{\sigma}} = 10^{0.13076} = 1.351.$$

The two components are similar in magnitude: that is, after adjustment for the effect of latitude, the amount of variation among mean house prices in different towns is about the same as that among individual houses in the same town. The mean house price typically varies from town to town by a factor of 1.381, after adjustment for the effect of latitude. Strictly speaking, a town typically lies this far *above or below* the relationship with latitude, so that the typical *difference* between two towns at the same latitude is a factor of $10^{\sqrt{2\times0.01963}} = 1.5781$. Within each town, houses typically vary in price by a factor of 1.351. The mixed model analysis gave the estimated effect of latitude as -0.08147 (Section 1.8): that is, it is estimated that $\log_{10}$(house price, pounds) decreases by 0.08147 for each degree of latitude northward. This is equivalent to a reduction in house prices by a factor of $10^{-0.08147} = 0.8290$ for each degree north, or an increase by a factor of $10^{0.08147} = 1/0.8290 = 1.2063$ for each degree south. So, returning to the logarithmic scale:

- the typical variation between towns is equivalent to the effect of $0.14011/0.08147 = 1.720$ degrees of latitude and

- the typical variation between houses in the same town is equivalent to the effect of $0.13076/0.08147 = 1.605$ degrees of latitude.

When the analysis is restricted to an equal number of houses from each town, the MSs in the anova (Section 1.9) are related to the variance components as shown in Table 3.8. Because three houses are considered from each town, the variance component σ_T^2 has the coefficient 3 in the 'town' stratum. These expected MSs are the same if

$$H_0 : \sigma_T^2 = 0 \tag{3.37}$$

is true, so it is appropriate to perform an F test of the significance of 'town' against the residual variance. This gives

$$F = \frac{0.07700}{0.01943} = 3.96, P(F_{9,22} > 3.96) = 0.0040.$$

Table 3.8 Expected mean squares of the terms in the anova of the effect of latitude on house prices in England.

Source of variation	Expected MS
town	$3\sigma_T^2 + \sigma^2$
Residual	σ^2

That is, the variation among towns is highly significant, even after adjusting for the effect of latitude.

In R, the output of the commands

```
summary(houseprice.model4)
anova(houseprice.model4)
```

given in Section 1.11 provides the information required to obtain these variance component estimates. The output from the function `summary()` includes the following information:

```
Random effects:
 Formula: ~1 | town
         (Intercept)   Residual
 StdDev:   0.1401131 0.1307764
```

The two SDs presented here are $\hat{\sigma}_T$ and $\hat{\sigma}$, respectively.

When the same model is fitted by SAS using PROC MIXED, the variance components are presented in the table headed 'Covariance parameter estimates' (Section 1.12). These values agree with those produced by GenStat.

3.19 Estimation of variance components for block effects in the split-plot experimental design

Variance components can also be estimated for the block effects in the split-plot experiment considered in Chapter 2 (Model 2.1). The expected MSs in the anova (Section 2.2) are as shown in Table 3.9. These expected MSs show that, as in the case of the hierarchical model of the three-stage assay process (Model 3.1, Sections 3.2–3.4), it is appropriate to perform an F test of the significance of each block-effect term against the block-effect term below. The values obtained are as shown in Table 3.10.

These tests show that neither the null hypothesis concerning variation among days,

$$H_0 : \sigma_\Delta^2 = 0 \tag{3.38}$$

nor that concerning variation among presentations,

$$H_0 : \sigma_\Pi^2 = 0 \tag{3.39}$$

Table 3.9 Expected mean squares of the terms in the anova of the perception of saltiness of commercial brands of ravioli by trained assessors.

Source of variation	Expected MS
day	$36\sigma_\Delta^2 + 9\sigma_\Pi^2 + \sigma^2$
day.presentation	$9\sigma_\Pi^2 + \sigma^2$
day.presentation.serving	σ^2

Table 3.10 F tests for the significance of the block-effect terms in the anova of the perception of saltiness of commercial brands of ravioli by trained assessors.

Source of variation	DF	MS	F	p
day	2	371.7	2.67	0.15
day.presentation	6	139.3	1.39	0.23
day.presentation.serving	64	100.3		

can be rejected. Nevertheless, we can, if we wish, proceed to estimate these variance components. This is done by rearranging the formulae for the expected MSs as follows:

$$\hat{\sigma}_\Delta^2 = \frac{MS_{day} - MS_{Residual,\ day.presentation\ stratum}}{36} = \frac{371.7 - 139.3}{36} = 6.4556$$

$$\hat{\sigma}_\Pi^2 = \frac{MS_{Residual,\ day.presentation\ stratum} - MS_{Residual,\ day.presentation.serving\ stratum}}{9}$$

$$= \frac{139.3 - 100.3}{9} = 4.3333$$

$$\hat{\sigma}^2 = MS_{Residual,\ day.presentation.serving\ stratum} = 100.3$$

Allowing for rounding error, these values agree with those given by the mixed-model analysis (Section 2.4). They show that even if there is variation among the true means of days and presentations, it is slight compared to the variation among servings within each presentation.

In R, the estimates of the variance components in this model are included in the output of the command

```
summary(ravioli.model3)
```

which is presented in Section 2.8. The estimates are as follows:

$\hat{\sigma}_\Delta^2 = [\text{StdDev (Intercept)}]^2$ in the '~1 | day' stratum $= 2.540342^2 = 6.4533$;

$\hat{\sigma}_\Pi^2 = [\text{StdDev (Intercept)}]^2$ in the '1 | presentation %in% day' stratum $= 2.082927^2 = 4.3386$;

$\hat{\sigma}^2 = [\text{StdDev Residual}]^2$ in the '1 | presentation %in% day' stratum $= 10.01351^2 = 100.27$.

Note that the residual SD is the square root of the variance component for serving within presentation. Allowing for rounding error, these values agree with those produced by GenStat.

When the same model is fitted by SAS using `PROC MIXED`, the variance components are presented in the table headed 'Covariance parameter estimates' (Section 2.9). Allowing for rounding error, these values agree with those produced by GenStat.

3.20 Summary

For each random-effect term in a mixed model, a variance component – the part of the total variance that is due to that term – can be estimated.

The estimation of variance components is illustrated in an experiment to study a three-stage assay process. The experiment has a hierarchical design: deliveries of a chemical paste were sampled, a sample of casks was taken from each delivery and repeated tests were performed on each cask. The 'strength' of the paste was measured.

Estimates of the variance components can be obtained from the MSs in the analysis of variance, together with the number of replications at each level in the hierarchy. They can also be obtained by mixed modelling.

The expected MSs justify the F tests that are performed.

Estimates of variance components can be used to design a future sampling strategy. They provide the basis for an estimate of the optimum number of casks to be sampled from each delivery, and the optimum number of tests to be performed on each cask.

Genetic (heritable) and environmental (non-heritable) variance components are routinely estimated in plant and animal breeding, and are compared in a ratio called *heritability*.

The use of variance components in this way is illustrated by a field experiment to evaluate doubled-haploid breeding lines derived from a cross between two inbred varieties of barley. The experiment has a randomized block design, but is unbalanced as not all lines are represented in both blocks. The effects of breeding lines, as well as those of blocks and the residual effects, are specified as random.

Analysis of variance can be performed on the experiment if it is 'padded' with missing values to make it balanced, but it can also be analysed in its original form using mixed modelling.

Three tests for the significance of a variance component are compared, namely

- the F test, which determines the significance of a term in a balanced experimental design;

- comparison of the estimate of the variance component with its SE, to give a z statistic;

- the change in deviance due to dropping the term in question from the random-effect model, keeping the fixed-effect model unchanged. This likelihood ratio test is interpreted as an χ^2 statistic with 1 degree of freedom, but the p-value associated with this statistic is halved, because it is not considered possible that the true value of the variance component is negative.

The F test is to be preferred in the circumstances in which it is valid. The comparison of the estimate with its SE should not be used as a formal test, but is useful as a preliminary screening device when many possible models are under consideration, as it does not require more than one model to be fitted. The deviance-based test is a good substitute for the F test.

The calculation of heritability and its use to predict the genetic advance under selection in a breeding programme are illustrated.

If the heritability is less than 1, the predicted genetic advance is less than the value given by a naive prediction from the mean of the selected breeding lines. This 'shrinkage' of predicted values is considered more fully in Chapter 5.

3.21   Exercises

3.1 In a research on artificial insemination of cows, a series of semen samples from six bulls was tested for the ability to conceive (Snedecor and Cochran, 1989, Section 13.7, pp. 245–247). The results are presented in Table 3.11.

(a) Arrange these data for analysis by GenStat, R or SAS.

(b) Analyse the data by the analysis of variance and by mixed modelling, making the assumption that the bulls studied have been chosen at random from a population of similar animals. Perform the best available test to determine whether the variation among bulls is significant.

The data are percentages based on slightly different numbers of tests: the assumption, made in the analyses specified in Part (b), that the variance among samples within each bull is constant is therefore not quite correct.

(c) Obtain diagnostic plots of the residuals and investigate whether there is evidence of a serious breach of the assumptions on which your analyses are based.

(d) Estimate the following variance components:

(i) among bulls

(ii) among samples within each bull.

Suppose that the results from this experiment are to be used to design an assay procedure to estimate the fertility of similar bulls in the future. Suppose also that the cost of including an additional bull in the assay is three times the cost of obtaining an additional sample from a bull already included.

(e) How many samples should be tested from each bull included in the assay?
N.B. The sums of squares and MSs in the anova of these data do not agree with those presented by Snedecor and Cochran.

3.2 An experiment using 36 samples of Portland cement is described by Davies and Goldsmith (1984, Section 6.75, pp. 154–158). The samples were 'gauged' (i.e. mixed with water and worked) by three gaugers, each one gauging 12 samples. After the samples had set, their compressive strength was tested by three breakers, each breaker testing four samples from

Table 3.11   Conception rates obtained from semen samples from six bulls.

Bull	Percentage of conceptions to services								
1	46	31	37	62	30				
2	70	59							
3	52	44	57	40	67	64	70		
4	47	21	70	46	14				
5	42	64	50	69	77	81	87		
6	35	68	59	38	57	76	57	29	60

Table 3.12   Compressive strength of samples of Portland cement (lb/sq.in.) mixed by three gaugers, measured by three breakers.

Gauger	Breaker					
	1		2		3	
1	5280	5520	4340	4400	4160	5180
	4760	5800	5020	6200	5320	4600
2	4420	5280	5340	4880	4180	4800
	5580	4900	4960	6200	4600	4480
3	5360	6160	5720	4760	4460	4930
	5680	5500	5620	5560	4680	5600

Source: Numerical data on the compressive strength of Portland cement, presented. Davies and Goldsmith (1984). Data reproduced by kind permission of Pearson Education Ltd.

each of the three gaugers. The results are presented in Table 3.12. (Data reproduced by the kind permission of Pearson Education Ltd.)

(a) Arrange these data for analysis by GenStat, R or SAS.

'Gauger' and 'Breaker' are both to be specified as random-effect terms.

(b) Justify these decisions.

(c) Analyse these data by the analysis of variance and by mixed modelling.

(d) Perform F tests for the significance of the following model terms:

  (i) Gauger

  (ii) Breaker

  (iii) Gauger × Breaker interaction.

(e) Obtain estimates of the variance components for the same model terms and for the residual term, from the anova. Confirm your answers from the mixed modelling results.

(f) According to the evidence from this experiment, which of these sources of variation needs to be taken into account when obtaining an estimate of the mean strength of samples of Portland cement?

The breakers in this experiment were human assistants who operated the testing machine, and the variation in the results that they produce is due to personal factors in their preliminary adjustment of the machine.

(g) Suppose that the cost of employing an additional breaker is 10 times the cost of getting the current breaker to test an additional sample. Determine the number of samples that each breaker should test in order to obtain the most accurate estimate possible of the strength of the cement, at a fixed cost.

3.3 Two inbred cultivars of wheat were hybridized, and the seed of 48 F_2-derived F_3 families (that is, families in the third progeny generation, each derived by inbreeding from a single plant in the second progeny generation) was obtained. The seed was sown in field plots in competition with ryegrass, in two replications in a randomized block design, with 'family' as the treatment factor. The mean grain yield per plant was determined in each plot. The results obtained are presented in Table 3.13. (Data reproduced by kind permission of S. Mokhtari.)

Table 3.13   Yield per plant (g) of F_3 families of wheat grown in competition with ryegrass in a randomized block design.

	A	B	C		A	B	C		A	B	C
1	block	family	yield	34	1	19	0.733	67	2	44	4.400
2	1	11	3.883	35	1	26	1.700	68	2	1	4.900
3	1	9	3.717	36	1	30	1.800	69	2	35	4.283
4	1	7	3.850	37	1	46	1.720	70	2	3	5.233
5	1	41	1.817	38	1	12	4.083	71	2	13	3.717
6	1	2	7.483	39	1	27	3.667	72	2	26	4.717
7	1	25	3.483	40	1	31	4.317	73	2	28	2.817
8	1	10	2.500	41	1	39	4.033	74	2	12	7.467
9	1	38	3.683	42	1	47	1.950	75	2	33	4.567
10	1	32	1.167	43	1	4	5.000	76	2	45	5.480
11	1	42		44	1	24	2.600	77	2	34	7.933
12	1	23	0.717	45	1	34	4.250	78	2	18	4.833
13	1	8	2.883	46	1	13	4.300	79	2	7	4.933
14	1	17	5.883	47	1	33	4.250	80	2	2	5.300
15	1	43	2.900	48	1	37	3.783	81	2	31	7.500
16	1	5	2.650	49	1	48	2.617	82	2	40	3.083
17	1	16	2.517	50	2	9	8.550	83	2	46	2.900
18	1	21	1.933	51	2	11	9.633	84	2	10	4.917
19	1	18	2.417	52	2	5	5.733	85	2	32	4.400
20	1	15	3.683	53	2	25	8.117	86	2	17	6.533
21	1	44	0.940	54	2	39	4.633	87	2	23	5.783
22	1	36	4.617	55	2	21	1.450	88	2	27	4.767
23	1	20	2.817	56	2	41	3.240	89	2	38	4.867
24	1	1	4.260	57	2	42		90	2	30	3.983
25	1	22	2.667	58	2	14	2.850	91	2	22	4.017
26	1	40	6.917	59	2	37	3.850	92	2	47	1.775
27	1	35	7.883	60	2	19	5.020	93	2	48	1.540
28	1	6	3.050	61	2	36	8.740	94	2	20	
29	1	14	2.533	62	2	43	1.300	95	2	24	8.033
30	1	29	5.400	63	2	16	4.600	96	2	4	4.020
31	1	45	3.150	64	2	29	9.100	97	2	15	3.183
32	1	3	2.150	65	2	6	4.600				
33	1	28	1.367	66	2	8	4.900				

Source: Data reproduced by kind permission of S. Mokhtari.

In the analysis of these data, 'family' will be specified as a random-effect term.

(a) Justify this decision.

(b) Analyse these data by the analysis of variance and by mixed modelling. Determine whether the variation among families is significant according to the F test.

(c) Obtain estimates of the variance components for the following model terms:

  (i) residual

  (ii) family

  (iii) block.

(d) Compare the estimate of the variance component for the term 'family' with its SE. Does their relative magnitude suggest that this term might reasonably be dropped from the model?

(e) Test the significance of the term 'family' by comparing the deviances obtained with and without the inclusion of this term in the model.

(f) Obtain diagnostic plots of the residuals from your analysis.

(g) Compare the results of the significance tests conducted in Parts (b) and (e) with each other, and with the informal evaluation conducted in Part (d). Consider how fully the assumptions of each test are likely to be met. Considered together, do these tests indicate that the term 'family' should be retained in the model?

(h) Estimate the heritability of yield in this population of families. (N.B. The estimate obtained using the methods described in this chapter is slightly biased downward, as some of the residual variance is due to genetic differences among plants of the same family.)

(i) Estimate the genetic advance that is expected if the highest-yielding 10% of the families are selected for further evaluation. Find the expected mean yield of the selected families and compare it with the observed mean yield of the same families. Account for the difference between these values.

(j) When these estimates of genetic advance and expected mean yield are obtained, what assumptions are being made about the future environment in which the families are evaluated?

These data have also been analysed and interpreted by Mokhtari *et al.* (2002).

References

Allard, R.W. (1960) *Principles of Plant Breeding*, John Wiley & Sons, Inc., New York, 485 pp.

Becker, W.A. (1975) *Manual of Quantitative Genetics*, Students Book Corporation, Pullman, WA.

Crainiceanu, C.M. and Ruppert, D. (2004) Likelihood ratio tests in linear mixed models with one variance component. *Journal of the Royal Statistical Society B*, **66**, 165–185.

Davies, O.L. and Goldsmith, P.L. (1984) *Statistical Methods in Research and Production with Special Reference to the Chemical Industry*, 4th edn, Longman, London, 478 pp.

Mokhtari, S., Galwey, N.W., Cousens, R.D. and Thurling, N. (2002) The genetic basis of variation among wheat F_3 lines in tolerance to competition by ryegrass (*Lolium rigidum*). *Euphytica*, **124**, 355–364.

Snedecor, G.W. and Cochran, W.G. (1967) *Statistical Methods*, 6th edn, Iowa State University Press, Ames, IA, 593 pp.

Snedecor, G.W. and Cochran, W.G. (1989) *Statistical Methods*, 8th edn, Iowa State University Press, Ames, IA, 503 pp.

Stram, D.O. and Lee, J.W. (1994) Variance components testing in the longitudinal mixed effects setting. *Biometrics*, **50**, 1171–1177.

4

Interval estimates for fixed-effect terms in mixed models

4.1 The concept of an interval estimate

We have seen in earlier chapters that the estimates of fixed effects produced by mixed-model analysis are similar to those produced by the more familiar analysis methods. Regression analysis gives coefficients that are interpreted as slopes and intercepts; analysis of variance gives treatment means. Mixed-model analysis, when the same terms are placed in the fixed-effect model, gives similar estimates, which can be interpreted in the same way (Chapter 1 – slopes and intercepts; Chapter 2 – treatment means). However, the apparent *precision* of these estimates, indicated by their standard errors (SEs), is affected by the decision to specify other terms in the model as random. In this chapter, we shall explore the precision with which fixed effects are estimated by mixed-model analysis. The single value so far given as the estimate of each model parameter (slope, intercept or mean) is referred to as a *point estimate*, and the precision of each point estimate is indicated by enclosing it in an *interval estimate*. This indicates the range within which the true value of the parameter can reasonably be supposed to lie.

Before determining the interval estimates of the various model parameters introduced so far, we need to set out the general principles to be followed when obtaining such an estimate. Consider a model parameter β. The point estimate of this parameter is referred to as $\widehat{\beta}$, and the basis for the calculation of the interval estimate is the SE of the point estimate, $\text{SE}_{\widehat{\beta}}$. To obtain this, $\widehat{\beta}$ is treated as an observation of a random variable $\widehat{B}$, the variance of which, $\sigma^2_{\widehat{\beta}}$, is a function of the variance components in the model fitted. In order to regard $\widehat{\beta}$ in this way, it is necessary to envisage an effectively infinite population of data sets, from which the actual data set studied has been sampled at random, each of which would yield a value $\widehat{\beta}$. In the case of a randomized experimental design, such a population is provided by the alternative distributions of the treatments over the experimental units (see Section 2.6). It is usually assumed that these

Introduction to Mixed Modelling: Beyond Regression and Analysis of Variance, Second Edition. N. W. Galwey.
© 2014 John Wiley & Sons, Ltd. Published 2014 by John Wiley & Sons, Ltd.
Companion website: http://www.wiley.com/go/beyond_regression

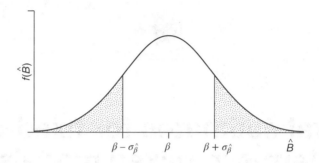

Figure 4.1   Distribution of the random variable $\hat{B}$, of which the parameter point estimate $\hat{\beta}$ is an observation.

Figure 4.2   An interval estimate of the parameter β, based on the standard error of the point estimate.

values of $\hat{\beta}$ are normally distributed around the true value β: that is,

$$\hat{B} \sim N(\beta, \sigma_{\hat{\beta}}^2). \tag{4.1}$$

These ideas are illustrated in Figure 4.1.

An estimate of the variance $\sigma_{\hat{\beta}}^2$, referred to as $\hat{\sigma}_{\hat{\beta}}^2$, is obtained using methods presented in Sections 4.2 and 4.3. The SE of the point estimate of the parameter is then given by the square root of this variance estimate: that is,

$$SE_{\hat{\beta}} = \hat{\sigma}_{\hat{\beta}}. \tag{4.2}$$

This value is added to and subtracted from the point estimate to obtain the boundaries of the interval estimate, as illustrated in Figure 4.2. This *number line* represents the range of possible values for the true parameter value β. The shaded areas indicate values outside the range in which the true value β can reasonably be supposed to lie. The boundaries of this region are symmetrical on either side of the point estimate $\hat{\beta}$.

The concept of an interval estimate can be applied not only to an individual parameter, but also to a mathematical function that combines the parameters. In particular, it can be applied to the difference between two parameters.

4.2 Standard errors for regression coefficients in a mixed-model analysis

In the study of the relationship between latitude and house prices in Chapter 1, estimates of the fixed effects (constant, i.e. intercept, and effect of latitude, i.e. slope) obtained by three different methods were presented, together with their SEs (Table 1.6, Section 1.8): they are

Table 4.1 Comparison of the parameter estimates, and their SEs, obtained from different methods of analysis of the effect of latitude on house prices in England.

Term	Method of analysis					
	Regression analysis ignoring towns (Model 1.1)		Regression analysis on town means		Mixed-model analysis (Model 1.3)	
	Estimate	SE$_{Estimate}$	Estimate	SE$_{Estimate}$	Estimate	SE$_{Estimate}$
Constant	9.68	1.00	9.44	1.87	9.497	1.9248
Latitude	−0.0852	0.0188	−0.0804	0.0353	−0.08147	0.036272

Table 4.2 Comparison of the parameter estimates, and their SEs, obtained from different methods of analysis of the effect of latitude on house prices in England, using data 'trimmed' to three houses per town.

Term	Method of analysis					
	Regression analysis ignoring towns (Model 1.1)		Regression analysis on town means		Mixed-model analysis (Model 1.3)	
	Estimate	SE$_{Estimate}$	Estimate	SE$_{Estimate}$	Estimate	SE$_{Estimate}$
Constant	9.04	1.37	9.04	2.01	9.038	2.0068
Latitude	−0.0726	0.0259	−0.0726	0.0378	−0.07260	0.037835

repeated in Table 4.1. As already noted, the point estimates of the effect of latitude obtained by the three methods are very similar, but the SE from the regression analysis ignoring towns is considerably smaller than those from the other two analyses and is misleading. The SEs from the other two analyses are both reasonable: the slight difference between them arises from the unequal numbers of houses sampled in different towns. Similar arguments apply to the estimates of the constant and their SEs.

The method of calculation of SEs of estimates in a mixed-model analysis is generally complicated, but when the number of observations per group is constant it is relatively straightforward. In order to compare the calculations for the three models in this simpler case, the estimates and their SEs from the data set 'trimmed' to three houses per town, which were presented in Section 1.9, are collated in Table 4.2. The SEs from the regression analysis on town means and from the mixed-model analysis are now the same, allowing for rounding error.

The methods for calculating these SEs are as follows. To obtain the SE of the effect of latitude from the regression analysis ignoring towns, we define

x_i = the latitude of the ith house,

n = the number of houses represented in the dataset,

$$\bar{x} = \frac{\sum_{i=1}^{n} x_i}{n} \text{ (the mean value of latitude),} \tag{4.3}$$

$$S_{XX} = \sum_{i=1}^{n} (x_i - \bar{x})^2 \text{(the \textit{corrected sum of squares} of latitude),} \tag{4.4}$$

and

$$\hat{\sigma}^2 = \text{the estimate of } \sigma^2 \text{ from Model 1.2} = MS_{\text{Resid}} \text{ from the anova of this model.}$$

We refer to the true effect of latitude as β_1, and the point estimate of this effect as $\hat{\beta}_1$. The estimate of the variance of $\hat{\beta}_1$ is then given by

$$\hat{\sigma}^2_{\hat{\beta}_1} = \frac{\hat{\sigma}^2}{S_{XX}}, \tag{4.5}$$

and

$$SE_{\hat{\beta}_1} = \hat{\sigma}_{\hat{\beta}_1} = \sqrt{\frac{\hat{\sigma}^2}{S_{XX}}}. \tag{4.6}$$

In the present case,

$$n = 33,$$

$$\bar{x} = \frac{53.7947 + 53.7947 + \cdots + 51.7871}{33} = \frac{1749.8337}{33} = 53.0253,$$

$$S_{XX} = (53.7947 - 53.0253)^2 + (53.7947 - 53.0253)^2 + \cdots$$
$$+ (51.7871 - 53.0253)^2 = 53.7878,$$

$$\hat{\sigma}^2 = 0.03615,$$

$$\hat{\sigma}^2_{\hat{\beta}_1} = \frac{0.03615}{53.7878} = 0.000672$$

and

$$SE_{\hat{\beta}_1} = \sqrt{\frac{0.03615}{53.7878}} = 0.02592$$

as stated above.

To obtain the SE of the effect of latitude from the regression analysis on town means, the same method of calculation is used, but town means replace the individual house values, and the value of MS_{Resid} obtained from the town means is used as $\hat{\sigma}^2$. Thus we obtain

$$n_{\text{town means}} = 11,$$

$$\bar{x}_{\text{town means}} = \frac{53.7947 + 53.2591 + \cdots + 51.7871}{11} = \frac{583.2779}{11} = 53.0253,$$

$$S_{XX,\text{town means}} = (53.7947 - 53.0253)^2 + (53.2591 - 53.0253)^2 + \cdots + (51.7871 - 53.0253)^2$$
$$= 17.92927,$$

and

$$\hat{\sigma}^2_{\text{town means}} = 0.02567.$$

We modify Equation 4.6 to give

$$SE_{\hat{\beta}_1,\text{town means}} = \hat{\sigma}_{\hat{\beta}_1,\text{town means}} = \sqrt{\frac{\hat{\sigma}^2_{\text{town means}}}{S_{XX,\text{town means}}}}. \tag{4.7}$$

Substituting the numerical values into Equation 4.7, we obtain

$$SE_{\widehat{\beta}_1, \text{town means}} = \sqrt{\frac{0.02567}{17.92927}} = 0.03784$$

as stated above.

The SE of the effect of latitude from the mixed-model analysis is numerically identical to that from the regression analysis on town means, but can be viewed somewhat differently. It is not based on a single variance estimate: it is calculated using both the among-towns and the within-towns variance components. To see how these are combined, we need to consider the relationship between variance components and mean squares (MSs) in this analysis, as discussed in Section 3.18. In Table 3.8 we noted that

$$\text{Expected MS}_{\text{town}} = r\sigma_T^2 + \sigma^2$$

where

r = number of houses sampled in each town = 3.

Now the value of MS_{town} is closely related to the variance among the town means: specifically,

$$\sigma_{\text{town means}}^2 = \frac{\text{Expected MS}_{\text{town}}}{r} = \sigma_T^2 + \frac{\sigma^2}{r}$$

Replacing the expected MSs by the observed MSs in this formula, and the true variance components by their estimates, we obtain

$$\widehat{\sigma}_{\text{town means}}^2 = \frac{\text{MS}_{\text{town}}}{r} = \widehat{\sigma}_T^2 + \frac{\widehat{\sigma}^2}{r}. \tag{4.8}$$

Substituting Equation 4.8 into Equation 4.7 we obtain

$$SE_{\widehat{\beta}_1, \text{town means}} = \sqrt{\frac{\widehat{\sigma}_T^2 + \frac{\widehat{\sigma}^2}{r}}{S_{XX, \text{town means}}}}. \tag{4.9}$$

This version of the formula makes explicit the dependence of the SE on both σ_T^2 and σ^2. Both variance components tend to reduce the precision of the estimate $\widehat{\beta}_1$. The reduction in precision due to σ^2 can be overcome by increasing the value of r – that is, the variation among houses within each town will have less effect if more houses are sampled from each town. However, this strategy does not reduce the impact of σ_T^2, which can be overcome only by sampling more towns.

The estimate of the constant, $\widehat{\beta}_0$, is of less interest than that of the slope, $\widehat{\beta}_1$, but the SEs of the different estimates are presented here for completeness. The SE of $\widehat{\beta}_0$ from the regression analysis ignoring towns is given by

$$SE_{\widehat{\beta}_0} = \sqrt{\left(\frac{1}{n} + \frac{\bar{x}^2}{S_{XX}}\right)\widehat{\sigma}^2}. \tag{4.10}$$

The numerical value is given by

$$SE_{\widehat{\beta}_0} = \sqrt{\left(\frac{1}{33} + \frac{53.0253^2}{53.7878}\right) \times 0.03615} = 1.375.$$

The corresponding value from the regression analysis on the town means is given by

$$
SE_{\hat{\beta}_0} = \sqrt{\left(\frac{1}{n_{\text{town means}}} + \frac{\bar{x}^2_{\text{town means}}}{S_{XX,\text{town means}}} \right) \hat{\sigma}^2}
$$

$$
SE_{\hat{\beta}_0} = \sqrt{\left(\frac{1}{11} + \frac{53.0253^2}{17.92927} \right) \times 0.02567} = 2.007. \qquad (4.11)
$$

The SE of the constant from the mixed model analysis is given by

$$
SE_{\hat{\beta}_0} = \sqrt{\left(\frac{1}{n_{\text{town means}}} + \frac{\bar{x}^2_{\text{town means}}}{S_{XX,\text{town means}}} \right) \left(\hat{\sigma}^2_T + \frac{\hat{\sigma}^2}{r} \right)}, \qquad (4.12)
$$

which gives the same numerical value as Equation 4.11. As in the case of $\hat{\beta}_1$, the precision of the estimate is reduced by both σ^2_T and σ^2. As before, the reduction in precision due to σ^2 can be overcome by increasing the value of r, but that due to σ^2_T can be overcome only by sampling more towns.

4.3　Standard errors for differences between treatment means in the split-plot design

We will next consider the precision of the treatment effects in the split-plot experiment considered in Chapter 2, and how this is related to the variance components identified in that experiment. The tables of means and the accompanying SEs from this experiment (Section 2.2) are repeated here, for ease of reference.

Tables of means

Variate: saltiness

Grand mean 29.28

brand	A	B	C	D
	21.78	19.22	43.23	32.91

assessor	ALV	ANA	FAB	GUI	HER	MJS	MOI
	35.27	4.83	47.52	37.86	29.14	36.01	21.53

assessor	NOR	PER
	33.41	18.00

brand	assessor	ALV	ANA	FAB	GUI	HER	MJS
A		30.44	4.46	40.84	23.01	28.21	24.50
B		32.67	0.00	18.56	30.44	17.82	11.14
C		39.35	11.88	74.99	60.13	40.09	57.17
D		38.61	2.97	55.68	37.86	30.44	51.23

brand	assessor	MOI	NOR	PER
A		15.59	15.59	13.36
B		13.36	33.41	15.59
C		30.44	54.20	20.79
D		26.73	30.44	22.27

Standard errors of differences of means

Table	brand	assessor	brand assessor
rep.	27	12	3
s.e.d.	3.213	4.088	8.351
d.f.	6	64	66.70

Except when comparing means with the same level(s) of

brand		8.176
d.f.		64

In a designed experiment, the natural focus of estimation is on *comparisons* between means, rather than individual means. The SEs of these differences have to take account of natural variation among main plots as well as among sub-plots in the same main plot. In the context of the experiment on the sensory evaluation of ravioli, this means that they must take account of σ_{Π}^2, the variance component representing variation among presentations, as well as σ^2, the component representing variation among servings in the same presentation. The resulting formulae are presented in Table 4.3. The numerical values in this table agree with those in the GenStat output, allowing for rounding error.

The considerations that determine which variance component(s) contribute to the precision of each comparison are similar to those that determine which stratum of the anova should be used to test the significance of each treatment term (Section 2.2). The formulae in the table show that the precision of comparisons between brands is affected not only by σ^2, but also by σ_{Π}^2, because these comparisons always require the comparison of different presentations. However, the precision of comparisons between assessors is affected only by σ^2, because these comparisons are based entirely on the comparison of servings in the same presentation. The precision of comparisons between brand.assessor means depends on the particular pair of brand.assessor combinations being compared. If they involve the same brand, the comparison between them is based entirely on the comparison of servings in the same presentation, and its precision depends on σ^2 only. However, if they involve different brands, the comparison between them requires comparison of different presentations, and its precision depends on both σ^2 and σ_{Π}^2. For example, suppose that the treatment 'Brand C, Assessor FAB' is to be compared with 'Brand C, Assessor ANA'. Both treatments involve the same brand, so $SE_{Difference} = 8.176$. On each day, this pair of treatments occurred in a single presentation. The values from the same presentation are compared, and the natural variation among presentations does not reduce the precision of the comparison. However, suppose that 'Brand C, Assessor FAB' is to be compared with 'Brand B, Assessor ANA'. These two treatments involve different brands, so $SE_{Difference} = 8.351$. On each day, these two treatments occurred in different presentations, and the natural variation among presentations reduces the precision of the comparison. The value of $SE_{Difference}$ is accordingly larger.

Table 4.3   Standard errors of differences for the treatment terms in the split-plot experiment to compare the perception of saltiness of commercial brands of ravioli by trained assessors.

Treatment term	Type of comparison	SE$_{\text{difference}}$	
		Formula	Numerical value
brand	–	$\sqrt{\dfrac{2}{ra}(a\hat{\sigma}_{\Pi}^2 + \hat{\sigma}^2)} =$ $\sqrt{\dfrac{2}{ra}\text{MS}_{\text{Residual, day.presentation stratum}}}$	$\sqrt{\dfrac{2}{3\times9}(9 \times 4.3 + 100.3)} =$ $\sqrt{\dfrac{2}{3\times9} \times 139.3} = 3.212$
assessor	–	$\sqrt{\dfrac{2}{rb}\hat{\sigma}^2} =$ $\sqrt{\dfrac{2}{rb}\text{MS}_{\text{Residual, day.presentation.serving stratum}}}$	$\sqrt{\dfrac{2}{3\times4} \times 100.3} = 4.089$
brand.assessor	Between means with different levels of brand	$\sqrt{\dfrac{2}{r}(\hat{\sigma}_{\Pi}^2 + \hat{\sigma}^2)}$	$\sqrt{\dfrac{2}{3}(4.3 + 100.3)} = 8.351$
	Between means with the same level of brand	$\sqrt{\dfrac{2}{r}\hat{\sigma}^2} =$ $\sqrt{\dfrac{2}{r}\text{MS}_{\text{Residual, day.presentation.serving stratum}}}$	$\sqrt{\dfrac{2}{3} \times 4.3} = 8.177$

where
b = number of brands,
a = number of assessors,
r = number of replications (days).

4.4   A significance test for the difference between treatment means

The point estimate of the difference between any pair of means, $\hat{\delta}$, can be compared with its standard error, SE$_{\hat{\delta}}$, to determine whether it is significant: that is, to assess the strength of the evidence against the null hypothesis

$$H_0 : \delta = 0 \tag{4.13}$$

where δ is the difference between the true means. The method for doing so is as follows. The point estimate, the true value and the SE are used to specify a random variable,

$$T = \frac{\hat{\delta} - \delta}{\text{SE}_{\hat{\delta}}}. \tag{4.14}$$

This variable follows a standard probability distribution, the t distribution – a statement that can be expressed in symbolic shorthand as

$$T \sim t. \tag{4.15}$$

The t distribution is compared with the standard normal distribution (Distribution 3.29, introduced in Section 3.11) in Figure 4.3. The two distributions are similar in shape, but the precise

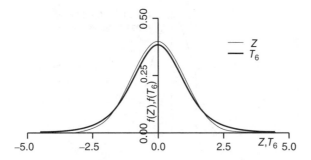

Figure 4.3   Comparison of the t distribution (DF = 6) with the standard normal distribution.

shape of the t distribution depends on the degrees of freedom (DF) of $SE_{\hat\delta}$, that is, the precision with which the variance of $\hat\delta$, $\sigma^2_{\hat\delta}$, is estimated. If the DF were infinite, $\sigma^2_{\hat\delta}$ would be known exactly, and T would follow the standard normal distribution exactly. In the case of comparisons between brands, $SE_{\hat\delta}$ has six DF, which is the DF for the residual MS against which the term *brand* was tested in the anova (Section 2.2). This rather small value gives a rather imprecise estimate of $\sigma^2_{\hat\delta}$, and a t distribution that is rather broader and flatter than the standard normal distribution, as shown in the figure.

In order to test H_0 using the t distribution, the values of $\hat\delta$ and $SE_{\hat\delta}$ obtained from the data and the value of δ obtained from H_0 (Equation 4.13) are substituted into Equation 4.14 to give an observed value of T, referred to as t. A parameter α is then specified, which defines a *critical region* of the t distribution, comprising an area of $\alpha/2$ in each of its tails. For a T variable with v DF, the *critical values*, at the boundaries of the critical region, are $\pm t_{v,\alpha/2}$. These ideas are illustrated in Figure 4.4 for $\alpha = 0.05$ and $\alpha = 0.01$, in a t distribution with six DF. The hatched areas comprise the 5% critical region of the distribution, an area totalling 0.05 in the two tails of the distribution. The cross-hatched areas comprise the 1% critical region, specified in the same way. The evidence against H_0 is assessed by determining whether the observed value t lies within the critical region for each value of α considered: that is, whether

$$|t| > t_{v,\alpha/2}. \tag{4.16}$$

For example, consider the comparison between the mean values of saltiness for brands C and D. This gives

$$t = \frac{(\text{Mean}_{\text{Brand C}} - \text{Mean}_{\text{Brand D}}) - 0}{SE_{\text{difference between brands}}} = \frac{(43.23 - 32.91) - 0}{3.213} = \frac{10.32}{3.213} = 3.21.$$

This value exceeds $t_{6,0.025} = 2.45$, but not $t_{6,0.005} = 3.71$: that is, it lies in the critical region for $\alpha = 0.05$, but not in that for $\alpha = 0.01$, and the difference between these brands is significant at the 5% level, but not at the 1% level. A *two-tailed* test is used here because the decision to subtract the mean saltiness of brand D from that of brand C is arbitrary. If the value for brand C is subtracted from that for brand D, the value $t = -3.21$ is obtained. This also lies in the critical region for $\alpha = 0.05$ but not in that for $\alpha = 0.01$, and the conclusion reached is the same. For a fuller introduction to the t distribution and its use for the comparison of means, see, for example, Moore and McCabe (2012, Sections 7.1 and 7.2, pp. 403–467).

Significance tests for comparisons between assessors, or between brand.assessor combinations, are made in the same way, using the appropriate value of DF. The DF for a comparison

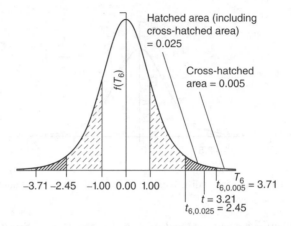

Figure 4.4 The t distribution (DF = 6), showing critical values for significance tests and the corresponding critical regions.

between assessors is determined in the same way as that for comparisons between brands: that is, it is given by the DF for the residual term against which 'assessor' is tested, namely

$$\text{DF}_{\text{Residual, day.presentation.serving stratum}} = 64.$$

However, the DF for a comparison between brand.assessor combinations depends on the particular comparison considered. For a comparison that involves the same brand, the DF is that for the term against which 'brand.assessor' is tested: again, $\text{DF}_{\text{Residual, day.presentation.serving stratum}}$. However, for a comparison that involves different brands, the precision of which depends on two variance components, the determination of the value of DF is less straightforward. It is obtained by the method of Satterthwaite (1946), which was applied to the split-plot situation by Taylor (1950). For an alternative reference on this topic see Steel and Torrie (1981, Section 16.2, pp. 381–382). Steel and Torrie's approach leads directly to a critical value for t, and does not obtain the DF explicitly. We will not examine the method of Satterthwaite in detail, but will note that the value of DF obtained is always:

- greater than the minimum $\text{DF}_{\text{Residual}}$ in any of the strata where effects contributing to the table are estimated: in the present case,

$$\min(\text{DF}_{\text{Residual, day.presentation stratum}}, \ \text{DF}_{\text{Residual, day.presentation.serving stratum}}) = \min(6, 64) = 6,$$

  and

- less than the sum of $\text{DF}_{\text{Residual}}$ in those strata: in the present case,

$$\text{DF}_{\text{Residual, day.presentation stratum}} + \text{DF}_{\text{Residual, day.presentation.serving stratum}} = 6 + 64 = 70.$$

The value given in the GenStat output, 66.70, meets these criteria. As noted earlier (Section 1.2), the value of DF for a model term represents the number of independent pieces of information to which that term is equivalent. Normally, therefore, values of DF are integers. However, in this case, a non-integer value of DF is specified. As an example of a case in which this value is required, consider the comparison between 'Brand D, Assessor HER' and 'Brand A, Assessor GUI'. This gives the value

$$t = \frac{40.09 - 23.01}{8.351} = 2.0453.$$

The critical values of a t distribution with non-integer DF cannot be looked up in standard statistical tables. However, some software systems, including GenStat, are able to cope with this situation. The significance of this t statistic is given by the following GenStat statement:

```
PRINT CUT(2.0453; 66.70)
```

The function name CUT indicates that the Cumulative probability (the p-value) for the Upper tail of the t distribution is to be obtained. The output of this statement is as follows:

```
CUT((2.045;66.7))
0.02239
```

That is,

$$P(T_{66.70} > 2.0453) = 0.02239.$$

But again, a two-tailed test is appropriate, so the p-value we require is

$$P(T_{66.70} > 2.0453 \text{ or } T_{66.70} < -2.0453) = P(|T_{66.70}| > 2.0453) = 2 \times P(T_{66.70} > 2.0453)$$
$$= 2 \times 0.02239 = 0.04478.$$

The probability of obtaining so extreme a t value by chance is 0.04478, indicating that the difference between these means is significant at the 5% level, but not at the 1% level.

If no software that can deal correctly with non-integer values of DF is available, the p-value required can be estimated by interpolation between the values given by the integers on either side of the actual value of DF, in this case 66 and 67. Thus

$$P(T_{66} > 2.0453) = 0.02241$$

and

$$P(T_{67} > 2.0453) = 0.02238.$$

The p-value given by the actual DF is intermediate between those given by the integer DF values, as it should be.

4.5   The least significant difference (LSD) between treatment means

As an alternative to the SE of the difference between two means, the *least significant difference* (LSD) can be calculated. This is the value of $\hat{\delta}$ that gives a value of t on the boundary of the critical region, that is, the smallest value of t that satisfies the inequality in Equation 4.16, and is given by

$$\mathrm{LSD}_{\hat{\delta}} = t_{v,\alpha/2} \cdot \mathrm{SE}_{\hat{\delta}}. \tag{4.17}$$

Conversely, the range of values of δ that *do not* produce a value of t in the critical region defines a $100 \times (1 - \alpha)\%$ *confidence interval* for δ. The boundaries of this confidence interval are given by

$$CI = \hat{\delta} \pm LSD_{\hat{\delta}}.$$

For example, if we substitute into Equation 4.17 the 5% critical value of t, namely

$$t_{6,0.05/2} = 2.45,$$

and the SE for comparisons between brands, namely

$$SE_{\hat{\delta}} = 3.213,$$

we obtain the LSD between brands,

$$LSD_{\hat{\delta}} = 2.45 \times 3.213 = 7.87.$$

The $100 \times (1 - 0.05) = 95\%$ confidence interval for the difference between Brand C and Brand D is

$$(43.23 - 32.91) \pm 7.87 = 10.32 \pm 7.87.$$

The concept of the LSD is illustrated graphically in Figure 4.5. This number line represents the range of possible values for the true difference δ between the treatment means, with zero at the centre. The stippled areas indicate values outside the range $\pm SE_{\hat{\delta}}$. Such values, if observed, provide at least tentative evidence against the null hypothesis (Equation 4.13). The hatched areas represent values outside the range $\pm LSD$ ($\alpha = 0.05$). Such values provide significant evidence against H_0, at least at the 5% level. This range is wider than the range $\pm SE_{\hat{\delta}}$ – that is, treatment means must differ by more than $SE(\hat{\delta})$ in order to be significantly different. The cross-hatched areas represent values outside the range $\pm LSD$ ($\alpha = 0.01$). Such values provide significant evidence against H_0, at least at the 1% level. An observed value, $\hat{\delta}$, significant at the 5% level but not at the 1% level, is represented on the number line.

The following GenStat statement specifies that LSDs, rather than SEs, are to accompany the table of means from an analysis of variance:

```
ANOVA [FPROB = yes; PSE = lsd; LSDLEVEL = 5] saltiness
```

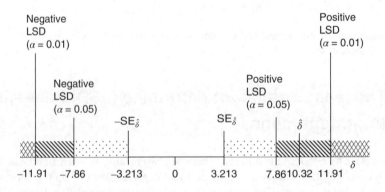

Figure 4.5 Comparison of least significant differences based on two critical regions ($\alpha = 0.05$ and $\alpha = 0.01$) and the SE of the difference.

The option setting 'PSE = lsd' specifies that SEs are to be replaced by LSDs, and the option setting 'LSDLEVEL = 5' specifies that the LSDs are to be determined at the 5% level of significance: that is, that the parameter α is to be set to 0.05. The output produced by this statement is the same as that presented in Section 4.3, except for the table of LSDs, which is as follows:

Least significant differences of means (5% level)

Table	brand	assessor	brand assessor
rep.	27	12	3
l.s.d.	7.861	8.167	16.670
d.f.	6	64	66.70

Except when comparing means with the same level(s) of

brand		16.334
d.f.		64

If the option setting 'LSDLEVEL = 5' is changed to 'LSDLEVEL = 1', the table of LSDs is as follows:

Least significant differences of means (1% level)

Table	brand	assessor	brand assessor
rep.	27	12	3
l.s.d.	11.910	10.853	22.144
d.f.	6	64	66.70

Except when comparing means with the same level(s) of

brand		21.706
d.f.		64

These numerical values are marked in Figure 4.5. Note that each LSD at the 1% level is larger than the corresponding LSD at the 5% level. As was indicated by the corresponding significance test (Section 4.4), the observed difference between Brands C and D, $\hat{\delta} = 10.32$, is significant at the 5% level, but not at the 1% level.

In the case of a balanced experimental design such as this split-plot design, for which GenStat's analysis of variance directives (BLOCKSTRUCTURE, TREATMENTSTRUCTURE and ANOVA) can be used, these directives give a clear, simple and complete statement of the point and interval estimates of treatment means. However, we will also examine how the same information is displayed in the output from the mixed modelling directives (VCOMPONENTS and REML), which can be applied to data sets that are not balanced. In the present case, the appropriate statements are

```
VCOMPONENTS [FIXED = brand * assessor] \
    RANDOM = day / presentation / serving
REML [PRINT = model, components, Wald, means] saltiness
```

The first part of the output from the REML statement was presented in Section 2.4. The tables of means are as follows:

Table of predicted means for Constant

29.28   Standard error: 1.855

Table of predicted means for brand

brand	A	B	C	D
	21.78	19.22	43.23	32.91

Standard error of differences: 3.213.

Table of predicted means for assessor

assessor	ALV	ANA	FAB	GUI	HER	MJS	MOI	NOR
	35.27	4.83	47.52	37.86	29.14	36.01	21.53	33.41

assessor	PER
	18.00

Standard error of differences: 4.088.

Table of predicted means for brand.assessor

assessor brand	ALV	ANA	FAB	GUI	HER	MJS	MOI
A	30.44	4.46	40.84	23.01	28.21	24.50	15.59
B	32.67	0.00	18.56	30.44	17.82	11.14	13.36
C	39.35	11.88	74.99	60.13	40.09	57.17	30.44
D	38.61	2.97	55.68	37.86	30.44	51.23	26.73

assessor brand	NOR	PER
A	15.59	13.36
B	33.41	15.59
C	54.20	20.79
D	30.44	22.27

Standard errors of differences.
Average	8.311
Maximum	8.351
Minimum	8.176

Average variance of differences: 69.08.

Standard error of differences for same level of factor:
	brand	assessor
Average	8.176	8.351
Maximum	8.176	8.351
Minimum	8.176	8.351

The means for the treatment terms are the same as those given by the corresponding anova, allowing for rounding error. The SEs are the same as those given by the anova, with the addition of an SE for the constant (equivalent to the grand mean from the ANOVA statement). However, the way in which the two SEs for the brand.assessor table are presented is not quite as neat as in the output of the ANOVA statement. The output first notes that the value of $SE_{Difference}$ varies according to the means being compared, and presents the average, maximum and min-imum values. The average value presented does not apply to any individual comparison. The average variance of the differences is also presented: this is almost, but not quite, the same as the square of the average value of $SE_{Difference}$ $(8.311^2 = 69.07)$. If there were a wide dis-crepancy between the two, the average variance might be considered preferable as a measure of precision, due to the additive property of variances (for an example of which see Sections 5.2 and 5.3). This information might be the best available from an unbalanced data set, but in the present case we can interpret the various values of $SE_{Difference}$ in relation to the two treat-ment factors. The next part of the output does this, noting the values of $SE_{Difference}$ that occur when the means compared have the same level of one or other of the treatment factors. As in the output of the ANOVA statement, these values show that comparisons that involve the same brand are made with greater precision. In the analysis of a balanced experimental design, as in the present case, there is only one value for each factor (8.176 for comparisons involving the same brand and 8.351 for comparisons involving the same assessor), but the output allows for the possibility of a range of values by presenting average, maximum and minimum values of $SE_{Difference}$.

4.6   Standard errors for treatment means in designed experiments: A difference in approach between analysis of variance and mixed-model analysis

Investigators often wish to present SEs of means rather than of differences between means. Although this may appear to be a simpler idea, it raises issues that do not occur in relation to SEs of differences. We will therefore consider the calculation of SEs of means in the context of a simpler experimental design, namely the randomized block design.

   In an experiment to test the effect of different manufacturing specifications (a range of combinations of particle size and compacting pressure) on the tensile strength of iron powder sinters, described by Bennett and Franklin (1954, Section 8.52, pp. 518–519), each combina-tion of manufacturing specifications was tested in each of three furnaces. Thus the experiment had a randomized block design, with furnaces as the blocks and manufacturing specifications as the treatments. The data are presented in the spreadsheet in Table 4.4. (Data reproduced by the kind permission of Wiley and Sons, Inc.)

   The manufacturing specifications comprised all combinations of two levels of compacting pressure (25 and 50) and six particle-size ranges (A–F). Thus the treatment effects could be partitioned into main effects of pressure and size, and pressure.size interaction effects. How-ever, it is simpler, and sufficient for the present purpose, to regard the 12 pressure.size combi-nations as a single treatment factor. Thus the model to be fitted to these data (Model 4.18) is

$$y_{ij} = \mu + \phi_i + \varepsilon_{ij} + \gamma_{k|ij} \tag{4.18}$$

Table 4.4 Tensile strengths of iron powder sinters manufactured to a range of specifications, investigated in an experiment with a randomized block design.

	A	B	C		A	B	C
1	furnace!	mfng_spec!	tnsl_strngth	20	2	50A	21.3
2	1	25A	11.3	21	2	50B	21.4
3	1	25B	12.2	22	2	50C	22.0
4	1	25C	12.9	23	2	50D	24.1
5	1	25D	12.1	24	2	50E	25.5
6	1	25E	16.9	25	2	50F	22.1
7	1	25F	14.3	26	3	25A	10.0
8	1	50A	21.1	27	3	25B	9.9
9	1	50B	21.1	28	3	25C	11.3
10	1	50C	21.7	29	3	25D	13.3
11	1	50D	24.4	30	3	25E	12.4
12	1	50E	23.6	31	3	25F	13.8
13	1	50F	23.5	32	3	50A	18.8
14	2	25A	11.9	33	3	50B	19.5
15	2	25B	10.4	34	3	50C	21.6
16	2	25C	12.4	35	3	50D	23.8
17	2	25D	13.9	36	3	50E	23.3
18	2	25E	14.9	37	3	50F	20.5
19	2	25F	15.0				

Source: Data reproduced by kind permission of Wiley and Sons, Inc.

where

y_{ij} = the value of tensile strength from the jth sample in the ith furnace (the ijth furnace.sample combination),

μ = the grand mean value of tensile strength,

ϕ_i = the effect of the ith furnace,

ε_{ij} = the effect of the jth sample in the ith furnace and

$\gamma_{k|ij}$ = the effect of the kth manufacturing specification, being the specification tested in the ijth furnace.sample combination.

It is assumed that the block effects are random, that is, that the ϕ_i are independent values of a random variable Φ, such that

$$\Phi \sim N(0, \sigma_\Phi^2), \tag{4.19}$$

and that the ε_{ij} are independent values of a random variable E, such that

$$E \sim N(0, \sigma^2), \tag{4.20}$$

The following GenStat statements read the data and analyse them according to Model 4.1:

```
IMPORT 'IMM edn 2\\Ch 4\\iron sinters.xlsx'
BLOCKSTRUCTURE furnace
TREATMENTSTRUCTURE mfng_spec
ANOVA [FPROB = yes; PSE = differences, means] tnsl_strngth
```

The output of the ANOVA statement is as follows:

Analysis of variance

Variate: tnsl_strngth

Source of variation	d.f.	s.s.	m.s.	v.r.	F pr.
furnace stratum	2	15.6817	7.8408	10.38	
furnace.*units* stratum					
mfng_spec	11	886.4367	80.5852	106.72	<0.001
Residual	22	16.6117	0.7551		
Total	35	918.7300			

Message: the following units have large residuals.

furnace 1 *units* 4	−1.5	s.e. 0.7
furnace 1 *units* 5	1.7	s.e. 0.7
furnace 3 *units* 5	−1.4	s.e. 0.7

Tables of means

Variate: tnsl_strngth

Grand mean 17.4

mfng_spec	25A	25B	25C	25D	25E	25F	50A
	11.1	10.8	12.2	13.1	14.7	14.4	20.4

mfng_spec	50B	50C	50D	50E	50F
	20.7	21.8	24.1	24.1	22.0

Standard errors of means

Table	mfng_spec
rep.	3
d.f.	22
e.s.e.	0.50

Standard errors of differences of means

Table	mfng_spec
rep.	3
d.f.	22
s.e.d.	0.71

The F test for the term 'mnfctrng_spec' shows that the effect of the manufacturing specification on tensile strength is highly significant. The null hypothesis that there is no real variation among furnaces,

$$H_0 : \sigma_\Phi^2 = 0, \qquad (4.21)$$

is not explicitly tested in the GenStat output. However, the F value for this term, 10.38, provides such a test, namely

$$P(F_{2,22} > 10.38) = 0.0006687.$$

Thus there is highly significant evidence against this null hypothesis.

By default, GenStat presents only the SEs of differences between the treatment means. However, the option setting 'PSE = differences, means' in the ANOVA statement has caused the SEs both of differences and of means to be produced.

Before comparing the SEs of differences and means, we will obtain the corresponding mixed-model analysis of these data. This is produced by the following GenStat statements:

```
VCOMPONENTS [FIXED = mfng_spec] RANDOM = furnace
REML [PRINT = model, components, means, wald; PSE = differences] \
   tnsl_strngth
```

The REML statement produces the following output:

REML variance components analysis

Response variate:	tnsl_strngth
Fixed model:	Constant + mfng_spec
Random model:	furnace
Number of units:	36

Residual term has been added to model.

Sparse algorithm with average information (AI) optimization.

Estimated variance components

Random term	component	s.e.
furnace	0.5905	0.6537

Residual variance model

Term	Model(order)	Parameter	Estimate	s.e.
Residual	Identity	Sigma2	0.755	0.2277

Tests for fixed effects

Sequentially adding terms to fixed model.

Fixed term	Wald statistic	n.d.f.	F statistic	d.d.f.	F pr.
mfng_spec	1173.97	11	106.72	22.0	<0.001

n.d.f., numerator degrees of freedom; d.d.f., denominator degrees of freedom.

Dropping individual terms from full fixed model.

Fixed term	Wald statistic	n.d.f.	F statistic	d.d.f.	F pr.
mfng_spec	1173.97	11	106.72	22.0	<0.001

Message: denominator DF for approximate F-tests are calculated using algebraic derivatives ignoring fixed/boundary/singular variance parameters.

Table of predicted means for Constant

17.45   Standard error: 0.467

Table of predicted means for mfng_spec

mfng_spec	25A	25B	25C	25D	25E	25F	50A
	11.07	10.83	12.20	13.10	14.73	14.37	20.40

mfng_spec	50B	50C	50D	50E	50F
	20.67	21.77	24.10	24.13	22.03

Standard error of differences: 0.7095.

We need to check whether the estimates of variance components from this analysis are consistent with the MSs in the preceding anova. The relationships among these values are as shown in Table 4.5.

It is not possible to specify in the same REML statement that SEs both of differences and of means are to be printed. However, the following additional statement will produce the SEs of means:

```
REML [PRINT = means; PSE = estimates] tnsl_strngth
```

The output of this statement is as follows:

Table of predicted means for Constant

17.45   Standard error: 0.467

Table of predicted means for mfng_spec

mtng_spec	25A	25B	25C	25D	25E	25F	50A
	11.07	10.83	12.20	13.10	14.73	14.37	20.40

mfng_spec	50B	50C	50D	50E	50F
	20.67	21.77	24.10	24.13	22.03

Standard error: 0.6697.

The two types of SEs produced by each of the two methods of analysis are collated in Table 4.6. The formulae and the numerical values for the SE of differences agree between the two methods of analysis, but those for the SE of means do not. The SE of means from the analysis of variance depends on the residual variance only, whereas that from the mixed-model analysis depends also on the variation among blocks (furnaces). Since each treatment mean is based on information from all the blocks, the latter is the more natural view of the matter. The more variation there is among blocks, the less precise must be the treatment mean, unless it is intended

Table 4.5 Relationships between estimated variance components and $MS_{Residual}$ values in the randomized block experiment on the effects of manufacturing specifications on the tensile strength of iron sinters.

g = number of treatments (manufacturing specifications) = 12.
The slight discrepancies in the numerical relationships are due to rounding error.

Algebraic relationship	Numerical relationship
$g\hat{\sigma}_{\Phi}^2 + \hat{\sigma}^2 = MS_{Residual, \text{ furnace stratum}}$	$12 \times 0.5905 + 0.755 \approx 7.8408$
$\hat{\sigma}^2 = MS_{Residual, \text{ furnace.*units*stratum}}$	$0.755 \approx 0.7551$

Table 4.6 Comparison of SEs obtained by different methods of analysis of the randomized block experiment on the effects of manufacturing specifications on the tensile strength of iron sinters.

r = number of replications (furnaces) = 3.

Method of analysis	SE of differences between means	SE of means
Analysis of variance	$\sqrt{\frac{2}{r}}\hat{\sigma} = \sqrt{\frac{2}{3} \times 0.755} = 0.709$	$\sqrt{\frac{1}{r}}\hat{\sigma} = \sqrt{\frac{1}{3} \times 0.755} = 0.502$
Mixed-model analysis	$\sqrt{\frac{2}{r}}\hat{\sigma} = \sqrt{\frac{2}{3} \times 0.755} = 0.7095$	$\sqrt{\frac{1}{r}(\hat{\sigma}_{\Phi}^2 + \hat{\sigma}^2)} =$ $\sqrt{\frac{1}{3} \times (0.5905 + 0.755)} = 0.6697$

to be *conditional on* the blocks, that is, to apply only to future observations on the same blocks. This is an unrealistic restriction, since in most cases observations will never be made again on the blocks in question. But for the purpose of a *comparison* between treatment means, variation among blocks does not matter: this is why σ_{Φ}^2 does not contribute to the SE of differences in either method of analysis. And even when the SE of means is presented in preference to the SE of differences, its true function is generally to assist the interpretation of comparisons between treatment means. It is in tacit recognition of this that σ_{Φ}^2 is omitted from the calculation of the SE of means in the output produced by GenStat's analysis of variance directives. Mixed-model analysis is a tool for purists.

If the block term is specified as a fixed-effect term in mixed-model analysis, the SEs of means are the same as those obtained from the analysis of variance, and it appears that the discrepancy between the two methods is resolved. However, the inadequacy of this approach is revealed when we return to consideration of the split-plot design. In this case, specifying the main-plot term as a fixed-effect term is not an option: to do so would result in entirely the wrong analysis. For example, in the case of the experiment considered in Chapter 2, if 'presentation' were included in the model in the TREATMENTSTRUCTURE statement, all terms would be tested against $MS_{Resid, \text{ day.presentation.serving stratum}}$, which would over-estimate the precision of comparisons between brands. The question of which variance components should be taken into account when calculating SEs for estimates from mixed modelling is considered more fully by Welham *et al.* (2004).

4.7 Use of R to obtain SEs of means in a designed experiment

The following commands import the data presented in Table 4.4 into R, fit Model 4.18 and produce the resulting anova:

```
rm(list = ls())
ironsinters <- read.table(
   "IMM edn 2\\Ch 4\\iron sinters.txt",
   header=TRUE)
attach(ironsinters)
furnace <- factor(furnace)
mfng_spec <- factor(mfng_spec)
ironsinters.aov <- aov(tnsl_strngth ~ mfng_spec + Error(furnace))
summary(ironsinters.aov)
```

The output of these commands is as follows:

```
Error: furnace
          Df Sum Sq Mean Sq F value Pr(>F)
Residuals  2  15.68   7.841

Error: Within
          Df Sum Sq Mean Sq F value Pr(>F)
mfng_spec 11  886.4   80.59   106.7 <2e-16 ***
Residuals 22   16.6    0.76
---
Signif. codes:  0 '***' 0.001 '**' 0.01 '*' 0.05 '.' 0.1 ' ' 1
```

The values are the same as those given by GenStat (Section 4.6), allowing for rounding. The following command produces the table of treatment means from this analysis:

```
model.tables(ironsinters.aov, type = "means", se = TRUE)
```

The output of this command is as follows:

```
Tables of means
Grand mean

17.45

 mfng_spec
mfng_spec
   25A    25B    25C    25D    25E    25F    50A    50B    50C
50D    50E    50F
11.067 10.833 12.200 13.100 14.733 14.367 20.400 20.667 21.767
24.100 24.133 22.033
```

```
Warning message:
In model.tables.aovlist(ironsinters.aov, type = "means", se = TRUE)
   SEs for type 'means' are not yet implemented
```

The mean values are the same as those given by GenStat. However, note that R is not able to give SEs for these means, though a message indicates that this may become possible in future.

The following commands use mixed modelling to fit Model 4.18 and summarize the results:

```
library(nlme)
ironsinters.lme.a <- lme(tnsl_strngth ~ mfng_spec,
    random = ~ 1|furnace)
anova(ironsinters.lme.a)
ironsinters.lme.b <- lme(tnsl_strngth ~ 0 + mfng_spec,
    random = ~ 1|furnace)
summary(ironsinters.lme.b)
```

The models fitted by the two lme() functions are equivalent, but differently parameterized, the first so as to produce the most informative anova, the second so as to produce more easily interpreted parameter estimates. The term '0' in the model fitted by the second lme() function prevents the fitting of an intercept, so that the first factor in the model is represented by means, rather than by departures from such an intercept. The output of the anova() function is as follows:

```
            numDF denDF    F-value p-value
(Intercept)     1    22 1398.0772  <.0001
mfng_spec      11    22  106.7246  <.0001
```

The *F* value for 'mfng_spec' and the DF are the same as those obtained by GenStat. Part of the output of the function summary() is as follows:

```
Fixed effects: tnsl_strngth ~ 0 + mfng_spec
                    Value Std.Error DF  t-value p-value
mfng_spec25A 11.06667 0.6697153 22 16.52444        0
mfng_spec25B 10.83333 0.6697153 22 16.17603        0
mfng_spec25C 12.20000 0.6697153 22 18.21670        0
mfng_spec25D 13.10000 0.6697153 22 19.56055        0
mfng_spec25E 14.73333 0.6697153 22 21.99940        0
mfng_spec25F 14.36667 0.6697153 22 21.45190        0
mfng_spec50A 20.40000 0.6697153 22 30.46071        0
mfng_spec50B 20.66667 0.6697153 22 30.85889        0
mfng_spec50C 21.76667 0.6697153 22 32.50138        0
mfng_spec50D 24.10000 0.6697153 22 35.98544        0
mfng_spec50E 24.13333 0.6697153 22 36.03522        0
mfng_spec50F 22.03333 0.6697153 22 32.89955        0
```

These treatment means and their common SE agree with those produced by GenStat. The t tests relate to the null hypothesis that the true mean for the treatment in question is zero, which is not of interest.

4.8 Use of SAS to obtain SEs of means in a designed experiment

The following SAS statements import the data presented in Table 4.4 into SAS, fit Model 4.18 and produce the resulting anova:

```
PROC IMPORT OUT = sinters DBMS = EXCELCS REPLACE
    DATAFILE = "&pathname.\IMM edn 2\Ch 4\iron sinters.xlsx";
    SHEET = "for SAS";
RUN;

ODS RTF;
PROC ANOVA;
    CLASS furnace mfng_spec;
    MODEL tnsl_strngth = furnace mfng_spec;
    MEANS mfng_spec;
RUN;
ODS RTF CLOSE;
```

Part of the output of PROC ANOVA is as follows:

Source	DF	Sum of squares	Mean square	F value	Pr > F
Model	13	902.1183333	69.3937179	91.90	<0.0001
Error	22	16.6116667	0.7550758		
Corrected total	35	918.7300000			

R-square	Coeff Var	Root MSE	tnsl_strngth mean
0.981919	4.979662	0.868951	17.45000

Source	DF	Anova SS	Mean square	F value	Pr > F
furnace	2	15.6816667	7.8408333	10.38	0.0007
mfng_spec	11	886.4366667	80.5851515	106.72	<0.0001

Level of mfng_spec	N	tnsl_strngth	
		Mean	Std Dev
25A	3	11.0666667	0.97125349
25B	3	10.8333333	1.20968315
25C	3	12.2000000	0.81853528
25D	3	13.1000000	0.91651514
25E	3	14.7333333	2.25462488
25F	3	14.3666667	0.60277138
50A	3	20.4000000	1.38924440
50B	3	20.6666667	1.02143690
50C	3	21.7666667	0.20816660
50D	3	24.1000000	0.30000000
50E	3	24.1333333	1.19303534
50F	3	22.0333333	1.50111070

The values in the anova tables are equivalent to those given by GenStat, and the treatment means agree with those given by GenStat. The values of Std Dev are simply calculated from the observations on each treatment. Consequently they vary from treatment to treatment, and contain a component due to variation among blocks (furnaces), although furnace has not been specified as a random-effect term. The SEs of the treatment means are obtained from these values by the formula

$$SE_{\text{Treatment mean}} = \frac{\text{Std Dev}}{\sqrt{r}} \qquad (4.22)$$

where

r = No. of observations contributing to the treatment mean.

For example, the SE of the mean for treatment 25A is

$$SE_{\text{Treatment mean}} = \frac{0.97125349}{\sqrt{3}} = 0.5608.$$

In the present simple experimental design, the values of Std Dev given by SAS are connected to the value of $SE_{\text{Treatment mean}}$ given by the GenStat REML statement by the formula

$$SE_{\text{Treatment mean}} = \sqrt{\frac{\text{mean}([\text{Std Dev}]^2)}{r}}. \qquad (4.23)$$

The following statements use PROC MIXED, and hence mixed modelling, to fit Model 4.18:

```
ODS RTF;
PROC MIXED ASYCOV NOBOUND;
   CLASS furnace mfng_spec;
   MODEL tnsl_strngth = mfng_spec /DDFM = KR HTYPE = 1;
```

```
    RANDOM furnace;
    LSMEANS mfng_spec;
RUN;
ODS RTF CLOSE;
```

Part of the output of PROC MIXED is as follows:

| Convergence criteria met. |

Covariance parameter estimates	
Cov Parm	**Estimate**
furnace	0.5905
Residual	0.7551

Asymptotic covariance matrix of estimates			
Row	**Cov Parm**	**CovP1**	**CovP2**
1	furnace	0.4273	−0.00432
2	Residual	−0.00432	0.05183

Least squares means						
Effect	**mfng_spec**	**Estimate**	**Standard error**	**DF**	**t value**	**Pr > |t|**
mfng_spec	25A	11.0667	0.6697	7.7	16.52	<0.0001
mfng_spec	25B	10.8333	0.6697	7.7	16.18	<0.0001
mfng_spec	25C	12.2000	0.6697	7.7	18.22	<0.0001
mfng_spec	25D	13.1000	0.6697	7.7	19.56	<0.0001
mfng_spec	25E	14.7333	0.6697	7.7	22.00	<0.0001
mfng_spec	25F	14.3667	0.6697	7.7	21.45	<0.0001
mfng_spec	50A	20.4000	0.6697	7.7	30.46	<0.0001
mfng_spec	50B	20.6667	0.6697	7.7	30.86	<0.0001
mfng_spec	50C	21.7667	0.6697	7.7	32.50	<0.0001
mfng_spec	50D	24.1000	0.6697	7.7	35.99	<0.0001
mfng_spec	50E	24.1333	0.6697	7.7	36.04	<0.0001
mfng_spec	50F	22.0333	0.6697	7.7	32.90	<0.0001

The variance component estimates agree with those produced by GenStat, as do the treatment means and their common SE.

4.9 Summary

Each parameter estimate (slope, intercept or mean) is a *point estimate* and can be enclosed in an *interval estimate*.

In order to obtain an interval estimate for a parameter β, the parameter estimate $\hat{\beta}$ is regarded as an observation of a random variable, usually with distribution

$$\hat{B} \sim N\left(\beta, \sigma_{\hat{\beta}}^2\right).$$

An estimate of $\sigma_{\hat{\beta}}^2$, designated $\hat{\sigma}_{\hat{\beta}}^2$, can be obtained from the data, and the SE of $\hat{\beta}$ is given by its square root,

$$SE_{\hat{\beta}} = \hat{\sigma}_{\hat{\beta}}.$$

An interval estimate for the true parameter value β is given by $\hat{\beta} \pm SE_{\hat{\beta}}$.

When several observations of a response variable Y are taken at each value of an explanatory variable X, and β is the slope of the regression line relating Y to X, an appropriate estimate of $SE_{\hat{\beta}}$ can be obtained either by performing a regression analysis on the group means, or by a mixed-model analysis.

If the data are analysed as if the observations within each group were mutually independent, $SE_{\hat{\beta}}$ will be under-estimated, that is, the precision of $\hat{\beta}$ will be over-estimated.

The value of $SE_{\hat{\beta}}$ depends on the variance component estimates for the terms that are relevant to the estimation of β. Larger variance component estimates give a larger value of $SE_{\hat{\beta}}$.

In a designed experiment, the emphasis is on comparison between treatments. Therefore, SEs of differences between means are often presented, rather than SEs of the means themselves.

In a split-plot experiment, several types of comparison can be made. The variance components that contribute to $SE_{\text{difference}}$ for each type are shown in Table 4.7.

An interval estimate for the difference between the true means, δ, is given by

$$\hat{\delta} \pm SE_{\hat{\delta}}.$$

The significance of the difference between treatment means can be determined by a t test.

The DF of the t statistic depends on the precision of the variance estimate $\hat{\sigma}_{\hat{\delta}}^2$, that is, the number of pieces of information that effectively contribute to the term in question. The value is given by DF_{Residual} in the appropriate stratum of the anova, except in the case of a comparison

Table 4.7   Variance components that contribute to each type of comparison between treatments in a split-plot design.

Type of comparison*	Variance components that contribute to $SE_{\text{difference}}$
Between levels of Factor A	Main-plot residual, sub-plot residual
Between levels of Factor B	Sub-plot residual only
Between A.B combinations with different levels of Factor A	Main-plot residual, sub-plot residual
Between A.B combinations with the same level of Factor A	Sub-plot residual only

* Factor A = the treatment factor that varies only among main plots, not within each main plot.
Factor B = the treatment factor that varies among sub-plots within each main plot.

between A.B combinations with different levels of Factor A, when information from two strata is combined and a non-integer value of DF must be calculated.

The difference between two means that gives a t value that is just significant at a particular level (say, $\alpha = 0.05$, the 5% significance level) is called the *least significant difference*.

The LSD provides the basis for an alternative interval estimate for δ, the $100 \times (1 - \alpha)\%$ confidence interval (i.e. if $\alpha = 0.05$, the 95% confidence interval). This is given by

$$CI = \hat{\delta} \pm LSD_{\hat{\delta}}.$$

In any model term, the variance components that contribute to SE_{mean} are not necessarily the same as those that contribute to $SE_{difference}$. For example, in a randomized block design, when the treatment means are considered and compared using GenStat's mixed modelling directives, $\hat{\sigma}^2_{block}$ contributes to SE_{mean} but not to $SE_{difference}$.

When the treatment means are presented using GenStat's analysis of variance directives, $\hat{\sigma}^2_{block}$ does not contribute to either SE_{mean} or $SE_{difference}$. This is only valid if the means are intended to be conditional upon the blocks observed, which is unrealistic. However, SE_{mean} is usually presented in order to assist the comparison of means, and it is in tacit recognition of this that $\hat{\sigma}^2_{block}$ is omitted from both SEs.

4.10 Exercises

4.1 Refer to your results from Exercise 1.1, in which effects on the speed of greyhounds were modelled. Obtain SEs for the estimates of the constant and of the effect of age that you obtained from your mixed model.

4.2 Refer to your results from Exercise 2.1, in which effects on the yield of oats were modelled. Examine the output produced by the analysis of variance to determine the effects of variety and nitrogen level in a split-plot design.

(a) Obtain the SE for the difference between the mean yields for the following factor levels or combinations of levels:

(i) nitrogen Level 1 vs. nitrogen Level 3

(ii) variety 'Victory' at nitrogen Level 1 vs. variety 'Victory' at nitrogen Level 3

(iii) variety 'Victory' at nitrogen Level 1 vs. variety 'Golden Rain' at nitrogen Level 3.

(b) Obtain the LSD at the 5% level between variety means. Are there any two varieties that are significantly different according to this criterion?

(c) Obtain the SE for nitrogen-level means on the following bases:

(i) conditional on the blocks and main plots

(ii) taking into account the random effects of blocks and main plots.

4.3 (a) If you are using the software R to solve these exercises, use the method given in Section 2.8 to obtain SEs for the following results from the split-plot experiment to compare commercial brands of ravioli:

(i) the difference between the means for treatments 'Brand C, Assessor FAB' and 'Brand C, Assessor ANA'

(ii) the difference between the means for Brand C and Brand B

(iii) the difference between the means for Assessor ANA and Assessor FAB

  (iv) the mean for treatment 'Brand B, Assessor ANA'

  (v) the mean for Brand B

  (vi) the mean for Assessor ANA.

(b) For each result in Part (a), and for the difference between the means for treatments 'Brand C, Assessor FAB' and 'Brand B, Assessor ANA' considered in Section 2.8, state which other results have the same SE.

References

Bennett, C.A. and Franklin, N.L. (1954) *Statistical Analysis in Chemistry and the Chemical Industry*, John Wiley & Sons, Inc., New York, 724 pp.

Moore, D.S. and McCabe, G.P. (2012) *Introduction to the Practice of Statistics*, 7th edn, Freeman, New York, 694 pp.

Taylor, J. (1950) The comparison of pairs of treatments in split-plot experiments. *Biometrika*, **37**, 443–444.

Satterthwaite, F.E. (1946) An approximate distribution of estimates of variance components. *Biometrics Bulletin*, **2**, 110–114.

Steel, R.G.D. and Torrie, J.H. (1981) *Principles and Procedures of Statistics: A Biometrical Approach*, 2nd edn, McGraw-Hill.

Welham, S., Cullis, B., Gogel, B. *et al.* (2004) Prediction in linear mixed models. *Australia and New Zealand Journal of Statistics*, **46**, 325–347.

5

Estimation of random effects in mixed models: Best Linear Unbiased Predictors (BLUPs)

5.1 The difference between the estimates of fixed and random effects

In ordinary regression analysis and analysis of variance, the random-effect terms are always *nuisance variables*: residual variation or block effects. The effects of individual levels of such terms are not of interest. However, we have seen that in a mixed model, effects that are of intrinsic interest may be specified as random – for example, the effects of the individual breeding lines in the barley field trial discussed in Chapter 3. The decision to specify a term as random causes a fundamental change in the way in which the effect of each level of that term is estimated. We will illustrate this change and its consequences in the context of the barley breeding lines. However, the concepts introduced and the arguments presented apply equally to any situation in which replicated, quantitative evaluations are available for the comparison of members of some population – for example, new chemical entities to be evaluated as potential medicines by a pharmaceutical company, or candidates for admission to a university on the basis of their examination scores.

It is easy to illustrate the relationship between the estimates of fixed and random effects in data that are grouped by a single factor – for example, a fully randomized design leading to a one-way anova. The data from the barley field trial are classified by two factors, line and block. Fortunately, however, the effects of blocks are negligible (Section 3.10). We will therefore treat this experiment as having a single-factor design, reducing Model 3.24 from

$$y_{ij} = \mu + \delta_i + \epsilon_{ij} + \phi_{k|ij}$$

to Model 5.1, namely

$$y_j = \mu + \phi_{k|j} + \epsilon_j \tag{5.1}$$

Introduction to Mixed Modelling: Beyond Regression and Analysis of Variance, Second Edition. N. W. Galwey.
© 2014 John Wiley & Sons, Ltd. Published 2014 by John Wiley & Sons, Ltd.
Companion website: http://www.wiley.com/go/beyond_regression

where the block effect δ_i is omitted and

y_j = the grain yield of the jth plot,

μ = the grand mean (overall mean) value of grain yield,

δ_i = the effect of the ith block,

$\phi_{k|j}$ = the effect of the kth breeding line, being the line sown in the jth plot,

ϵ_j = the residual effect of the jth plot.

We can then compare an analysis in which the variation among lines is specified as a fixed-effect term with one in which it is specified as a random-effect term. In the analysis of designed experiments, it is not normal practice to omit features of the design from the model fitted: it is justified in the present case solely in the interests of clarity.

The variation among lines in the original, 'unpadded' data set can be modelled as a fixed-effect term using GenStat's analysis of variance directives, as follows:

```
IMPORT 'IMM edn 2\\Ch 3\\barley progeny.xlsx'; SHEET = 'Sheet1'
BLOCKSTRUCTURE
TREATMENTSTRUCTURE line
ANOVA [FPROBABILITY = yes] yield_g_m2
```

The same model can be fitted using the mixed-modelling directives, with no random-effects model specified, as follows:

```
VCOMPONENTS [FIXED = line]
REML [PRINT = model, components, deviance, means; \
    PTERMS = 'constant' + line] yield_g_m2
```

In order to specify 'line' as a random-effects term, it is moved to the random-effects model in the VCOMPONENTS statement; thus:

```
VCOMPONENTS RANDOM = line
REML [PRINT = model, components, deviance, means, effects; \
    PTERMS = 'constant'+ line; METHOD = Fisher] yield_g_m2
```

In order to obtain standard errors (SEs) of the differences between means, the option setting 'METHOD = Fisher' must be used (see Sections 2.5 and 11.10).

The breeding-line means obtained from these two analyses are compared in Table 5.1. They are not the same: in the case of a high-yielding line such as Line 7, the random-effect mean is lower than the fixed-effect mean, whereas for a low-yielding line such as Line 16, the opposite is the case. The fixed-effect means are the simple means of the observations for the line in question. For example, the mean for Line 7 is

$$\frac{907.38 + 820.08}{2} = 863.73,$$

whereas that for Line 3, which occurs only in Block 1, is the single plot value 873.04. Each of these means is taken as an estimate of the true mean yield of the breeding line in question. But the plant breeder knows that if he/she selects the highest-yielding lines this year for further evaluation, he/she is selecting on both genetic and environmental variation. This year's environmental component will not contribute to the selected lines' yield next year, and consequently, their mean yield will generally be somewhat lower. We have already seen how this

Table 5.1 Comparison between the estimates of mean yields of breeding lines of barley obtained when breeding line is specified as either a fixed- or a random-effect model term.

Line	Fixed-effect mean	Random-effect mean	Line	Fixed-effect mean	Random-effect mean
1	654.5	639.9	43	757.6	724.8
2	483.3	510.2	44	695.4	673.6
3	873.0	782.5	45	547.8	552.2
4	719.1	674.9	46	675.6	657.3
5	799.0	730.7	47	192.4	306.9
6	802.4	761.6	48	715.6	690.2
7	863.7	812.1	49	826.9	781.8
8	468.2	486.7	50	803.4	762.4
9	681.0	661.7	51	719.3	693.3
10	760.8	727.4	52	603.0	593.8
11	580.8	579.4	53	559.1	563.2
12	436.6	477.5	54	834.4	787.9
13	508.1	519.5	55	555.3	560.5
14	922.7	860.6	56	479.1	495.7
15	600.1	591.7	57	182.0	299.6
16	225.4	286.9	58	478.5	495.2
17	755.5	723.1	59	594.0	590.2
18	315.8	361.3	60	428.8	454.3
19	523.2	532.0	61	542.0	547.4
20	632.5	614.4	62	730.5	682.9
21	514.6	532.1	63	659.4	644.0
22	340.7	410.5	64	664.8	648.4
23	711.6	686.9	65	766.1	731.7
24	567.2	568.2	66	767.0	732.5
25	428.6	472.0	67	554.9	558.0
26	609.8	603.2	68	355.7	394.2
27	747.1	716.2	69	423.1	449.6
28	679.9	647.5	70	950.5	883.5
29	769.4	734.4	71	582.7	580.9
30	353.3	419.3	72	212.2	276.1
31	196.4	263.1	73	355.3	420.7
32	753.1	721.0	74	520.4	529.6
33	808.5	766.7	75	617.3	603.8
34	656.3	641.4	76	344.5	413.2
35	464.5	483.6	77	357.9	422.6
36	927.5	864.5	78	260.1	315.5
37	724.0	697.1	79	378.4	412.8
38	591.0	587.8	80	483.0	498.9
39	179.3	249.0	81	309.7	356.3
40	489.6	504.3	82	251.3	348.1
41	429.9	455.2	83	280.4	368.4
42	673.0	655.2			

leads to an expected genetic advance under selection that is smaller than the difference between the mean of the selected lines and that of the full set of lines (Section 3.15). Mixed-model analysis provides a way of building the pessimism of the plant breeder more fully into the formal analysis of the data, giving a similar adjustment to the mean of each individual breeding line. This adjusted mean is the random-effect mean.

5.2   The method for estimation of random effects: The best linear unbiased predictor (BLUP) or 'shrunk estimate'

The adjustment to obtain the random-effect mean is made as follows. Following Model 5.1, the true mean of the kth breeding line is represented by

$$\mu_k = \mu + \phi_k. \tag{5.2}$$

In the fixed-effect means in Table 5.1, this value is estimated by

$$\hat{\mu}_k = \frac{\sum_{j=1}^{r_k} y_{kj}}{r_k} \tag{5.3}$$

where
y_{kj} = the jth observation of the kth breeding line,
r_k = the number of observations of the kth breeding line.

The overall mean of the population of breeding lines, μ, is estimated by mixed modelling as about

$$\hat{\mu} = 572.6$$

(Chapter 3 – GenStat: Section 3.10, Constant = 572.6; R: Section 3.16, Intercept = 572.47; SAS: Section 3.17, Intercept = 572.65). Note that this is not quite the same as the mean of all the observations (=581.6) or the mean of the line means (=569.1). Although we are treating this estimate as given, and using it to explain the estimation of the effects of individual lines, these two estimation steps are really interconnected in a single process. Something more will be said about the estimation of μ in a moment.
Rearranging Equation 5.2, we obtain

$$\phi_k = \mu_k - \mu. \tag{5.4}$$

Similarly, an estimate of ϕ_k is given by

$$\hat{\phi}_k = \hat{\mu}_k - \hat{\mu}. \tag{5.5}$$

This ordinary estimate of the difference between a treatment mean and the overall mean is called the *Best Linear Unbiased Estimate* (*BLUE*). To allow for the expectation that high-yielding lines in the present trial will perform less well in a future trial – and that low-yielding lines will perform better – the BLUE can be replaced by a 'shrunk estimate' called the *Best Linear Unbiased Predictor* (*BLUP*). The formula for the required shrinkage is

$$\mathrm{BLUP}_k = \mathrm{BLUE}_k \cdot \text{shrinkage factor}_k = (\hat{\mu}_k - \hat{\mu}) \cdot \left(\frac{\hat{\sigma}_G^2}{\hat{\sigma}_G^2 + \frac{\hat{\sigma}_E^2}{r_k}} \right). \tag{5.6}$$

The estimated variance components in this equation, $\hat{\sigma}_G^2$ and $\hat{\sigma}_E^2$, are obtained by mixed modelling as described in Section 3.10. Note that the shrinkage factor is the same as the estimated heritability defined in Equation 3.33, except that the average number of replications per line is replaced by the actual number for the line under consideration. The relationship in Equation 5.6, combined with the constraint

$$\sum_{k=1}^{p} \text{BLUP}_k = 0, \tag{5.7}$$

where
p = number of breeding lines,

determines the value of $\hat{\mu}$ as well as those of the BLUPs. For Line 7, substituting the values given by Model 5.1, we obtain:

$$\text{BLUP}_7 = (863.7 - 572.5) \cdot \left(\frac{30667}{30667 + \frac{13226}{2}} \right) = 239.54$$

(Note that the variance component estimates are slightly different from those obtained in Section 3.10, because the block term has been dropped in Model 5.1.) A new estimate of the mean for the kth breeding line is then given by

$$\hat{\mu}_k' = \hat{\mu} + \text{BLUP}_k. \tag{5.8}$$

For Line 7,

$$\hat{\mu}_7' = 572.5 + 239.54 = 812.04.$$

As Line 7 is relatively high-yielding, its shrunk mean, 812.04, is lower than its unadjusted mean, 863.7.

The original estimates (BLUE_k and $\hat{\mu}_k$) are compared with the shrunk estimates (BLUP_k and $\hat{\mu}_k'$) for an arbitrary subset of the breeding lines (about a quarter of the total) in Figure 5.1. Note that BLUP_k is shrunk towards zero relative to BLUE_k, and $\hat{\mu}_k'$ is correspondingly shrunk towards the estimate of the overall mean. In practice, it is usually the values at one extreme that are of interest – for example, the high-yielding breeding lines. Values far from the grand mean are shrunk more than those close to the mean. As should be expected, the amount of shrinkage specified by Equation 5.6 is large when

- the genetic variance, $\hat{\sigma}_G^2$, is small

- the environmental variance, $\hat{\sigma}_E^2$, is large

- the number of replications of the breeding line under consideration, r_k, is small.

Provided that the number of replications is constant over breeding lines, the shrinkage of BLUPs does not change the ranking of the means. However, if the number of replications is unequal, *crossovers* may occur, as in the present case, where Line 3 is estimated to be higher yielding than Line 7 on the basis of their BLUEs, but lower yielding on the basis of their BLUPs.

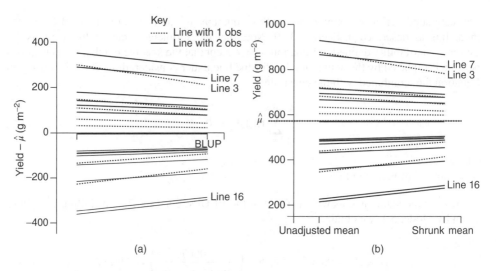

Figure 5.1   (a,b) Comparison between BLUEs and BLUPs, and between unadjusted and shrunk means, for yields of breeding lines of barley.

5.3   The relationship between the shrinkage of BLUPs and regression towards the mean

The relationship between the shrunk and unadjusted means can also be illustrated by a scatter diagram, as shown in Figure 5.2. Each point represents a breeding line. The shrinkage towards the overall mean is indicated by the fact that the points representing breeding lines that have an estimated yield above $\hat{\mu}$ lie below the line

shrunk mean = unadjusted mean,

whereas those representing breeding lines with an estimated yield below $\hat{\mu}$ lie above this line. That is, the points lie approximately along a line that is flatter than this line. Moreover, points based on a single observation lie on a flatter line than those based on two observations. The crossover of Lines 3 and 7 is indicated by the fact that the point representing Line 3 lies below, but to the right of, that representing Line 7.

The flattening of the line of points on this scatter diagram is reminiscent of the commonly observed phenomenon of *regression towards the mean*, and this is no accident. An exploration of the connection between the two phenomena will help to clarify the distinction between the BLUE and the BLUP, and the sense in which each can be regarded as a 'best' statistic.

Suppose that the values of σ_G^2 and σ_E^2 for a population of breeding lines are known with considerable precision from experiments like that just described. Then suppose that the yield of a large number of new breeding lines, drawn at random from the same underlying population, is measured in an experiment with a single replication. Though much is known about the population, not much is known about the new lines individually – only the information from a single observation on each. If another experiment were performed on the same large sample of breeding lines, a second observation would be obtained on each. The relationship between the present and future observations (designated Y_{obs} and Y_{new}, respectively) would

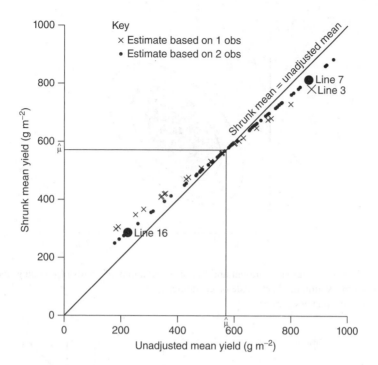

Figure 5.2 An alternative representation of the comparison between unadjusted and shrunk mean yields of breeding lines of barley.

then be as shown in Figure 5.3. Each point represents the value of the present observation on an individual breeding line, and a possible value for the future observation. The figure shows that the performance of each breeding line in the future experiment can be predicted from the available information, but not very accurately. High-yielding lines in the present experiment will generally give a high yield in the future experiment, and low-yielding lines a low future yield, but this relationship, which is due to the genetic value of each line, is blurred by the environmental component of each observation, present and future. The distribution of points in the figure – the probability distribution – is summarized by the ellipses. These are contour lines, each indicating a path along which the density of points (the probability density) is constant – a high density on the inner ellipse, a low density on the outer one. As all these density contours are the same shape, and concentric, a single contour is sufficient to indicate the general shape of the distribution, and this convention will be used in subsequent figures.

The criterion for the conversion of a BLUE to a BLUP, given in Equation 5.6, can also be represented graphically on this probability distribution, as follows. Consider Figure 5.4. The variance of the observed values in the present experiment is given by

$$\text{var}(Y_{\text{obs}}) = \sigma_G^2 + \sigma_E^2, \tag{5.9}$$

and in a future experiment $\text{var}(Y_{\text{new}})$ will be the same. The genetic component of each existing observation contributes to new observations on the same breeding line, but the environmental component does not: hence the covariance between Y_{obs} and Y_{new} is given by

$$\text{cov}(Y_{\text{obs}}, Y_{\text{new}}) = \sigma_G^2. \tag{5.10}$$

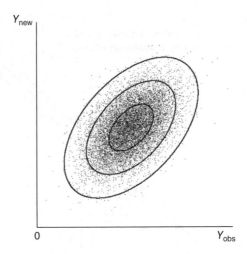

Figure 5.3 Joint distribution of present and future observations of yields of breeding lines of barley, when a single observation has been made on each line.

For explanation of conventions, see text.

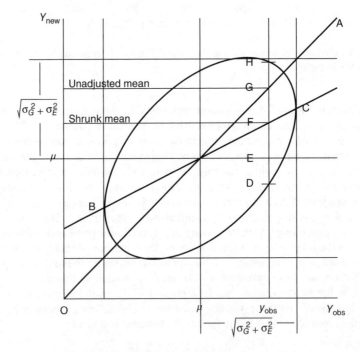

Figure 5.4 Summary of the relationship between present and future observations of yields of breeding lines of barley, when a single observation has been made on each line, showing the criterion for the conversion of a BLUE to a BLUP.

For an explanation of the annotations on this figure, see the text.

These variance and covariance values specify the bivariate-Normal distribution of Y_{obs} and Y_{new}, and the ellipse in the figure represents the 1-standard-deviation probability-density contour of this distribution. The simplest prediction of the future yield of any breeding line is its yield in the present experiment. This prediction is given by the equation

$$Y_{new} = Y_{obs},$$

which is represented by the line OA on the figure. For a breeding line that gives a yield y_{obs}, the point G is the corresponding position on OA, and the vertical coordinate of this point gives the prediction of the line's future yield. This is the unadjusted mean (based on a single observation in the present case), and the corresponding BLUE is given by its difference from the overall mean μ, that is, the distance EG. But the figure shows that this will not be the mean future yield of all the lines that give a yield of y_{obs} in the present experiment. The distances DF and FH are equal: hence the distribution of Y_{new}, *among those lines for which*

$$Y_{obs} = y_{obs},$$

is symmetrical about the point F, and the vertical coordinate of this point is the mean value of this distribution. This is the shrunk mean, and the corresponding BLUP is given by its difference from μ, that is, the distance EF. The point F lies on the line BC, which connects the left-most and right-most points on the ellipse, and which is the line of best fit obtained when Y_{obs} is treated as the explanatory variable, and Y_{new} as the response variable, in a regression analysis. For values of Y_{obs} above μ, BC lies below OA and the expected value of Y_{new} is less than Y_{obs}. Conversely, for values of Y_{obs} below μ, BC lies above OA and the expected value of Y_{new} is greater than Y_{obs}. This is the phenomenon of regression towards the mean noted by Francis Galton, which gave regression analysis its name and which was mistakenly interpreted as indicating that, over time, a biological population would converge to mediocrity unless steps were taken to prevent this. The connection between the shrinkage of BLUPs and regression towards the mean is also noted – together with connections to many other areas of statistics – in the classic paper by Robinson (1991, Section 5.2).

Now consider the situation when prediction is based not on a single observation of each breeding line, Y_{obs}, but on the mean of r observations, designated $\overline{Y}_{obs}$. This case is illustrated, together with the previous one, in Figure 5.5. We note that

$$\mathrm{var}(\overline{Y}_{obs}) = \sigma_G^2 + \frac{\sigma_E^2}{r}, \tag{5.11}$$

less than $\mathrm{var}(Y_{obs})$, due to the greater reliability of the mean of several observations, and the consequent smaller contribution of environmental variation. However, the variance of individual new observations, Y_{new}, is unchanged. Consequently, the ellipse is steeper, and so is the regression line $B'C'$: it is closer to the line OA. There is less regression towards the mean. Specifically, the regression line is given by

$$(Y_{new} - \mu) = (\overline{Y}_{obs} - \mu) \cdot \left(\frac{\sigma_G^2}{\sigma_G^2 + \frac{\sigma_E^2}{r}} \right). \tag{5.12}$$

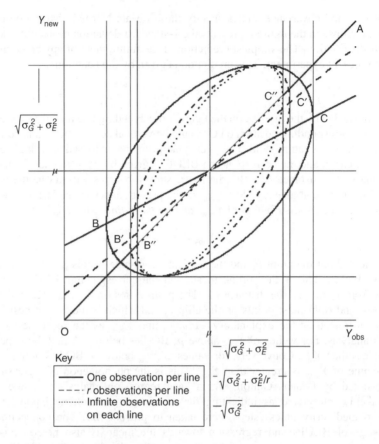

Figure 5.5 Summary of the relationship between present and future observations of yields of breeding lines of barley when r observations have been made on each line, showing the reduced discrepancy between BLUE and BLUP.

Comparison of this equation with Equation 5.6 shows that the amount of regression towards the mean is precisely equivalent to the shrinkage of the BLUP relative to the BLUE.

Finally, consider the situation when the prediction is based on a large – effectively infinite – number of observations, also illustrated in Figure 5.5. The effect of environmental variation on $\overline{Y}_{obs}$ is eliminated and $\mathrm{var}(\overline{Y}_{obs})$ is consequently reduced to σ_G^2, making the ellipse so steep that the regression line B″C″ is superimposed on the line OA. $\overline{Y}_{obs}$ now needs no adjustment to give the expected value of Y_{new}: there is no regression towards the mean, and no shrinkage of the BLUE is required to obtain the BLUP. If the plant breeder could attain this situation (which would require unlimited experimental resources), he/she would have no need of pessimism in his/her predictions: they would be based on full knowledge of the genetic potential of each line and would require no adjustment to indicate its future *mean* performance (though future individual observations would still vary as much as ever).

Except in this limiting case, the BLUE and the BLUP are different, so they cannot both be 'best' for the same purpose. For what purpose should each be preferred? The answer to this question lies in a comparison of the following two equations:

$$E(Y_{\text{new}}|\mu_k) = \mu_k = E(\hat{\mu} + \text{BLUE}_k) \tag{5.13}$$

$$E(Y_{\text{new}}|\bar{y}_k, \hat{\mu}, \hat{\sigma}_G^2, \hat{\sigma}_E^2) = \mu + (\bar{y}_k - \mu) \cdot \left(\frac{\sigma_G^2}{\sigma_G^2 + \frac{\sigma_E^2}{r_k}} \right) = E(\hat{\mu} + \text{BLUP}_k). \tag{5.14}$$

Translated into words, these equations state the following:

- Equation 5.13. The expected value of a new observation on breeding line k, given that the true mean value for this breeding line is the unknown value μ_k, is also the expected value of the unadjusted mean of the sample of observations on this breeding line.

- Equation 5.14. The expected value of a new observation on breeding line k, given the unadjusted mean for this line, *together with the specified information about the population of lines to which it belongs* ($\hat{\mu}, \hat{\sigma}_G^2$ and $\hat{\sigma}_E^2$), is also the expected value of the shrunk mean.

Thus if a factor is specified as fixed, each level sampled is considered to tell us only about itself, but if it is specified as random, each level is considered to provide some information concerning the other levels in the sample. Our best prediction about the future performance of Line k as a barley variety (or new chemical entity k as a medicine, or examination candidate k as a university student) therefore depends partly on our reason for taking an interest in it. If we have chosen Line k because of its relative performance among the set of factor levels under consideration – for example, because it is high-yielding relative to the other lines studied – then we should be guided by the BLUP. However, if we have chosen it for reasons that have nothing to do with its relative yield, for example, its brewing quality or its resistance to some disease – in short, if we have chosen it 'because it is Line k' – then we should ask ourselves whether the comparative performance of other lines is relevant, and if we decide that it is irrelevant, we should be guided by the BLUE. Unfortunately there is no simple, objective criterion for deciding whether the other levels of the factor under consideration are relevant. As we have seen (Section 1.6), it is not necessary that the levels comprise a representative sample from a population: it is sufficient that they form an exchangeable set, and more will be said about this situation below (Section 5.6). But how should we decide when this weaker criterion is met? 'There is nothing in (the mathematics) that requires the component problems (i.e. the factor levels) to have some sensible relation to each other' (Efron and Morris, 1977). This is *Stein's Paradox*, the BLUP being closely related to the *James–Stein estimator*. Efron and Morris give a fairly accessible account of this difficult issue, and Robinson (1991, Sections 5.3 and 6.1) robustly asserts that 'estimates of the characteristics of butterflies in Brazil, ball bearings in Birmingham and Brussels sprouts in Belgium ought not to be related to each other'.

However, sometimes common sense will suggest that factor levels should be regarded as exchangeable even though we have a special interest in one particular level. For example, suppose that we are interested in the performance of examination candidate k not because of his or her position in the distribution of students, but because we are his or her anxious parents. If Candidate k's performance is excellent on this occasion, we will of course hope for good results in the future. But we may prudently remember that there is an element of chance in these things, note his/her position in the overall distribution, resolve not to set our hopes too high, and specify 'candidate' as a random-effect term in our model. We can make our assumption of exchangeability more robust by specifying that only levels of the same type as Level k should be considered – barley lines that share the brewing quality or disease resistance, chemical entities with a similar molecular structure or university students in the same socio-economic group. Or we may specify that all levels should be considered, but with an adjustment for such additional variables. This was our approach when we decided that the house prices in every town in the sample discussed in Chapter 1 were relevant to our prediction of house prices in Durham, but only after taking into account the latitude of each town, by including 'latitude' as a fixed-effect model term.

5.4　Use of R for the estimation of fixed and random effects

The following R commands import the original, 'unpadded' data on the barley breeding lines, fit the model omitting the block term and specifying 'line' as a fixed-effect term and print the results:

```
rm(list = ls())
barleyprogeny.unbalanced <- read.table(
   "IMM edn 2\\Ch 3\\barley progeny.txt",
   header=TRUE)
attach(barleyprogeny.unbalanced)
fline <- factor(line)
fblock <- factor(block)
barleyprogeny.model1lm <- lm(yield_g_m2 ~ fline)
summary(barleyprogeny.model1lm)
```

The output of the function `summary()` is as follows:

```
Call:
lm(formula = yield_g_m2 ~ fline)

Residuals:
    Min      1Q  Median      3Q     Max
-203.71  -43.41    0.00   43.41  203.71

Coefficients:
             Estimate Std. Error t value Pr(>|t|)
(Intercept)   654.485     81.442   8.036 4.75e-11 ***
fline2       -171.155    141.062  -1.213 0.229839
fline3        218.555    141.062   1.549 0.126645
fline4         64.655    141.062   0.458 0.648389
fline5        144.505    141.062   1.024 0.309825
fline6        147.870    115.177   1.284 0.204212
fline7        209.245    115.177   1.817 0.074339 .
fline8       -186.300    115.177  -1.618 0.111102
```

```
fline9           26.470      115.177     0.230 0.819026
fline10         106.270      115.177     0.923 0.359938
.
.
.
fline82        -403.185      141.062    -2.858 0.005878 **
fline83        -374.045      141.062    -2.652 0.010271 *
---
Signif. codes:  0 '***' 0.001 '**' 0.01 '*' 0.05 '.' 0.1 ' ' 1

Residual standard error: 115.2 on 59 degrees of freedom
Multiple R-squared: 0.8733,      Adjusted R-squared: 0.6973
F-statistic: 4.961 on 82 and 59 DF,  p-value: 6.685e-10
```

The estimated mean for Line 1 is the intercept (654.485), and the effect of every other line is estimated relative to the value for Line 1: for example, for Line 7,

$$\text{fixed-effect mean} = 654.485 + 209.245 = 863.730.$$

The following commands fit the model omitting the block term and specifying 'line' as a random-effect term, and present the results:

```
library(nlme)
barleyprogeny.model1lme <- lme(yield_g_m2 ~ 1,
    data = barleyprogeny.unbalanced, random = ~ 1|fline)
summary(barleyprogeny.model1lme)
coef(barleyprogeny.model1lme)
```

The function `coef()` displays the estimates of the random effects in the model specified as its argument – in the present case, the shrunk means for the breeding lines. The coefficients for the fixed-effect terms – in this case, the intercept only – are included in the output of the function `summary()`.

The output of these commands is as follows:

```
Linear mixed-effects model fit by REML
 Data: barleyprogeny.unbalanced
       AIC      BIC    logLik
  1878.384 1887.23 -936.192

Random effects:
 Formula: ~1 | fline
         (Intercept) Residual
StdDev:     175.1198 115.0031
```

```
Fixed effects: yield_g_m2 ~ 1
                Value Std.Error DF  t-value p-value
(Intercept) 572.4734  21.67055 83 26.41711       0

Standardized Within-Group Residuals:
        Min           Q1          Med          Q3          Max
-1.62399201 -0.53822597 -0.02744873  0.46482593  2.05367544

Number of Observations: 142
Number of Groups: 83

     (Intercept)
1       639.9374
2       510.1906
3       782.4734
4       674.9465
5       730.7361
6       761.5776
7       812.0656
8       486.6841
9       661.7120
10      727.3568
.
.

.
82      348.0758
83      368.4353
```

The BLUP for each line can be obtained by rearranging Equation 5.8 to give

$$\text{BLUP}_k = \widehat{\mu}'_k - \widehat{\mu}. \tag{5.15}$$

For Line 7,

$$\text{BLUP}_7 = 812.0656 - 572.4734 = 239.5922.$$

This is almost, but not exactly, the same as the value given by GenStat (239.54 – Section 5.2).

5.5   Use of SAS for the estimation of random effects

The following SAS statements import the original, 'unpadded' data on the barley breeding lines, fit the model omitting the block term and specifying 'line' as a fixed-effect term and present the results:

```
PROC IMPORT OUT = barley DBMS = EXCELCS REPLACE
   DATAFILE = "&pathname.\IMM edn 2\Ch 3\barley progeny.xlsx";
   SHEET = "sheet1 SAS";
RUN;

ODS RTF;
```

```
PROC GLM;
   CLASS line;
   MODEL yield_g_m2 = line /SOLUTION;
RUN;
ODS RTF CLOSE;
```

Part of the output of `PROC GLM` is as follows:

Source	DF	Sum of squares	Mean square	F value	Pr > F
Model	82	5396990.540	65816.958	4.96	<0.0001
Error	59	782674.772	13265.674		
Corrected total	141	6179665.312			

R-square	Coeff Var	Root MSE	yield_g_m2 mean
0.873347	19.80432	115.1767	581.5735

Source	DF	Type I SS	Mean square	F value	Pr > F
line	82	5396990.540	65816.958	4.96	<0.0001

Source	DF	Type III SS	Mean square	F value	Pr > F
Line	82	5396990.540	65816.958	4.96	<0.0001

Parameter	Estimate		Standard error	t value	Pr > |t|
Intercept	280.4400000	B	115.1767082	2.43	0.0179
line 1	374.0450000	B	141.0620826	2.65	0.0103
line 2	202.8900000	B	162.8844627	1.25	0.2178
line 3	592.6000000	B	162.8844627	3.64	0.0006
line 4	438.7000000	B	162.8844627	2.69	0.0092
line 5	518.5500000	B	162.8844627	3.18	0.0023
linc 6	521.9150000	B	141.0620826	3.70	0.0005
line 7	583.2900000	B	141.0620826	4.13	0.0001
line 8	187.7450000	B	141.0620826	1.33	0.1883
line 9	400.5150000	B	141.0620826	2.84	0.0062
line 10	480.3150000	B	141.0620826	3.40	0.0012

$\vdots$

line 82	−29.1400000	B	162.8844627	−0.18	0.8586
line 83	0.0000000	B	.	.	.

Note: The $X'X$ matrix has been found to be singular, and a generalized inverse was used to solve the normal equations. Terms whose estimates are followed by the letter 'B' are not uniquely estimable.

The estimated mean for Line 83 is the intercept (280.44), and the effect of every other line is estimated relative to the value for Line 83: for example, for Line 7,

$$\text{fixed-effect mean} = 280.44 + 583.29 = 863.73.$$

The note reflects the fact that the choice of Line 83 as the factor level relative to which other effects are defined is arbitrary (cf. the arbitrary choice of two towns relative to which the effects of other towns are defined, described in Section 1.4). The following statements fit the model omitting the block term and specifying 'line' as a random-effect term, and present the results for Lines 1–4:

```
ODS RTF;
PROC MIXED ASYCOV NOBOUND;
   CLASS line;
   MODEL yield_g_m2 = /DDFM = KR HTYPE = 1 SOLUTION;
   RANDOM line /SOLUTION;
   ESTIMATE 'Shrunk mean, Line 1' INTERCEPT 1 | line 1;
   ESTIMATE 'Shrunk mean, Line 2' INTERCEPT 1 | line 0 1;
   ESTIMATE 'Shrunk mean, Line 3' INTERCEPT 1 | line 0 0 1;
   ESTIMATE 'Shrunk mean, Line 4' INTERCEPT 1 | line 0 0 0 1;
RUN;
ODS RTF CLOSE;
```

The SOLUTION option in the RANDOM statement indicates that the random-effect parameter estimate for each line – that is, its BLUP – is to be printed. The ESTIMATE statements obtain and print the shrunk means for each line, as a weighted sum of model parameters. Thus

shrunk mean for Line 1 = 1 × intercept + 1 × effect of Line 1,

shrunk mean for Line 2 = 1 × intercept + 0 × effect of Line 1 + 1 × effect of Line 2,

and so on. Similar, increasingly verbose statement could be used to obtain the shrunk means for the other lines. The shrunk means for all 83 lines can be obtained by replacing the ESTIMATE statements by the following:

```
%MACRO loop();
%DO i=1 %TO 83;
   ESTIMATE   "Shrunk mean, Line &i"
      INTERCEPT 1| line %DO j=1 %TO &i-1; 0 %END; 1;
%END;
%MEND loop;
OPTIONS MPRINT;
%loop;
```

However, a detailed explanation of the 'macro' syntax used in these statements is beyond the scope of the present account.

Part of the output of PROC MIXED is as follows:

Convergence criteria met.

Covariance parameter estimates	
Cov Parm	**Estimate**
Line	30655
Residual	13230

Asymptotic covariance matrix of estimates			
Row	**Cov Parm**	**CovP1**	**CovP2**
1	Line	38832845	−3532116
2	Residual	−3532116	5863037

Fit statistics	
−2 Res Log Likelihood	1872.4
AIC (smaller is better)	1876.4
AICC (smaller is better)	1876.5
BIC (smaller is better)	1881.2

AIC, Akaike Information Criterion; BIC, Bayesian Information Criterion.

Null model likelihood ratio test		
DF	**Chi-square**	**Pr > ChiSq**
1	39.72	<0.0001

Solution for fixed effects					
Effect	**Estimate**	**Standard error**	**DF**	**t value**	**Pr > $|t|$**
Intercept	572.48	21.6738	81.9	26.41	<0.0001

Solution for random effects						
Effect	**line**	**Estimate**	**Std Err Pred**	**DF**	**t value**	**Pr > $|t|$**
line	1	67.4540	76.8380	106	0.88	0.3820
line	2	−62.2705	99.1142	127	−0.63	0.5310
Line	3	209.95	99.1142	127	2.12	0.0361
Line	4	102.45	99.1142	127	1.03	0.3033
Line	5	158.23	99.1142	127	1.60	0.1129
Line	6	189.08	76.8380	106	2.46	0.0155
Line	7	239.56	76.8380	106	3.12	0.0023
Line	8	−85.7798	76.8380	106	−1.12	0.2668

Solution for random effects						
Effect	line	Estimate	Std Err Pred	DF	t value	Pr > |t|
Line	9	89.2259	76.8380	106	1.16	0.2482
Line	10	154.86	76.8380	106	2.02	0.0464

⋮   ⋮

Line	82	−224.35	99.1142	127	−2.26	0.0253
Line	83	−204.00	99.1142	127	−2.06	0.0416

Estimates					
Label	Estimate	Standard error	DF	t value	Pr > |t|
Shrunk mean, Line 1	639.93	74.8350	92.7	8.55	<0.0001
Shrunk mean, Line 2	510.20	98.1859	121	5.20	<0.0001
Shrunk mean, Line 3	782.43	98.1859	121	7.97	<0.0001
Shrunk mean, Line 4	674.92	98.1859	121	6.87	<0.0001
Shrunk mean, Line 5	730.70	98.1859	121	7.44	<0.0001
Shrunk mean, Line 6	761.55	74.8350	92.7	10.18	<0.0001
Shrunk mean, Line 7	812.04	74.8350	92.7	10.85	<0.0001
Shrunk mean, Line 8	486.70	74.8350	92.7	6.50	<0.0001
Shrunk mean, Line 9	661.70	74.8350	92.7	8.84	<0.0001
Shrunk mean, Line 10	727.34	74.8350	92.7	9.72	<0.0001

⋮   ⋮

Shrunk mean, Line 82	348.12	98.1859	121	3.55	0.0006
Shrunk mean, Line 83	368.48	98.1859	121	3.75	0.0003

The column headed 'Estimate' in the table headed 'Solution for Random Effects' gives the BLUPs, and the corresponding column in the table headed 'Estimates' gives the shrunk means. These agree fairly closely with those produced by GenStat.

5.6   The Bayesian interpretation of BLUPs: Justification of a random-effect term without invoking an underlying infinite population

When the specification of a factor as a random-effect term is justified without invoking an underlying infinite population of levels, using the concept of exchangeability (Sections 1.6 and 2.6), an alternative justification can be given for the use of shrunk estimates of random effects. This is the *Bayesian* approach, in which the evidence about a parameter value (e.g. the mean of a factor level or the slope of a regression line) obtained from the data is combined with a prior belief concerning the value in order to obtain a posterior belief. We will examine this approach in general terms before considering its use to obtain BLUPs.

In Bayesian statistics, degrees of belief are represented by probability distributions, giving higher probability to values that are considered more credible. This interpretation of the concept of probability as a measure of belief is philosophically different from the *frequentist* interpretation, which considers the probability of an event to be the proportion of occasions on which that event occurs 'in the long run': a lucid account of the relationship between the two interpretations is given by Hacking (2001, Chapter 11, pp. 127–139, Chapter 16, pp. 189–200 and Chapter 21, pp. 256–260). The frequentist interpretation can be applied to estimates of a parameter, which will vary from one sample to the next, but cannot be applied to the true value, which is unknown but does not vary. However, the Bayesian interpretation can be applied to the true parameter value, because we have a belief about it – namely that it lies somewhere near our estimate. Fortunately the two concepts of probability obey the same laws, and can, with caution, be used together. In particular, the relationship between our belief concerning the true value of a parameter (in the present case, the mean yield of a particular breeding line) and an estimate of the same parameter (in the present case, the mean value from that breeding line in the field trial) is as follows.

We define Δ, the true value and D, the estimate of Δ from the data, both regarded as random variables. It can be shown that for any given value of Δ, say δ, and any given value of D, say d,

$$P(\Delta = \delta | D = d) = \frac{P(D = d | \Delta = \delta) \cdot P(\Delta = \delta)}{\sum_\delta P(D = d | \Delta = \delta) \cdot P(\Delta = \delta)} \qquad (5.16)$$

where $P(\Delta = \delta | D = d)$ means 'The probability that Δ equals δ given that D equals d'. The symbol $\sum_\delta$ indicates summation over all possible values of δ: if δ is a continuous variable, this summation becomes an integration. This is Bayes' Theorem: its proof is straightforward, and is given, for example, by Hacking (2001, Chapter 7, pp. 69–71). It shows that in order to decide on our belief concerning the true value given the evidence (the *posterior probability distribution* of Δ, $P(\Delta = \delta | D = d)$), we must specify our belief before we see the evidence (the *prior distribution*, $P(\Delta = \delta)$).

The probability of our estimate given any particular true value, $P(D = d | \Delta = \delta)$, is given by the distribution

$$D | \delta, \sigma_D^2 \sim N(\delta, \sigma_D^2). \qquad (5.17)$$

That is, if D were estimated repeatedly, the estimates would be clustered around the true value δ with a certain variance, σ_D^2. An estimate of this variance is provided by SE_d^2. It is convenient also to specify $P(\Delta = \delta)$ using a normally distributed variable,

$$\Delta | \mu_\Delta, \sigma_\Delta^2 \sim N(\mu_\Delta, \sigma_\Delta^2) \qquad (5.18)$$

the values of μ_Δ and σ_Δ^2 being chosen to give a distribution that correctly describes our prior belief. We next define two *weights*, namely

$$w_{\text{prior}} = \frac{1}{\sigma_\Delta^2} \qquad (5.19)$$

and

$$w_D = \frac{1}{\sigma_D^2}. \qquad (5.20)$$

It can then be shown that our posterior belief concerning the true value, $P(\Delta = \delta | D = d)$, is expressed by the distribution

$$\Delta | \mu_\Delta, \sigma_\Delta^2, d, \sigma_D^2 \sim N \left(\frac{w_{\text{prior}} \mu_\Delta + w_D d}{w_{\text{prior}} + w_D}, \frac{1}{w_{\text{prior}} + w_D} \right). \tag{5.21}$$

That is:

- the mean of the posterior distribution is a weighted mean of the prior mean and the estimate from the data and

- the variance of the posterior distribution is inversely related to the weights, and hence directly related to the variances of the component variables.

The mean of the posterior distribution is always intermediate between those of the component variables, and its variance is always smaller than that of either component variable: that is, precision is always gained by combining the two sources of information.

It can be shown that if all values of Δ from $-\infty$ to $+\infty$ were considered equally probable before inspecting the data, then on the basis of the data, Δ would have the distribution

$$\Delta | d, \sigma_D^2 \sim N(d, \sigma_D^2) \tag{5.22}$$

Thus the posterior distribution (Distribution 5.21) is equivalent to the distribution of a new variable,

$$\Delta | \mu_\Delta, \sigma_\Delta^2, d, \sigma_D^2 = \frac{w_{\text{prior}}(\Delta | \mu_\Delta, \sigma_\Delta^2) + w_D(\Delta | d, \sigma_D^2)}{w_{\text{prior}} + w_D}, \tag{5.23}$$

that is, a weighted mean of $\Delta | \mu_\Delta, \sigma_\Delta^2$ (which has Distribution 5.18) and $\Delta | d, \sigma_D^2$ (which has Distribution 5.22), the weights being inversely proportional to the variances of the component variables. Thus the Bayesian approach can be informally summarized as

posterior belief = prior belief + belief on the basis of the data alone.

In the Bayesian approach to statistical analysis, prior belief may come from expert opinion. For example, an expert in barley breeding may believe that the true mean yield of lines derived from the cross 'Chebec' $\times$ 'Harrington' is usually between 352.4 and 703.6 g/m^2, that is, 572.5 ± 175.1 g/m^2 (implausibly precise values: the reason for choosing them will become apparent shortly). This belief can be expressed by the statements

$$\mu_\Delta = 572,$$

$$\sigma_\Delta^2 = 175.1^2,$$

and hence

$$\Delta | \mu_\Delta, \sigma_\Delta^2 \sim N(572.5, 175.1^2)$$

(from which the units of measurement, g/m^2, have been dropped for convenience). The expert will therefore be sceptical about the high mean of Line 7, $d = 863.7$ g/m^2. The SE of this estimate is

$$\sqrt{\frac{\hat{\sigma}_E^2}{r_k}} = \sqrt{\frac{13226}{2}} = 81.3 \, \text{g/m}^2,$$

so approximately,

$$\sigma_D^2 = 81.3^2.$$

The expert's posterior belief concerning the true mean yield of this breeding line is obtained by combining his/her prior belief and the evidence, using the Equations and Distribution 5.19–5.21:

$$w_{\text{prior}} = \frac{1}{175.1^2} = 0.00003261$$

$$w_D = \frac{1}{81.3^2} = 0.00015122$$

$$\Delta | \mu_\Delta, \sigma_\Delta^2, d, \sigma_D^2 \sim$$

$$N \left(\frac{0.00003261 \times 572.5 + 0.00015122 \times 863.7}{0.00003261 + 0.00015122}, \frac{1}{0.00003261 + 0.00015122} \right)$$

$$\Delta | \mu_\Delta, \sigma_\Delta^2, d, \sigma_D^2 \sim N(812.04, 73.76^2).$$

But in the context of mixed modelling, we are interested in a prior distribution derived not from expert opinion but from a set of estimates of which the estimate under consideration is a member: in the present case, the estimated mean yields of the sample of breeding lines. The mean of all these estimates, and the variance of the estimates around this value, then provide the parameters of the prior distribution. This is sometimes known as the *empirical Bayesian approach* (see also Section 8.4). The first parameter of Distribution 5.21 can be rearranged as

$$\text{posterior mean}(\Delta | \mu_\Delta, \sigma_\Delta^2, d, \sigma_d^2) = \mu_\Delta + (d - \mu_\Delta) \cdot \frac{\sigma_\Delta^2}{\sigma_\Delta^2 + \sigma_d^2}, \tag{5.24}$$

and this expression bears a strong resemblance to the formula for the shrunk mean given in Equations 5.6 and 5.8: indeed, when the prior distribution is obtained in this way, we can re-write Equation 5.24 as

$$\text{posterior mean}(\Delta | \mu_\Delta, \sigma_\Delta^2, d, \sigma_d^2) = \mu_\Delta + \text{BLUP}(\Delta). \tag{5.25}$$

where

$$\text{BLUP}(\Delta) = (d - \mu_\Delta) \cdot \frac{\sigma_\Delta^2}{\sigma_\Delta^2 + \sigma_d^2} \tag{5.26}$$

In the present case, we substitute

$$\mu_\Delta = \hat{\mu} = 527.5$$

$$\sigma_\Delta^2 = \hat{\sigma}_G^2 = 30667 = 175.1^2$$

$$d = \hat{\mu}_k = 863.7$$

$$\sigma_D^2 = \frac{\hat{\sigma}_E^2}{r_k} = \frac{13226}{2}$$

and obtain

$$\text{posterior mean}(\Delta | \mu_\Delta, \sigma_\Delta^2, d, \sigma_d^2) = 527.5 + (863.7 - 527.5) \cdot \frac{30667}{30667 + \frac{13226}{2}} = 812.04,$$

which is the same as the value obtained from the prior distribution based on expert opinion, as the values specified for μ_Δ and σ_Δ^2 are the same. The SE of this shrunk mean is given by

$$\sqrt{\text{posterior var}(\Delta | \mu_\Delta, \sigma_\Delta^2, d, \sigma_d^2)} = \sqrt{\frac{1}{w_{\text{prior}} + w_D}} = \sqrt{\frac{1}{0.00003261 + 0.00015122}} = 73.76.$$

In this Bayesian approach to BLUPs, the collective features of the exchangeable set of factor levels (in this case μ_Δ and σ_Δ^2) are regarded as prior knowledge, to be taken into account when interpreting the data on each individual level (in this case the values of D). If we decide that the factor levels do not form an exchangeable set, but are so disparate that each is irrelevant to the estimation of the effect of the others (see the discussion of Stein's paradox in Section 5.3), we can express this irrelevance by specifying $\sigma_\Delta^2 = \infty$ instead of using the variance component for the factor under consideration. The prior distribution is then uniform from $-\infty$ to $+\infty$, and the posterior mean is the unshrunk mean for each factor level: that is, from a Bayesian point of view, a fixed-effect term is simply a random-effect term with a flat prior distribution.

It should be noted that the BLUP and its SE are only strictly valid when d and the estimate of σ_Δ^2 are independent, whereas in practice they are almost always obtained from the same information. This does not matter much when the number of factor levels is large, as in the present case: the contribution of each breeding line to $\hat{\sigma}_G^2$ is small, and hence the correlation (lack of independence) between $\hat{\mu}_k$ and $\hat{\sigma}_G^2$ is negligible. But when the number of factor levels is small, as in the sample of English towns considered in Chapter 1, this correlation starts to matter, and the BLUPs for individual towns, and their SEs, should not be taken too literally. In this context, it is helpful to distinguish two purposes for which a model term may be specified as random:

- to obtain BLUPs for the effects of the term itself or

- to ensure that it is taken into account when determining the precision of effect estimates in other terms.

For example, in the present case, our primary interest is in the yields of the individual breeding lines, and we will still be interested in shrinking our estimates of these as a defence against over-optimism, whether or not we feel justified in regarding them as a representative sample from a much larger population. For this purpose, exchangeability is enough, though we require a moderately large set of factor levels in order to obtain trustworthy BLUPs. In connection with the James–Stein estimator, Efron and Morris (1977) advise that as few as nine levels are tolerable and 15 or 20 are ample, and it seems reasonable to extend these judgements to BLUPs. However, in the study of the relationship between latitude and house prices, our main motive for specifying 'town' as a random-effect term was to ensure that this term contributed to the SE for the estimate of the effect of latitude, and for this purpose it is helpful to envisage a much larger population of towns for which this estimate will be valid, though a small sample of towns is enough to give us a usable SE. We took relatively little interest in the mean prices in individual towns, whether represented by BLUEs or by BLUPs. In the case of a randomized experimental design, the contribution of block and residual effects to the SE of a treatment effect is justified by the large number of possible permutations of these effects over the treatments, from which the actual permutation was randomly selected (see Sections 2.6 and 4.1). But it is probably not helpful to envisage the random permutation of English towns over their latitudes.

5.7 Summary

In ordinary regression analysis and analysis of variance, random-effect terms are always regarded as nuisance variables – residual variation or block effects – but this is not always appropriate. Effects that are of intrinsic interest may be specified as random.

The decision to specify a term as random causes a fundamental change in the way in which the effect of each of its levels is estimated.

This is illustrated in the field experiment to evaluate breeding lines of barley (Sections 3.8–3.17), but the same concepts and arguments apply equally to any situation in which replicated, quantitative evaluations are available for the comparison of members of some population, for example:

- new chemical entities to be evaluated as potential medicines

- candidates for admission to a university on the basis of their examination scores.

We saw earlier (Sections 3.15 and 3.20) that the predicted genetic advance due to selection among the barley breeding lines is less than the value given by a naïve prediction from the mean of the selected lines. Mixed-model analysis gives a similar adjustment to the mean of each individual breeding line.

When a term is specified as random, the ordinary difference between the mean for each level and the grand mean (the BLUE) is replaced by a 'shrunk estimate' (the BLUP).

The shrinkage of the BLUP, relative to the BLUE, is large when:

- the variance component for the term in question is small

- the residual variance is large

- the number of replications of the factor level under consideration is small.

Crossovers may occur as a result of the shrinkage, so that a level of a random-effect term that is ranked higher than another on the basis of the BLUEs is ranked lower on the basis of the BLUPs.

The shrinkage of the BLUP is equivalent to the phenomenon of regression towards the mean.

This is illustrated by considering the present observations of each level of a random-effect term as predictions of future observations. The line of best fit relating present observations to future observations (the regression line) is flatter than a line of unit slope passing through the origin.

The degree of flattening of the line of best fit (the amount of regression towards the mean) is equivalent to the shrinkage of the BLUP relative to the BLUE.

The discrepancy between the BLUE and the BLUP tells us that our best prediction about the future performance of any factor level of a random-effect term depends partly on our reason for taking an interest in it, as follows:

- if we have chosen the level in question because of its relative performance among the levels under consideration, we should be guided by the BLUP

- however, if we have chosen it for reasons unrelated to its relative performance, and consider the comparative performance of other levels to be irrelevant, we should be guided by the BLUE.

When the levels of a random-effect term are regarded as an exchangeable set rather than a sample from an infinite population, the BLUP can be justified by a *Bayesian* argument. This can be informally summarized as

posterior belief = prior belief + belief on the basis of data alone.

In the '*empirical Bayesian*' approach, the prior probability distribution of belief is specified by the grand mean and the variance component for the term in question. The estimated mean of the level under consideration (e.g. the mean for a particular barley breeding line) and its variance are combined with this prior distribution to obtain the posterior distribution. The mean of this posterior distribution is the shrunk mean for the level in question (grand mean + BLUP). The square root of the variance of the posterior distribution gives the SE of this shrunk mean.

It is helpful to distinguish two purposes for which a model term may be specified as random:

- to obtain BLUPs for the effects of the term itself. For this purpose, the criterion of exchangeability is sufficient.

- to ensure that the term is taken into account when determining the precision of effect estimates in other terms. For this purpose, it may be helpful to envisage a much larger population from which the levels of the term are sampled. In the case of a treatment effect in a randomized experimental design, the contribution of block and residual effects to the SE is justified by the large number of possible permutations of the treatments over these effects.

In order to achieve trustworthy BLUPs, a fairly large set of levels (at least 9 and ideally 15 or more) must be studied, so that the estimated variance component for the term in question and the estimated mean of each level are nearly independent.

5.8   Exercises

5.1 Return to the data set concerning the yield of F_3 wheat families in the presence of ryegrass, introduced in Exercise 3.3.

   (a) Using mixed modelling, obtain an estimate of the mean yield of each of the F_3 families, specifying 'family' as a fixed-effect term.

   (b) Using mixed modelling and specifying 'family' as a random-effect term, obtain the following:

   (i) an estimate of the overall mean of the population of F_3 families

   (ii) the BLUP for the effect of each family.

   (c) From the output of the analysis performed in Section (b), obtain the following:

   (i) an estimate of the variance component for 'family'

   (ii) an estimate of the residual variance component.

   Obtain also the number of observations of each family.

(d) From the information obtained in Parts (a)–(c), compare the relationship between the BLUPs and the estimates of family means obtained specifying 'family' as a fixed-effect term with that given in Equation 5.6.

(e) Obtain the shrunk mean for each family, and plot the shrunk means against the means obtained when 'family' is specified as a fixed-effect term (the unadjusted means). The point representing one of the families deviates from the general relationship between these two types of mean.

(f) What is the distinguishing feature of this family?

5.2 In many types of plant, exposure to low temperature at an early stage of development causes flowering to occur more rapidly: this phenomenon is called *vernalization*. An inbred line of chickpea with a strong vernalization response and a line with little or no vernalization response were crossed, and the F_1 hybrid progeny were self-fertilized to produce the F_2 generation. Each F_2 plant was self-fertilized to produce an F_3 family. The seed of each F_3 family was divided into two batches. Germinating seeds of one were vernalized by exposure to low temperature ($4\,^{\circ}C$) for 4 weeks. The other batch provided a control. All F_3 seeds were then sown, and allowed to grow. The plants were arranged in groups of four: within each group the plants were of the same family and had received the same low-temperature treatment. Generally there were 12 plants (i.e. three groups of four) in each family and each exposed to low-temperature treatment, but in some families fewer or more plants were available. The number of days from sowing to flowering was recorded for each plant. The first and last few rows of the spreadsheet holding the data are presented in Table 5.2: the complete data set is held in the file 'chickpea vernalisation.xlsx', available from this book's website, at the web address given in the Preface. (Data reproduced by the kind permission of S. Abbo, Field Crops and Genetics, The Hebrew University of Jerusalem.)

(a) Divide the data between two spreadsheets, one holding only the results from the vernalized plants, the other, only those from the control plants.

(b) Analyse the results from the vernalized plants by mixed modelling, specifying 'family', 'group' and 'plant' as random-effect terms.

(c) Obtain an estimate of the component of variance for each of the following terms:

  (i) family

  (ii) group within family

  (iii) plant within group.

  Which term in your mixed model represents residual variation?

(d) Estimate the heritability of time to flowering in vernalized plants from this population of families. (N.B. The estimate obtained using the methods described in Chapter 3 is slightly biased downwards, as some of the residual variance is due to genetic differences among plants of the same family.)

(e) Obtain the unadjusted mean and the shrunk mean, and the BLUP, for the number of days from sowing to flowering in each family.

(f) Extend Equation 5.6 to the present situation, in which two components of variance contribute to the shrinkage of the BLUPs. Use the values obtained above to check your equation.

(g) Repeat the steps indicated in Sections (b), (c) and (d) of this exercise for the control plants. Comment on the difference between the estimates of variance components and heritability obtained from the vernalized plants and the control plants.

Table 5.2   Time from sowing to flowering of F_3 chickpea plants with and without exposure to a vernalizing stimulus.

	A	B	C	D	E
1	plant_group	plant	family	low_T	days_to_flower
2	60	1	88	vernalized	63
3	60	2	88	vernalized	64
4	60	3	88	vernalized	61
5	60	4	88	vernalized	70
6	1	1	14	vernalized	75
7	1	2	14	vernalized	
8	1	3	14	vernalized	
9	1	4	14	vernalized	
10	103	1	29	control	78
11	103	2	29	control	82
12	103	3	29	control	82
13	103	4	29	control	88
⋮					⋮
1092	41	3	44	vernalized	70
1093	41	4	44	vernalized	64

Source: Data reproduced by kind permission of S. Abbo, Field Crops and Genetics, The Hebrew University of Jerusalem.

(h) Plot the shrunk mean for each family obtained from the control plants against the corresponding value obtained from the vernalized plants. Comment on the relationship between the two sets of means.

5.3 Return to the house price data analysed in Chapter 1.

(a) Fit the mixed model introduced in Chapter 1 to these data. Obtain the intercept and slope of the line of best fit relating log(house price) to latitude. Obtain the BLUP for the effect of each town. Do you consider this data set to be adequate to permit interpretation of the BLUPs?

(b) Obtain the fitted value of log(house price) for each town, that is, the value on the line of best fit at the latitude of each town.

(c) Hence obtain the shrunk mean value of log(house price) for each town.

(d) Obtain the BLUE for each town.

(e) Produce a figure similar to Figure 1.4, but with the line of best fit from the mixed model instead of the simple regression line. Add the shrunk means to this plot. Comment on their distribution relative to the simple means.

(f) Plot the shrunk means against the simple means. Identify any crossovers among the towns for these variables.

(g) Plot the shrunk means against the simple means. Again, identify any crossovers.

References

Efron, B. and Morris, C. (1977) Stein's paradox in statistics. *Scientific American*, **236**, 119–127.

Hacking, I. (2001) *An Introduction to Probability and Inductive Logic*, Cambridge University Press, Cambridge, 302 pp.

Robinson, G.K. (1991) That BLUP is a good thing: the estimation of random effects. *Statistical Science*, **6**, 15–51.

6

More advanced mixed models for more elaborate data sets

6.1 Features of the models introduced so far: A review

The mixed models introduced in the examples presented so far are reviewed in Table 6.1. These examples illustrate several features of the range of mixed models that can be specified. Each model term may be a variate, like 'latitude', in which each observation can have any numerical value, or a factor, like 'town', in which each observation must come from a specified set of levels. Each part of the model (the fixed-effect model and the random-effect model) may have more than one term, as in the random-effect model 'block + line'. Factors may be *crossed*, as in the model 'brand*assessor', which is equivalent to

$$\text{brand} + \text{assessor} + \text{brand.assessor} = \text{main effects of brand}$$

$$+ \text{ main effects of assessor} + \text{brand} \times \text{assessor interaction effects.}$$

Alternatively, they may be *nested*, as in the model 'day/presentation/serving', which is equivalent to

$$\text{day} + \text{day.presentation} + \text{day.presentation.serving} = \text{main effects of day}$$

$$+ \text{ effects of presentation within each day}$$

$$+ \text{ effects of serving within each day.presentation combination.}$$

Crossing and nesting of factors is specified using the notation of Wilkinson and Rogers (1973).

6.2 Further combinations of model features

More elaborate combinations of these features are permitted. More than two factors may be crossed, thus

$$a * b * c = a + b + c + a.b + a.c + b.c + a.b.c. \tag{6.1}$$

Introduction to Mixed Modelling: Beyond Regression and Analysis of Variance, Second Edition. N. W. Galwey.
© 2014 John Wiley & Sons, Ltd. Published 2014 by John Wiley & Sons, Ltd.
Companion website: http://www.wiley.com/go/beyond_regression

Table 6.1   Mixed models introduced in earlier chapters.

Example	Response variable	Fixed-effect model	Random-effect model	Design illustrated	Where discussed
Effects on house prices in England	logprice	latitude	town	Regression analysis on grouped data	Chapter 1; Sections 3.18 and 4.2
Sensory evaluation of ravioli	saltiness	brand* assessor	day/presentation/ serving	Split-plot design	Chapter 2; Sections 3.19 and 4.3–4.5
Effects on the strength of a chemical paste	strength	–	delivery/cask/ test	Hierarchical design	Sections 3.2–3.7
Genetic variation in a barley field trial	yield_g_m2	–	block + line	Randomized block design	Sections 3.8–3.17; Chapter 5
Tensile strength of iron powder sinters	tnsl_strngth	mfng_spec	furnace	Randomized block design	Sections 4.6–4.8

The three-way interaction term 'a.b.c' represents variation among the individual combinations of levels of 'a', 'b' and 'c' that is not adequately represented by the main effects or the two-way interaction terms. If factors 'a', 'b' and 'c' each have two levels, it is 'the difference of a difference of a difference' (Mead, 1988, Section 13.1, p. 348).

Factors and variates may be used in the same part of the mixed model (the random-effect model or the fixed-effect model), and factors may be crossed or nested with variates. For example, if 'f' is a factor and 'v' is a variate, the model

$$f * v = f + v + f.v \qquad (6.2)$$

or

$$f/v = f + f.v \qquad (6.3)$$

is permitted. These models represent a set of sloping lines relating the response variate to the explanatory variate 'v'. The term 'f' represents variation in the intercept of the line among the levels of this factor. The term 'v' represents the mean slope of the line, averaged over the levels of 'f'. In the crossed model 'f * v', the term 'f.v' represents variation among the slopes at different levels of 'f' around the average slope. In the nested model 'f/v', in which no main effect of 'v' is estimated, the term 'f.v' represents all the variation in the response variate that is related to 'v', whether common to all levels of 'f' or specific to individual levels. The meaning of each term in the model 'f * v' is illustrated in Figure 6.1.

Variates may be crossed with other variates. In this case, the interaction term is equivalent to the product of the two variates: that is, if v1 and v2 are variates, and

$$v12 = v1 \times v2,$$

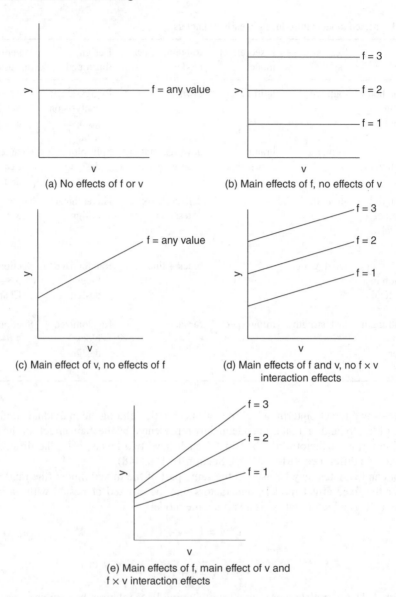

Figure 6.1   (a–e) Possible patterns in the response to a factor (f) and a variate (v).

then the model

$$v1 * v2 = v1 + v2 + v1.v2 \qquad (6.4)$$

is equivalent to

$$v1 + v2 + v12.$$

A fuller account of models that contain products and powers (squares, cubes, etc.) of explanatory variates, known as *polynomial models*, is given by Draper and Smith (1998, Section 12.1, pp. 251–253).

In all the examples considered so far, the significance of each fixed-effect model term is tested by comparing the mean square for that term with one other mean square. For example:

- in the regression analysis on grouped data, the effect of latitude is tested by comparing $MS_{latitude}$ with MS_{town} (Section 1.5).

- in the analysis of the split-plot design:

  - the effects of the brand of ravioli are tested by comparing MS_{brand} with $MS_{Residual}$ in the 'day.presentation' stratum (the main-plot stratum)

  - the effects of assessor are tested by comparing $MS_{assessor}$ with $MS_{Residual}$ in the 'day.presentation.serving' stratum (the sub-plot stratum) (Section 2.2).

This confinement of each fixed-effect term to a single stratum of an anova is a consequence of the *balance* of these data sets: it is not a universal feature of mixed models. If there were some towns (presumably large ones) within which houses had been sampled at more than one latitude, separate estimates of the effect of latitude among and within towns would be obtainable, though the information could be pooled over these strata. (See the discussions of within- and between-group estimates of effects in Sections 7.8 and 8.3.) If some assessors had been omitted from certain presentations of ravioli, then in order to obtain estimates of assessor effects based on all available information, between- and within-presentation information would have to be combined. In a mixed-model analysis, the pooling of information over strata is performed automatically by the computer software, but because of this issue, it is only in special cases that an equivalent analysis can be specified as an ordinary regression analysis or analysis of variance.

6.3 The choice of model terms to be specified as random

When constructing a mixed model, it is of course necessary to decide which terms should be allocated to the random-effects part. The first steps in this decision-making process are as follows:

- It is generally sensible to specify all variates as fixed-effect terms. To specify a variate as a random-effect term is to assume that the associated slope (the corresponding parameter estimate in the regression model) is one of an infinite population or an exchangeable set of possible slopes: this assumption is rarely if ever helpful.

- Factors should be classified as

  - Terms representing nuisance effects (block effects). All block terms should generally be specified as random: the problems that can arise if block effects are specified as fixed are considered in more detail in Section 6.5.

  - Terms representing effects of intrinsic interest (treatment effects). These may be fixed or random.

This leaves the question of whether a factor representing effects of intrinsic interest should be specified as fixed or random.

If a factor is specified as a random-effect term, it is assumed that the levels of this factor that have been studied are a random sample from an infinite population of possible levels, or else that they form an exchangeable set (see Sections 1.6, 5.3 and 5.6). Hence we need to decide whether the levels of a factor of interest comprise such a sample or such a set, and if so, whether we prefer to take account of this relationship between them, or regard them simply as levels of individual interest. The assumption that the factor levels studied comprise a random sample or an exchangeable set is stringent, but it should not necessarily be thought of as a radical step. On the contrary, the inferences made about other terms are more general when a factor is specified as random (see Section 1.6).

Quite often, the assumption that factor levels are a random sample turns out to be more reasonable than it might at first appear: see, for example, the justification for regarding block effects in a randomized experimental design as random (Section 2.6). Even if factor levels are not strictly randomly sampled from an infinite population, this may still be a reasonable modelling assumption. Factor levels often comprise a representative sample from a finite set, for example, 'English towns', 'barley breeding lines derived from the cross Chebec × Harrington'. Provided that the set of available levels (towns or breeding lines as the case may be) is several times larger than the set actually studied, such a factor can reasonably be regarded as a random-effects term. It may be appropriate to think of the levels studied as being representative not only of those actually available for study, but also of those levels that *might have* existed but that happen not to. This is more persuasive in the case of crop breeding lines than of towns. Even if no such population can be specified, it may still be valid to specify a factor as random, and to make inferences about its levels collectively, if we consider that they comprise an exchangeable set. This is appropriate if we consider that each level sampled not only tells us about itself, but also provides some information concerning the other levels in the sample: a possibility explored more fully in Sections 1.6 and 5.3.

One test sometimes used to determine whether a term should be regarded as random is to ask whether, if one of the levels chosen were replaced by a new level, the study would be essentially unchanged. The number of distinct breeding lines that could be produced from the cross Chebec × Harrington is effectively infinite, and it would clearly make no essential difference to the study of this population of lines (Sections 3.8–3.17) if one of those sampled were removed and replaced by a line not currently included. However, in the study of commercial brands of ravioli (Chapter 2), the brands included were deliberately selected, being of individual interest to manufacturers, retailers and consumers. To replace one of them with a newly chosen brand would effectively make it a different study. Even in such a case we might be tempted to assess each brand partly on the basis of information provided by all the others, though with only four brands in the study we would be reluctant to specify 'brand' as a random-effect term (see Section 5.6). When we took the decision to treat 'town' as a random-effect term in the study of the effect of latitude on house prices (Chapter 1), we effectively asserted that if, for example, Durham were replaced by York, we would still have a representative sample of towns and the study would be essentially unchanged. This test is related to the question of whether if we lost the house prices from Durham, we would consider it worthwhile to predict them from the rest of the data (see Section 1.6).

When the levels of a factor are deliberately chosen values of a continuous variable (e.g. different doses of a drug in a pharmacological study, or different levels of application of a chemical fertilizer in an agricultural experiment), the factor should be specified as a fixed-effect term. The levels of such a factor do not constitute a random sample: there is no underlying distribution that indicates the frequency with which each value is expected to occur. Nor are they exchangeable, as their position on the scale is expected to be related to their effect in a systematic way.

Decisions on whether a term is fixed or random need usually only be made for main effects. For interaction terms and some nested terms, the choice is usually determined by the following guidelines:

- *Guideline 1.* When two terms are crossed, if both terms are fixed, the interaction should usually be fixed: otherwise, it must be random. That is, in the model

$$A * B = A + B + A.B,$$

  - if A and B are both fixed, then A.B should usually be fixed
  - if either A or B is random, or both are random, then A.B must be random.

- *Guideline 2.* When one term is nested within another, the nested term may be either fixed or random if the term within which it is nested is fixed: otherwise, it must be random. That is, in the model

$$A/B = A + A.B,$$

  - if A is fixed, then A.B may be fixed or random
  - if A is random, then A.B must be random.

The first part of Guideline 1 does not apply to all situations: in Section 8.4 we will encounter a case in which A and B are fixed but the A.B interaction is specified as random.

6.4   Disagreement concerning the appropriate significance test when fixed- and random-effect terms interact: 'The great mixed-model muddle'

There is an inconsistency in the statistical literature concerning the method for the construction of the F test when a fixed-effect term and a random-effect term interact. We will here review the two methods that have been proposed, and consider which of them corresponds to the results obtained from mixed modelling.

Consider an experiment in which all combinations of two factors, A with n_A levels and B with n_B levels, are studied in r replications. A response variable Y is measured on each experimental unit. Thus the model to be fitted is

$$A * B \tag{6.5}$$

in the notation of Wilkinson and Rogers (1973), or in algebraic notation,

$$y_{ijk} = \mu + \alpha_i + \beta_j + (\alpha\beta)_{ij} + \varepsilon_{ijk} \tag{6.6}$$

where

y_{ijk} = the observation of the response variable on the kth replication of the ith level of A and the jth level of B,

μ = the grand mean (overall mean) value of the response variable,

α_i = the effect of the ith level of A,

β_j = the effect of the jth level of B,

$(\alpha\beta)_{ij}$ = the interaction between the ith level of A and the jth level of B and

ε_{ijk} = the residual effect, that is, the deviation of y_{ijk} from the value predicted on the basis of μ, α_i, β_j and $(\alpha\beta)_{ij}$.

It is assumed that the ε_{ijk} are independent values of a random variable E, such that

$$E \sim N(0, \sigma^2). \tag{6.7}$$

It is decided to specify A as a fixed-effect term and B as a random-effect term. The values β_j are then interpreted as independent values of a random variable B, such that

$$B \sim N(0, \sigma_B^2). \tag{6.8}$$

It follows from Rule 1 in Section 6.3 that the interaction term A.B is also a random-effect term: that is, the values $(\alpha\beta)_{ij}$ are interpreted as independent values of a random variable (AB), such that

$$(AB) \sim N(0, \sigma_{AB}^2). \tag{6.9}$$

The variance of the fixed effects of A is designated by κ_A.

The issue on which authorities differ is the expected mean squares (EMSs) that are implied by this model. The alternative views are illustrated by the accounts of Snedecor and Cochran (1989, Section 16.14, pp. 319–324) and Ridgman (1975, Chapter 8, pp. 136–137). These are presented in Table 6.2, together with the F tests that they imply. Snedecor and Cochran considered that the interaction component of variance, σ_{AB}^2, does not contribute to the EMS for the main effect of B. It follows that the main effect of B should be tested against the residual term, whereas the interpretation of Ridgman indicates that this term should be tested against the A.B interaction. The other F tests are the same on either interpretation.

We will now determine which of these interpretations is implemented in the software systems GenStat, R and SAS. The spreadsheet in Table 6.3 contains a small data set suitable for fitting Model 6.5.

The following GenStat statements read the data and analyse them according to Model 6.5, using analysis of variance methods:

Table 6.2 Expected mean squares obtained when a fixed- and a random-effect factor interact, according to two authorities.

A is the fixed-effect factor and B the random-effect factor. The notations of the earlier authors have been adapted to conform to that of the present account. The pairs of mean squares to be compared by F tests are indicated by angled lines. The lines representing tests of the main effects of B are solid, for greater prominence: those representing other tests are dotted.

Source of variation	DF	Expected mean square	
		Snedecor and Cochran, 1989	Ridgman, 1975
A	$n_A - 1$	$n_B r \kappa_A^2 + r\sigma_{AB}^2 + \sigma_E^2$	$n_B r \kappa_A^2 + r\sigma_{AB}^2 + \sigma_E^2$
B	$n_B - 1$	$n_A r\sigma_B^2 + \sigma_E^2$	$n_A r\sigma_B^2 + r\sigma_{AB}^2 + \sigma_E^2$
A.B	$(n_A - 1)(n_B - 1)$	$r\sigma_{AB}^2 + \sigma_E^2$	$r\sigma_{AB}^2 + \sigma_E^2$
Residual	$n_A n_B (r - 1)$	σ_E^2	σ_E^2

Table 6.3 Data from an experiment to test all combinations of factors A and B in two replications.

	A	B	C
1	A!	B!	y
2	1	1	30
3	1	1	19
4	1	2	51
5	1	2	56
6	1	3	65
7	1	3	53
8	2	1	49
9	2	1	41
10	2	2	72
11	2	2	79
12	2	3	74
13	2	3	67

	A	B	C
14	3	1	72
15	3	1	65
16	3	2	91
17	3	2	95
18	3	3	70
19	3	3	66
10	4	1	22
21	4	1	14
22	4	2	61
23	4	2	54
24	4	3	42
25	4	3	43

```
IMPORT 'IMM edn 2\\Ch 6\\fixed A, random B.xlsx'
BLOCKSTRUCTURE B + A.B
TREATMENTSTRUCTURE A
ANOVA [FPROBABILITY = yes; PRINT = aovtable] y
```

The output of the ANOVA statement is as follows:

Analysis of variance

Variate: y

Source of variation	d.f.	s.s.	m.s.	v.r.	F pr.
B stratum	2	3978.08	1989.04	13.66	
B.A stratum					
A	3	5179.46	1726.49	11.85	0.006
Residual	6	873.92	145.65	5.40	
B.A.*Units* stratum	12	323.50	26.96		
Total	23	10354.96			

The fixed-effect model term A is represented explicitly in this anova table, but some effort is needed to recognize the other terms from the model. The random-effect term B is the only term in the 'B' stratum of the table, the residual variation due to E is the only term in the 'B.A. * Units * ' stratum, and the term A.B is represented by the Residual term within the 'B.A' stratum. Following Ridgman's interpretation of the model, the F statistic for B is therefore given by

$$\frac{MS_B}{MS_{A.B}} = \frac{1989.04}{145.65} = 13.66,$$

which agrees with the value given by GenStat.

The following statements analyse the data using the same model, but with mixed-modelling methods:

```
VCOMPONENTS [FIXED = A] RANDOM = B + A.B
REML [PRINT = Wald, components] y
```

The output of the REML statement is as follows:

Estimated variance components

Random term	Component	s.e.
B	230.42	248.85
B.A	59.35	42.40

Residual variance model

Term	Model(order)	Parameter	Estimate	s.e.
Residual	Identity	Sigma2	26.96	11.01

Tests for fixed effects

Sequentially adding terms to fixed model.

Fixed term	Wald statistic	n.d.f.	F statistic	d.d.f.	F pr.
A	35.56	3	11.85	6.0	0.006

Dropping individual terms from full fixed model.

Fixed term	Wald statistic	n.d.f.	F statistic	d.d.f.	F pr.
A	35.56	3	11.85	6.0	0.006

Message: denominator degrees of freedom for approximate F tests are calculated using algebraic derivatives ignoring fixed/boundary/singular variance parameters.

According to Ridgman's formulae for the EMSs, σ_B^2 is estimated by

$$\frac{MS_B - MS_{A.B}}{n_A r} = \frac{1989.04 - 145.65}{4 \times 2} = 230.42,$$

and this agrees with the value given by the REML statement. Thus GenStat supports the interpretation of Ridgman, not that of Snedecor and Cochran.

The following commands import the data into R and analyse them according to Model 6.5 using analysis of variance methods:

```
rm(list = ls())
AxB <- read.table(
    "IMM edn 2\\Ch 6\\fixed A, random B.txt",
    header=TRUE)
attach(AxB)
fA <- factor(A)
fB <- factor(B)
AxB.modelaov <-
    aov(y ~ fA + Error(fB + fA:fB))
summary(AxB.modelaov)
```

The output of the function `summary()` is as follows:

```
Error: fB
            Df Sum Sq Mean Sq F value Pr(>F)
Residuals    2   3978    1989

Error: fB:fA
            Df Sum Sq Mean Sq F value  Pr(>F)
fA           3   5179  1726.5   11.85 0.00621 **
Residuals    6    874   145.7
---
Signif. codes:   0 '***' 0.001 '**' 0.01 '*' 0.05 '.' 0.1 ' ' 1

Error: Within
            Df Sum Sq Mean Sq F value Pr(>F)
Residuals 12  323.5   26.96
```

R does not present an F test of the significance of the main effect of 'B', and hence does not here distinguish between the interpretations of Snedecor and Cochran and of Ridgman.

The following commands analyse the data using mixed-modelling methods:

```
library(lme4)
AxB.modellmer <- lmer(y ~ fA + (1|fB) +(1|fA:fB),
    data = AxB)
summary(AxB.modellmer)
```

The output of these commands is as follows:

```
Linear mixed model fit by REML
Formula: y ~ fA + (1 | fB) + (1 | fA:fB)
    Data: AxB
```

```
   AIC   BIC logLik deviance REMLdev
 162.5 170.8 -74.27    170.9   148.5
Random effects:
 Groups    Name          Variance Std.Dev.
 fA:fB     (Intercept)    59.347   7.7037
 fB        (Intercept)   230.424  15.1797
 Residual                 26.958   5.1921
Number of obs: 24, groups: fA:fB, 12; fB, 3

Fixed effects:
            Estimate Std. Error t value
(Intercept)   45.667     10.053   4.542
fA2           18.000      6.968   2.583
fA3           30.833      6.968   4.425
fA4           -6.333      6.968  -0.909

Correlation of Fixed Effects:
     (Intr) fA2     fA3
fA2 -0.347
fA3 -0.347  0.500
fA4 -0.347  0.500   0.500
```

Again, the estimate of σ_B^2 agrees with that obtained from the EMSs according to Ridgman's formulae. Thus R also supports this interpretation, not that of Snedecor and Cochran.

SAS's PROC ANOVA does not distinguish between fixed- and random-effect terms. If a particular model term is to be tested against a term other than the residual, this must be specified explicitly in the SAS statements, and does not have automatic consequences for the testing of any other term. Hence PROC ANOVA embodies no assumptions concerning the appropriate significance test when fixed- and random-effect terms interact. However, PROC MIXED does make such assumptions. The following SAS statements analyse the data using this procedure:

```
PROC IMPORT OUT = fixarandb DBMS = EXCELCS REPLACE
    DATAFILE = "&pathname.\IMM edn 2\Ch 6\fixed A, random B.xlsx";
    SHEET = "for SAS";
RUN;

ODS RTF;
PROC MIXED ASYCOV NOBOUND;
    CLASS A B;
    MODEL y = A /DDFM = KR HTYPE = 1 SOLUTION;
    RANDOM B A*B;
RUN;
ODS RTF CLOSE;
```

Part of the output of PROC MIXED is as follows:

Convergence criteria met

Covariance parameter estimates	
Cov Parm	**Estimate**
B	230.42
A*B	59.3472
Residual	26.9583

Asymptotic covariance matrix of estimates				
Row	**Cov Parm**	**CovP1**	**CovP2**	**CovP3**
1	B	61927	−441.97	3.08E−11
2	A*B	−441.97	1798.18	−60.5626
3	Residual	3.08E−11	−60.5626	121.13

Type 1 tests of fixed effects				
Effect	**Num DF**	**Den DF**	**F value**	**Pr > F**
A	3	6	11.85	0.0062

Again, the estimate of σ_B^2 agrees with that obtained from the EMSs according to Ridgman's formulae. Thus SAS's PROC MIXED also supports this interpretation, not that of Snedecor and Cochran.

In order to justify the formula of Snedecor and Cochran, it is necessary to assume that for all j (i.e. for every level of B),

$$\sum_{i=1}^{n_A} (\alpha\beta)_{ij} = 0. \tag{6.10}$$

The *estimates* and *predictors* of effects in a random-effect model term are commonly constrained to sum to zero: we have seen this in the case of Best Linear Unbiased Predictors (BLUPs) in Equation 5.7 (Section 5.2), and it is true of the estimated residuals in any straightforward regression analysis. However, to assume, as in Equation 6.10, that the *true* effects in such a term sum to zero is much more questionable, and Nelder (1977) argued strongly against it, as follows:

It is common (though undesirable) in setting up a model with fixed effects, say $y_{ij} = \mu + \alpha_i + \beta_j$, to constrain both the αs and βs to sum to zero, a property borrowed from their estimates, a_i and b_j. When such constraints are carried over to random effects, however, inconsistencies immediately arise. Thus Kempthorne and Folks (1971) describe the mixed model for the two-way classification with interactions (A fixed and B random) in which the interaction effects $(\alpha\beta)_{ij}$ are first defined as independent $N(0, \sigma^2)$ variables and then have their B margin $\sum_{i=1}^{n_A} (\alpha\beta)_{ij}$ constrained to be zero. These mutually inconsistent properties are confusing to the beginner and indeed to the expert as well. They lead to unnecessary complexity in the rules for deriving EMSs (see e.g. Bennett and Franklin, 1954). They also lead to unrealistic hypotheses … whereby the variation in a margin is hypothesized to be zero though real interactions are supposed to be present in the body of the table.

(The notation in this quotation from Nelder has been changed for consistency with that used here.) Nelder's paper started a heated and long-running debate, in which he referred to the constraints that he considered unjustified, and their consequences, as 'the great mixed-model muddle' (Nelder, 1998). He later stated that this 'has now been resolved by the recognition of two simple principles' (Nelder, 2008). These are

1. *Marginality*, a concept which is outlined in Section 7.2. If the term A.B is random, then because B is marginal to A.B, the variance component σ^2_{AB} must contribute to the variation among the levels of B. (The same is true of A, but this is not in dispute.)

2. Constraints on parameters – or rather, the absence of constraints on their true values, as described in the quotation above.

The counter-arguments to Nelder's position are presented in discussions appended to the articles cited here, but the present author does not find them persuasive. Throughout this book, it is assumed that the formulae for the EMSs advocated by Nelder, presented by Ridgman and implicit in the F statistics and variance component estimates given by GenStat, R and SAS, are correct.

6.5 Arguments for specifying block effects as random

It was stated earlier (Section 6.3) that all block terms should normally be specified as random. However, this practice is by no means universal, especially when the degrees of freedom of a block term are low and its variance is consequently poorly estimated. The case for this recommendation must therefore be argued.

The usual analysis of the randomized block design is valid in a wider range of circumstances if the block effects can be regarded as random than if they are regarded as fixed. To illustrate this, consider a simple experiment with a single fixed-effect treatment factor, A, with n_A levels, arranged in a randomized block design with n_B blocks. Suppose it is thought likely that there will be block × treatment interaction effects. It is then proper to replicate the treatments within each block, so that any such interaction can be detected, and its significance and magnitude assessed. Suppose that r replications of each treatment are set up within each block, and that a response variable Y is measured on each experimental unit. The appropriate model for this design is then

$$A * block = A + block + A.block \tag{6.11}$$

in the notation of Wilkinson and Rogers (1973), or in algebraic notation,

$$y_{ijk} = \mu + \alpha_i + \beta_j + (\alpha\beta)_{ij} + \varepsilon_{ijk} \tag{6.12}$$

where

y_{ijk} = the observation of the response variable on the kth replication of the ith treatment in the jth block,

μ = the grand mean (overall mean) value of the response variable,

α_i = the effect of the ith treatment,

β_j = the effect of the jth block,

$(\alpha\beta)_{ij}$ = the interaction between the ith treatment and the jth block and

ε_{ijk} = the effect of the kth experimental unit that received the ith treatment in the jth block.

It is assumed that the ε_{ijk} are independent values of a random variable E, such that

$$E \sim N(0, \sigma^2). \tag{6.13}$$

Now compare the consequences of specifying 'block' as a fixed-effect term with those of specifying it as a random-effect term. The treatment factor A is a fixed-effect term, so if 'block' is specified as a fixed-effect term the interaction term 'A.block' will also be a fixed-effect term if we follow Guideline 1 in Section 6.3. The variance of the effects of B is then designated by κ_B^2, and that of the A.B interaction effects by κ_{AB}^2. On the other hand, if 'block' is specified as a random-effect term, 'A.block' will also be a random-effect term. The values β_j will then be interpreted as independent values of a random variable B, such that

$$B \sim N(0, \sigma_B^2), \tag{6.14}$$

and the values $(\alpha\beta)_{ij}$ as independent values of a random variable (AB), such that

$$(AB) \sim N(0, \sigma_{AB}^2). \tag{6.15}$$

The EMSs and F tests that follow from each decision are illustrated in Table 6.4. If 'block' is specified as a fixed-effect term, the main effects of blocks, the main effects of treatment and the block × treatment interaction term are all tested against the residual term, whereas if 'block' is specified as a random-effect term, the main effects are tested against the interaction term.

Now consider the case where there is no replication of each treatment within each block (i.e. where $r = 1$), or where the analysis is performed using the mean value of each treatment in each block instead of the raw data. The EMSs and F tests are then as shown in Table 6.5. If 'block' is specified as a fixed-effect term, no term in the anova can now be tested against any other term, whereas if 'block' is specified as a random-effect term, the main effects can still be tested against the interaction.

Table 6.4 Expected mean squares from a randomized block design in the presence of block × treatment interaction, with treatment replication within each block. κ_A^2 = variance of the fixed effects of A. κ_B^2 = variance of the fixed effects of B. κ_{AB}^2 = variance of the fixed effects of (AB). The pairs of mean squares to be compared by F tests are indicated by angled lines.

Source of variation	DF	Expected mean square	
		Block effects fixed	Block effects random
block	$n_B - 1$	$n_A r \kappa_B^2 + \sigma_E^2$	$n_A r \sigma_B^2 + r \sigma_{AB}^2 + \sigma_E^2$
A	$n_A - 1$	$n_B r \kappa_A^2 + \sigma_E^2$	$n_B r \kappa_A^2 + r \sigma_{AB}^2 + \sigma_E^2$
A.block	$(n_A - 1)(n_B - 1)$	$r \kappa_{AB}^2 + \sigma_E^2$	$r \sigma_{AB}^2 + \sigma_E^2$
Residual		σ_E^2	σ_E^2

Table 6.5 Expected mean squares from a randomized block design in the presence of block × treatment interaction, but with no treatment replication within each block. Conventions are as in Table 6.4.

Source of variation	DF	Expected mean square	
		Block effects fixed	Block effects random
block	$n_B - 1$	$n_A \kappa_B^2 + \sigma_E^2$	$n_A \sigma_B^2 + \sigma_{AB}^2 + \sigma_E^2$
A	$n_A - 1$	$n_B \kappa_A^2 + \sigma_E^2$	$n_B \kappa_A^2 + \sigma_{AB}^2 + \sigma_E^2$
A.block	$(n_A - 1)(n_B - 1)$	$\kappa_{AB}^2 + \sigma_E^2$	$\sigma_{AB}^2 + \sigma_E^2$

If the block effects are to be specified as fixed, then in order to perform significance tests it is now necessary to drop the interaction term 'A.block' from the model, simplifying it to

$$A + \text{block} \tag{6.16}$$

in the notation of Wilkinson and Rogers (1973), or in algebraic notation,

$$y_{ij} = \mu + \alpha_i + \beta_j + \varepsilon_{ij}. \tag{6.17}$$

where
  y_{ij} = the observation of the response variable on the ith treatment in the jth block,
  μ = the grand mean (overall mean) value of the response variable,
  α_i = the effect of the ith treatment and
  β_j = the effect of the jth block
  ε_{ij} = the effect of the experimental unit that received the ith treatment in the jth block.

The EMSs and F tests are then as shown in Table 6.6: the standard F tests for the randomized block design are now appropriate whether block effects are specified as fixed or random. However, if they are specified as random, the inclusion or omission of the interaction term makes no difference to the appropriate F test. The interpretation of the bottom stratum of the anova changes from 'A.block' (Table 6.5) to 'Residual' (Table 6.6), but arithmetically the analysis is unchanged. In summary, this argument shows that the standard interpretation of the randomized block design is appropriate whether or not block × treatment interaction is present, provided that the blocks studied are representative of the population for which inferences are to be made, or comprise an exchangeable set.

Even if it is possible to interpret an anova in which block effects are specified as fixed, the omission of the corresponding variance components from the calculation of significance tests and standard errors (SEs) may lead to over-estimation of the significance of other effects, and of the precision of their estimates. This can be illustrated by specifying all model terms in the ravioli split-plot experiment (except the residuals) as fixed-effect terms, using the following statements:

```
MODEL saltiness
FIT [FPROB = yes; PRINT = accumulated] \
   brand * assessor + day / presentation
```

Table 6.6 Expected mean squares from a randomized block design in the absence of block × treatment interaction.
Conventions are as in Table 6.4.

Source of variation	DF	Expected mean square	
		Block effects fixed	Block effects random
block	$n_B - 1$	$n_A \kappa_B^2 + \sigma_E^2$	$n_A \sigma_B^2 + \sigma_E^2$
A	$n_A - 1$	$n_B \kappa_A^2 + \sigma_E^2$	$n_B \kappa_A^2 + \sigma_E^2$
Residual	$(n_A - 1)(n_B - 1)$	σ_E^2	σ_E^2

This analysis cannot be specified using GenStat's analysis of variance directives (BLOCK-STRUCTURE, TREATMENTSTRUCTURE and ANOVA) because the variation represented by the term 'brand.assessor' overlaps with (is partially aliased with) that represented by the term 'day.presentation': therefore the analysis is specified using the regression analysis directives. Note that the term 'day.presentation.serving' is not included explicitly in the model: it is omitted so that it will be used as the residual term.

The anova produced by these statements is compared with the correct one in Table 6.7. The angled lines on this table indicate the mean squares that are compared in the F test of each treatment term. The mean squares in the two analyses are identical. However, when the block effects are specified as fixed, 'brand' is incorrectly tested against the same mean square as 'assessor' and 'brand.assessor'. This small mean square with its large degrees of freedom leads to an inflated value of F_{brand}, and an over-estimation of the significance of this term.

In other circumstances, a decision to specify block effects as fixed can lead to *under*-estimation of the significance of other effects and of the precision of their estimates. This is illustrated by the analysis of a balanced incomplete block design presented in Sections 9.1–9.4. In this design, each block contains only a sub-set of the treatments being compared, and the information concerning treatment effects is therefore distributed between two strata of the anova, the among-blocks and within-block strata. If the block effects are specified as fixed, the estimates of treatment effects are based on the within-blocks information only. The significance of the treatment term is then lower, and SE$_{\text{Difference}}$ for comparisons between treatment means is larger, than if the block effects are regarded as random. The case for specifying incomplete block effects as random, and hence recovering inter-block information, is presented more rigorously by Robinson (1991, Section 5.1).

The argument that block effects should be regarded as random may apply also to factor levels that do not obviously constitute blocks. For example, suppose that in an educational experiment, a new method is tested in some schools, while others are used as controls, and that the response variable measured is the attainment of individual children within each school. It is important to specify 'school' as a random-effect term: otherwise variation among schools will not contribute to the SE of the estimate of the effect of educational method, and the precision of this estimate will be exaggerated. As for the reluctance to specify block terms as random due to their low degrees of freedom and consequent poorly estimated variance components, this is less of a concern if attention is focused on the *effects* of treatments and the *differences* between their means, rather than the means themselves. This distinction has been more fully discussed earlier (Section 4.6).

Table 6.7 Comparison between the correct analysis of a split-plot experiment and an analysis with block effects regarded as fixed.

	Correct analysis				Analysis with block effects regarded as fixed			
Source of variation	DF	MS	F	p	Source of variation	DF	MS	F	p
day stratum									
day	2	371.7	2.67		day	2	371.7	3.71	0.0300
day.presentation stratum									
brand	3	3286.6	23.59	0.00101	brand	3	3286.6	32.78	5.86×10^{-13}
Residual	6	139.3	1.39		day.presentation	6	139.3	1.39	0.233
day.presentation.serving stratum									
assessor	8	1934.3	19.29	2.12×10^{-14}	assessor	8	1934.3	19.29	2.12×10^{-14}
brand.assessor	24	253.1	2.52	0.002	brand.assessor	24	253.1	2.52	0.00171
Residual	64	100.3			Residual	64	100.3		

6.6   Examples of the choice of fixed- and random-effect specification of terms

Some examples will illustrate how the guidelines for identifying random-effect terms, presented in Section 6.3, are applied. In the treatment of asthma, two classes of drug, corticosteroids and beta-agonists, may be prescribed singly or in combination. A clinical trial to compare the efficacy of several doses of each might have the design shown in Table 6.8. Each cell in this table represents a treatment, to which the same number of patients will be allocated.

The model to be fitted is

corticosteroid * beta-agonist.

A measure of lung function called the *forced expiratory volume* (*FEV*) will be used on each patient. The levels of both treatment factors, corticosteroid and beta-agonist, are values on numerical scales, so the main effects of both factors should be specified as fixed-effect terms. Therefore the interaction term 'corticosteroid.beta-agonist' should also be specified as fixed. It is likely that the interaction effects are not a set of unrelated values, but that they follow some fairly simple functional form. For example, there is some evidence that the effects of these two classes of drugs are *synergistic*, that is, they work better in combination than either does on its own. In this case, there may be more response to corticosteroid at higher doses of beta-agonist and vice versa. Such a relationship would vindicate the decision to specify the interaction effects as fixed.

In another clinical trial, the four doses of beta-agonist might all be given to each of the 12 patients, during successive time periods. In such a trial, the order in which the doses are given will be randomized, a different randomization being used in each patient, and there will be a 'wash-out period' between treatment periods, to eliminate possible carry-over effects. Suppose that two replicate observations of FEV are made on each patient in each time period. The experiment will then have the design illustrated in Figure 6.2.

The model to be fitted is

beta-agonist * patient.

The term *beta-agonist* is a fixed-effect term as before, but the patients studied can (if they have been properly selected) be considered as a representative sample from a population of asthma sufferers eligible to participate in the trial: hence it is natural to specify 'patient' as a random-effect term. It follows from Guideline 1 (Section 6.3) that the interaction term 'beta-agonist.patient' is also a random-effect term. If Patient 1 were replaced by a different patient, this patient would respond to any particular dose of beta-agonist in a different way

Table 6.8   Factorial design to study the efficacy of two drug types in the treatment of asthma.

Dose of corticosteroid	Dose of beta-agonist			
	None	Low	Medium	High
None				
Low				
High				

Patient 1	Medium dose		No dose		High dose		Low dose	
	Obs 1	Obs 2	Obs 1	Obs 2	Obs 1	Obs 2	Obs 1	Obs 2

Patient 2	No dose		Low dose		High dose		Medium dose	
	Obs 1	Obs 2	Obs 1	Obs 2	Obs 1	Obs 2	Obs 1	Obs 2

Patient 12	Medium dose		Low dose		No dose		High dose	
	Obs 1	Obs 2	Obs 1	Obs 2	Obs 1	Obs 2	Obs 1	Obs 2

Figure 6.2   The arrangement, in a factorial design, of an experiment to study the efficacy of a range of doses of a beta-agonist in the treatment of asthma.

(assuming that real beta-agonist.patient interaction effects were present): thus the interaction effects, as well as the main effects of patients, are a random sample from an underlying population. The mixed model specified is then as follows:

Fixed-effect model: beta-agonist

Random-effect model: patient + beta-agonist.patient.

This clinical trial effectively has as randomized block design, each patient comprising a block. The random-effect model can be expressed more succinctly as

patient/beta-agonist.

However, note that in the present case this does *not* mean

main effects of patient + effects of beta-agonist within each patient

but

main effects of patient + beta-agonist × patient interaction effects,

because the main effect term 'beta-agonist' is present in the fixed-effect model.

Note that it is assumed that the interaction effects at each level of beta-agonist in the same patient are independent values of the underlying random variable. This assumption is reasonable if the different levels of beta-agonist are unrelated to each other. However, if, as in the present case, there is some structure among the levels, it may be desirable to reflect this in the model. If so, it should be done both for the main effect term 'beta-agonist' and for the beta-agonist.patient interaction term. For example, if it is expected that there is a trend in FEV in response to a higher dose of beta-agonist, this can be represented by dividing 'beta-agonist' into three polynomial terms, namely

lin(beta-agonist) – the linear effect of the dose of beta-agonist

quad(beta-agonist) – the effect of $dose^2$

cub(beta-agonist) – the effect of $dose^3$.

Table 6.9 Specification of the terms to be compared by significance tests in the anova of an experiment to study the effect of a beta-agonist on asthma.

Source of variation	
Original specification	Alternative specification
lin(beta-agonist)	lin(beta-agonist)
quad(beta-agonist)	quad(beta-agonist)
cub(beta-agonist)	cub(beta-agonist)
patient	patient
lin(beta-agonist).patient	beta-agonist.patient
quad(beta-agonist).patient	
cub(beta-agonist).patient	

The model to be fitted is then

$$(\text{lin(beta-agonist)} + \text{quad(beta-agonist)} + \text{cub(beta-agonist)}) * \text{patient}.$$

It is partitioned between the fixed-effect and random-effect models as follows:

Fixed-effect model: lin(beta-agonist) + quad(beta-agonist) + cub(beta-agonist)

Random-effect model: patient + lin(beta-agonist).patient + quad(beta-agonist).patient + cub(beta-agonist).patient.

The significance of each fixed-effect term is tested against the corresponding random-effect term, as indicated by the angled lines in the 'Original specification' column of Table 6.9.

If it appears reasonable to assume that the variation due to the terms 'lin(beta-agonist) .patient', 'quad(beta-agonist).patient' and 'quad(beta-agonist).patient' are *homogeneous* – that is, that the variances due to these three terms are equal – then it may be decided to pool them, to give significance tests with more degrees of freedom in the denominator and more power, as shown in the 'Alternative specification' column of Table 6.9. Whether such pooling is permissible is a matter of judgement. The three interaction mean squares can be compared formally using a significance test, for example, Bartlett's test for homogeneity of variance (Snedecor and Cochran, 1989, Section 13.10, pp. 251–252), but it should be remembered that a failure to find significant differences does not prove that quantities are the same.

Alternatively, we may decide that the relationship between FEV and dose is likely to be nearly linear, and that four levels of dose are not adequate to detect any slight departures from linearity. We may then choose to divide 'beta-agonist' only into two terms, namely

lin(beta-agonist) – the linear effect of the dose of beta-agonist

dev(beta-agonist) – the deviation of the effects of individual doses of beta-agonist from the linear effect.

The model to be fitted is then

$$(\text{lin(beta-agonist)} + \text{dev(beta-agonist)}) * \text{patient}.$$

We have seen that in order to obtain a meaningful estimate of a linear trend over factor levels, it is necessary to specify the departures from this trend as random effects (Section 1.6).

Therefore dev(beta-agonist) should be specified as a random-effect term, and our new model is partitioned between the fixed-effect and random-effect models as follows:

Fixed-effect model: lin(beta-agonist)

Random-effect model: patient + dev(beta-agonist) + lin(beta-agonist).patient + dev(beta-agonist).patient.

However, the variance component for dev(beta-agonist), based on only four dose levels, will be poorly estimated, which may result in a large SEs for the linear trend. Therefore, if exploration suggests that this variance component is close to zero (see Section 3.13), we may choose to drop this term and the term dev(beta-agonist).patient from the model. But again, we should remember that failure to detect an effect does not prove that it is absent.

Next consider an ecological study in which soil samples are taken at several sites in each of the several areas of woodland. In each sample, the abundance of springtails (a small, very common type of insect) is measured. The model to be fitted is

area/site.

Suppose that it is decided to specify 'area' as a random-effect term: the areas sampled are to be considered as representative of a larger set of woodland areas. Following Guideline 2, the effect of site within area (the term 'area.site') must then also be specified as random. The choice of a nested model, with no main effect of site, means that Site 1 in Area 1 is presumed to have nothing in common with Site 1 in Area 2. If another area were substituted for Area 1, the effect of Site 1 within the new Area 1 would be different: thus the effect of site within area, like that of area, is sampled from an underlying population.

It may be argued that there *are* features in common between sites in different areas. For example, the abundance of springtails may be influenced by soil pH, and each area may comprise high- and low-pH sites. Such effects should be recognized by a main-effect term in the model, which would then become

area * site type.

The factor 'site type' might then be specified as a fixed-effect term: or, in the case of soil pH, if measurements on a numerical scale were available, it might be replaced by a variate, which would be a fixed-effect term. The model would then be

area * pH.

Although the main effect of any variate should usually be specified as a fixed-effect term, the interaction of a variate with a factor, or the effect of a variate nested within levels of a factor, may be either fixed or random. For example, consider a study in which repeated measurements, over time or some other continuous variable, are made on each experimental unit – for example, a study in which the degree of anxiety indicated by the behaviour of mice is scored daily for several days. The mean score is likely to vary between mice, but it is also likely that each mouse will become more or less anxious during the period of the experiment. However, this trend is also likely to vary between mice: some may become more anxious, others less. The model to be fitted is then

mouse/time.

It is often natural to specify the variation among experimental units at time zero (the main-effect term 'mouse' in the present case) as random. The variation among mice in the

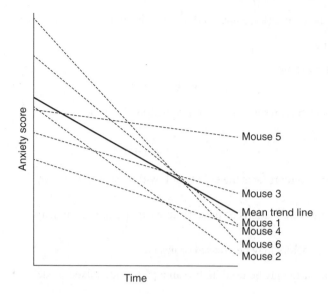

Figure 6.3 Possible pattern of results in an experiment to determine the effect of a factor (mouse) and a variate (time) on a response variate (anxiety score).

trend over the repeated measurements (the term 'mouse.time') is then also random, following Guideline 1. The resulting model comprises a set of regression lines, one for each experimental unit, with random variation among their slopes and intercepts – their coefficients.

It may be that there is some component of the trend over time that is expected to be common to all mice. For example, mice may generally become less anxious as they become used to the test situation. Such a common trend, around which the trend lines for individual mice vary, should be represented in the model by a main effect of time. The model to be fitted is then

$$\text{mouse} * \text{time}.$$

An example of a pattern of results that might be obtained from fitting this model is shown in Figure 6.3. A similar *random coefficient model* is considered in more detail later (Sections 7.5–7.7).

In the next chapter, we will apply the methods, principles and rules introduced in this chapter to real data sets and interpret the results.

6.7 Summary

The mixed models introduced in earlier chapters illustrate the following features of mixed models in general:

- each model term may be
  - a variate, in which each observation can have any numerical value, or
  - a factor, in which each must come from a specified set of levels.

- each part of the model (fixed and random) may have more than one term

- factors may be

  - crossed, for example

  $$a * b = a + b + a.b$$

  where term a.b represents the $a \times b$ interaction effects, or

  - nested, for example

  $$a/b = a + a.b$$

  where term a.b represents the effects of b within each level of a.

More elaborate combinations of these features are permitted, namely

- more than two factors may be crossed or nested

- factors and variates may be used in the same part of the mixed model

- factors may be crossed or nested with variates.

If 'f' is a factor and 'v' is a variate, the model 'f*v' or 'f/v' represents a set of sloping lines relating the response variate to the explanatory variate 'v'. The main-effect term 'v' represents the mean slope of the line, averaged over the levels of 'f', and is absent from the model 'f/v'.

Variates may be crossed with other variates. In this case, the interaction term is equivalent to the product of the two variates.

In the balanced data sets considered in earlier chapters, the significance of each fixed-effect model term is tested by comparing the mean square for that term with one other mean square. However, information about a term is often distributed over two or more strata defined by the random-effect terms. The information from these strata can be pooled when estimating fixed effects, but this is not straightforward, and in a mixed-model analysis it is generally done by computer software.

Rules are presented for deciding which terms should be allocated to the random-effects part of a mixed model, as follows:

- It is generally sensible to specify all variates as fixed-effect terms.

- Factors should be classified as

  - Terms representing nuisance effects – block effects, which should generally be specified as random. If block effects are regarded as fixed, significance tests may be mis-specified.

  - Terms representing effects of interest – treatment effects, which may be fixed or random.

To decide whether a treatment factor should be specified as random, we need to consider whether its levels are a random sample from an infinite population, or alternatively whether they form an exchangeable set. If so, we need to decide whether to take account of this relationship among them.

Even if factor levels are not strictly randomly sampled from an infinite population, this may still be a reasonable modelling assumption.

Although additional assumptions are made when a factor is regarded as a random-effect term, this should not necessarily be thought of as a radical step: on the contrary, the inferences made about other terms are more cautious than when the factor is regarded as fixed.

Decisions on whether a term is fixed or random need usually only be made for main effects. For interaction terms and some nested terms, the choice is determined by the following guidelines:

- *Guideline 1.* When two terms are crossed, if both terms are fixed the interaction should usually be fixed (but see Section 8.4 for an exception): otherwise, it must be random.

- *Guideline 2.* When one term is nested within another, the nested term may be either fixed or random if the term within which it is nested is fixed: otherwise, it must be random.

There is an inconsistency in the statistical literature concerning the method for the construction of the F test when a fixed-effect term (A) and a random-effect term (B) interact. Some authorities consider that B should be tested against the residual term, others that it should be tested against the A.B interaction term. This disagreement has been called *the great mixed-model muddle*. It is argued that the latter view is correct.

Examples of the application of the rules for specifying terms as fixed or random are given.

6.8 Exercises

6.1 In the experiment to compare four commercial brands of ravioli described in Chapter 2, it may be argued that 'assessor' should be specified as a random-effect factor.

   (a) Consider the case for this decision, and present the arguments for and against.

   (b) If 'brand' is specified as a fixed-effect term and 'assessor' as a random-effect term, how should 'brand.assessor' be specified?

   (c) Make these changes to the mixed model fitted to these data. Perform the new analysis and interpret the results. Explain the effect of the changes on the SEs of differences between brands.

   (d) Perform appropriate tests to determine whether the new random-effect terms are significant.

6.2 Return to the data set concerning the effect of oil type on the amount of wear suffered by piston rings, introduced in Exercise 2.2.

   (a) For each oil type, plot the value of wear against the ring number (1–4). Omit the values from the 'oil ring' from this plot. Is there evidence of a trend in the amount of wear from Ring 1 to Ring 4?

   (b) Repeat the mixed-model analysis performed on these data previously, but exclude the values from the 'oil ring'.

   (c) Now specify 'ring' as a variate instead of a factor. Fit this new model to the data by mixed modelling. Do the results confirm that there is a linear trend from Ring 1 to Ring 4? Does this trend vary significantly depending on the oil type?

   Note that when this change is made to the model, DF_{ring} is reduced from 3 to 1, and $DF_{oil.ring}$ from 6 to 2.

   (d) What source of variation is included in these terms in the previous model, but excluded in the present model?

Now suppose that the three types of oil tested are considered to comprise an exchangeable set.

(e) When this change is made, which parts of the expression 'oil*ring' should be regarded as fixed-effect terms, and which as random-effect terms?

(f) Fit this new model to the data by mixed modelling. Does the new model indicate that the linear trend from Ring 1 to Ring 4 varies significantly depending on the oil type?

References

Bennett, C.A. and Franklin, N.L. (1954) *Statistical Analysis in Chemistry and the Chemical Industry*, John Wiley & Sons, Inc., New York.

Draper, N.R. and Smith, H. (1998) *Applied Regression Analysis*, 3rd edn, John Wiley & Sons, Inc., New York, 706 pp.

Kempthorne, O. and Folkes, L. (1971) *Probability, Statistics and Data Analysis*, Iowa State University Press, Ames, IA.

Mead, R. (1988) *The Design of Experiments. Statistical Principles for Practical Application*, Cambridge University Press, Cambridge, 620 pp.

Nelder, J.A. (1977) A reformulation of linear models. *Journal of the Royal Statistical Society*, **140**, 48–77.

Nelder, J.A. (1998) The great mixed model muddle is alive and flourishing – alas! *Food Quality and Preference*, **9**, 157–159.

Nelder, J.A. (2008) What is the mixed model controversy? *International Statistical Review*, **76**, 134–139.

Ridgman, W.J. (1975) *Experimentation in Biology*, Blackie, Glasgow, 233 pp.

Robinson, G.K. (1991) That BLUP is a good thing: the estimation of random effects. *Statistical Science*, **6**, 15–51.

Snedecor, G.W. and Cochran, W.G. (1989) *Statistical Methods*, 8th edn, Iowa State University Press, Ames, IA, 503 pp.

Wilkinson, G.N. and Rogers, C.E. (1973) Symbolic description of factorial models for analysis of variance. *Applied Statistics*, **22**, 392–399.

7

Three case studies

7.1 Further development of mixed modelling concepts through the analysis of specific data sets

In this chapter, we shall apply the concepts introduced earlier to three data sets more elaborate than those considered so far. These were obtained from specialized investigations, the first into the causes of variation in bone mineral density (BMD) among human patients, the second into the possible value of lithium as a treatment for the degenerative neurological disease amyotrophic lateral sclerosis (ALS) and the third into the causes of variation in oil content in a grain crop. However, during the analysis of these data, new concepts and additional features of the mixed modelling process will be introduced which are widely applicable. In summary, these are as follows:

- *In the investigation of the causes of variation in BMD*

  - a fixed-effects model with several variates and factors
  - a polynomial model with quadratic and cross-product terms;
  - further interpretation of diagnostic plots of residuals, leading to the exclusion of outliers;
  - exploration and building of the fixed-effects model;
  - use of the marginality criterion to determine which terms to retain in a polynomial model;
  - predicted values from a model: their use as an aid to understanding the results of the modelling process;
  - graphical representation of predicted values;
  - the consequences of extrapolation of predictions outside the range of the data.

Introduction to Mixed Modelling: Beyond Regression and Analysis of Variance, Second Edition. N. W. Galwey.
© 2014 John Wiley & Sons, Ltd. Published 2014 by John Wiley & Sons, Ltd.
Companion website: http://www.wiley.com/go/beyond_regression

- *In the investigation of lithium as a treatment for ALS*

  – adjustment for a baseline value for each experimental subject when fitting the subject's trend over time;

  – intercepts and slopes of trend lines as random-effect terms;

  – allowance for possible covariance between these terms.

- *In the investigation of the causes of variation in oil content of a grain crop*

  – a random-effects model with several factors;

  – the marginality criterion applied to a linear trend and deviations from it;

  – exploration and building of the random-effects model;

  – the effect of one term within and among the levels of another: recognition of this distinction;

  – a (random factor) × (fixed variate) interaction term;

  – dependence of the magnitude of the variance component for the (random factor) × (fixed variate) term on the units of measurement of the variate;

  – comparison of the magnitudes of variance components;

  – graphical representation of the variation accounted for by the mixed model.

7.2   A fixed-effects model with several variates and factors

In an extensive study of osteoporosis, a disease in which the strength of the bones is abnormally low and the patient is at high risk of fractures, data were obtained from about 3600 individuals. One of the main observations taken was the BMD, a quantitative trait closely associated with osteoporosis, measured by an X-ray scan at various skeletal sites including the lumbar region of the spine. The main purpose of the study was to identify genetic factors associated with osteoporosis, and related individuals (parents and offspring, siblings, etc.) were therefore sampled. An extensive analysis of the genetic information in the data was reported by Ralston *et al.* (2005). Here we will focus on the possible relationships between BMD and other variables on which data were obtained, namely the sex, age, height and weight of the patient, and the hospital at which the scan was performed. These variables were observed in the expectation that adjustment for them would improve the precision with which the genetic influences on BMD could be estimated.

The variables to be considered here are listed in Table 7.1, and the first and last few rows of the spreadsheet holding the data are shown in Table 7.2. Each row represents an individual. (Data reproduced and used by kind permission of the FAMOS consortium.)

BMD is known to change with age, but this relationship is not expected to be linear: evidence from other studies, including studies over time on the same individual, indicates that a person's BMD reaches a peak at about age 40 and then declines. We will allow for this non-linearity by considering age^2, as well as age *per se*, for inclusion in the model fitted. Similarly, the relationship between BMD and height is not expected to be independent of that between BMD

Table 7.1 Variables recorded for each individual in a survey of osteoporosis in a human population.

Name	Type	Description
Gender	factor	1 = male; 2 = female
CentreNumber	factor	Number indicating the hospital (the centre) at which the individual was observed
Proband	factor	Whether the individual was a *proband*, that is, was recruited on the basis that they suffered from osteoporosis. (Other individuals were recruited on the basis that they were related to a proband.) Valid factor levels are 'No' and 'Yes'
EverHadFractures	factor	Whether the individual had ever had fractures. Valid factor levels are 'No', 'Unknown' and 'Yes'
Age	variate	Age in years at the time of measurement
Height	variate	Height in centimetres
Weight	variate	Weight in kilograms
BMD_Outlier	factor	Whether the individual is identified as having unusually high or low values of BMD for their age, height and weight. Valid factor levels are 'No' and 'Yes'
CalL2-L4BMD	variate	BMD, averaged over lumbar vertebrae 2, 3 and 4, calibrated to adjust for differences between the X-ray machines with which different individuals were measured

Table 7.2 Data from a survey of osteoporosis in a human population.

	A	B	C	D	E	F	G	H	I
1	Gender!	CentreNumber!	Proband!	EverHadFractures!	Age	Height	Weight	BMD_Outlier!	CalL2-L4BMD
2	2	02	Yes	No	62.00	150.00	53.10	No	0.772
3	1	02	No	Yes	61.00	177.00	76.10	No	0.970
4	2	02	No	No	40.00	157.50	56.90	No	1.238
5	1	02	No	Yes	39.00	182.00	73.00	No	0.950
⋮									⋮
3690	2	11	Yes	Yes	52.00	178.00	100.00	No	0.753
3691	1	11	No	Yes	36.00	185.00	98.60	No	1.099
3692	1	11	No	Yes	32.00	186.10	86.30	No	1.351

Source: Data reproduced and used by kind permission of the FAMOS consortium.

and weight: for example, a particular below-average value of BMD would be more extreme in a tall individual than in a short one. For this reason, height and weight are often combined into a single composite measure called *body mass index (BMI)*, thus

$$BMI = \frac{weight(kg)}{(height(cm))^2} \times 10000.$$

However, it may not be reasonable to assume that the interaction between height and weight is of the particular form specified by BMI. In order to give a good fit to a wide range of possible patterns of interaction, the following terms will be considered for inclusion in the model fitted:

height, weight, height^2, height $\times$ weight and weight^2. Because the probands have been selected on the basis that they suffer from osteoporosis, they are expected to give a distorted indication of the relationship between BMD and other variables. They will therefore be excluded from the modelling process. The following GenStat statements import the data and perform these preliminary manipulations:

```
IMPORT 'IMM edn 2\\Ch 7\\ost phenotypes.xlsx'
CALCULATE Agesq = Age ** 2
CALCULATE Htsq = Height ** 2
CALCULATE HtxWt = Height * Weight
CALCULATE Wtsq = Weight ** 2
RESTRICT CalL2_L4BMD; Proband .EQ. 1
```

The CALCULATE statements obtain the additional variables required, and the RESTRICT statement confines attention to those individuals for whom the value of the factor 'Proband' is 1, that is, 'No'. The restriction is applied directly only to the variate 'CalL2_L4BMD': when this variate is used in an analysis, the restriction will automatically be extended to the other variates and factors used.

It is believed that after allowing for the fact that different hospitals may have a preponderance of one or other sex, or may differ in the mean age of the patients scanned and so on, the remaining variation among hospitals can be regarded as a random variable – that is, either

- the hospitals can be regarded as a representative sample from a large population of hospitals available for study or

- at least they form an exchangeable set because, though we do expect the value of BMD to vary among hospitals, we have no idea which hospitals will give higher values than predicted from the fixed-effect terms, and which will give lower.

Therefore the initial model fitted (Model 7.1), comprising all the terms that are candidates for inclusion, is as follows:

Response variate:	CalL2_L4BMD
Fixed-effect model:	Gender + Age + Agesq + Height + Weight + Htsq + HtxWt + Wtsq
Random-effect model:	CentreNumber.

Note that all the fixed effect terms in this model, unlike those in the examples considered so far, vary both within and among levels of the random-effect term 'CentreNumber' (hospital). The fixed effects are therefore estimated mostly *within* each hospital, but they nevertheless explain part of the variation *among* hospitals. The following statements fit this model and display diagnostic plots of the residuals, but do not produce any other output:

```
VCOMPONENTS \
   [FIXED = Gender + Age + Agesq + \
   Height + Weight + Htsq + HtxWt + Wtsq] \
   RANDOM = CentreNumber
REML [PRINT = *] CalL2_L4BMD
VPLOT [GRAPHICS=high] fittedvalues, normal, halfnormal, histogram
```

The diagnostic plots are presented in Figure 7.1.

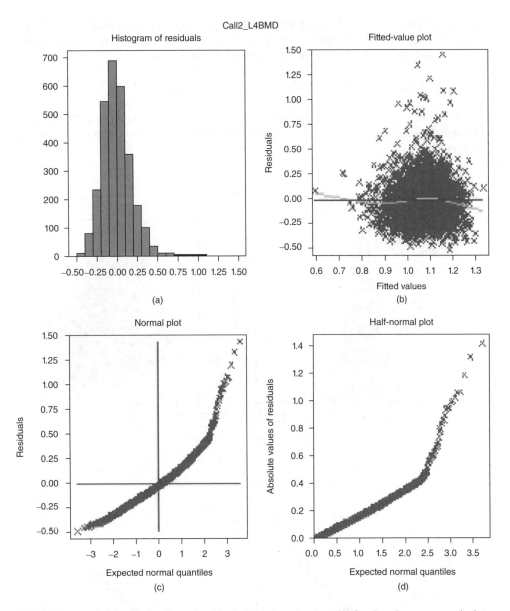

Figure 7.1   (a–d) Diagnostic plots of residuals from the mixed model fitted to the osteoporosis data, with exclusion of probands.

The residuals are clearly not normally and homogeneously distributed. The histogram has a long upper tail, and the fitted-value plot shows that this is due to a group of very large positive residual values associated with intermediate fitted values. This group of values also causes a kink in the distribution of the points in the normal and half-normal plots. Individuals with outlying values of BMD, assessed on the basis of all the skeletal sites, have already been identified in these data. It is believed that many of these observations represent regions of dense bone where previous fractures have healed, and are therefore unrepresentative of the overall composition of the patient's bones. The effect of excluding these values from the analysis is next explored. To do this, the RESTRICT statement is changed to

```
RESTRICT CalL2_L4BMD; Proband .eq. 1 .and. BMD_Outlier .eq. 1
```

and the VCOMPONENTS and REML statements are re-executed. The diagnostic plots are now as shown in Figure 7.2. The exclusion of the outliers has nearly, though not completely, eliminated the problem.

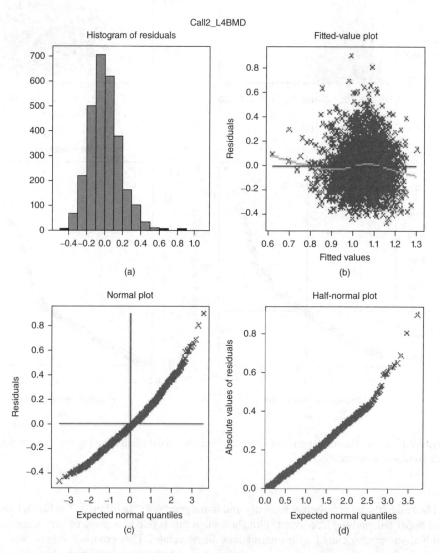

Figure 7.2 (a–d) Diagnostic plots of residuals from the mixed model fitted to the osteoporosis data, with exclusion of probands and BMD outliers.

The REML statement is then run with option settings that specify the usual items of output, as follows:

```
REML [PRINT = model, components, Wald] CalL2_L4BMD
```

The output from this statement is as follows:

REML variance components analysis

Response variate:	CalL2_L4BMD
Fixed model:	Constant + Gender + Age + Agesq + Height + Weight + Htsq + HtxWt + Wtsq
Random model:	CentreNumber
Number of units:	2801 (117 units excluded due to zero weights or missing values)

Residual term has been added to model.

Sparse algorithm with average information (AI) optimization.
All covariates centred.
Analysis is subject to the restriction on CalL2_L4BMD.

Estimated variance components

Random term	Component	s.e.
CentreNumber	0.00065	0.00041

Residual variance model

Term	Model(order)	Parameter	Estimate	s.e.
Residual	Identity	Sigma2	0.0286	0.00077

Tests for fixed effects

Sequentially adding terms to fixed model.

Fixed term	Wald statistic	n.d.f.	F statistic	d.d.f.	F pr.
Gender	67.33	1	67.33	2748.0	<0.001
Age	253.06	1	253.06	2789.5	<0.001
Agesq	14.32	1	14.32	2789.1	<0.001
Height	72.35	1	72.35	2788.3	<0.001
Weight	109.91	1	109.91	2786.4	<0.001
Htsq	33.48	1	33.48	2787.9	<0.001
HtxWt	7.74	1	7.74	2786.7	0.005
Wtsq	10.49	1	10.49	2788.4	0.001

Dropping individual terms from full fixed model.

Fixed term	Wald statistic	n.d.f.	F statistic	d.d.f.	F pr.
Gender	2.70	1	2.70	2748.0	0.101
Age	2.76	1	2.76	2789.5	0.097
Agesq	0.50	1	0.50	2789.1	0.479
Height	47.75	1	47.75	2788.3	<0.001
Weight	7.85	1	7.85	2786.4	0.005
Htsq	45.51	1	45.51	2787.9	<0.001
HtxWt	17.19	1	17.19	2786.7	<0.001
Wtsq	10.49	1	10.49	2788.4	0.001

Message: denominator degrees of freedom for approximate F tests are calculated using algebraic derivatives ignoring fixed/boundary/singular variance parameters.

The variance component for CentreNumber is somewhat larger than its standard error (SE), suggesting that there is real variation among the centres, though this component is much smaller than the residual variance among individuals. The F tests indicate that each successive term added to the fixed-effect model accounts for a highly significant amount of variation. However, in the fixed-effect part of a mixed model (as in an ordinary multiple regression model), the significance of a term depends on the order in which it is added to the model, except in certain special cases. Therefore, when all terms have been added, the effect of dropping each in turn is tested. In most cases, the term is still significant, though sometimes less so than when it was added in sequence. For example, Gender gives an F value of 67.33 when it is added as the first term in the model, but only 2.70 when it is dropped, all the other terms being retained. This is because much of the variation accounted for by Gender can alternatively be accounted for by other terms: males tend to be taller and heavier than females. The term 'Agesq' is non-significant when all other terms are present in the model, and can safely be dropped. The term 'Age' is also not quite significant when all other terms are present, but it is not necessarily safe to drop it. This is because it has been found to be non-significant only in a model that includes 'Agesq', and it is a general rule of statistical modelling that when one term in the model is a power of another, then if the higher-power term ('Agesq' in the present case) is included in a model, the lower-power term ('Age' in this case) must be retained. This is an instance of the *marginality criterion*, an important concept which merits a brief digression.

The marginality criterion states that when a *higher-order term*, namely an interaction or nested term in a factorial model, or a power greater than 1 or a product term in a polynomial model, is retained in the model, then the corresponding lower-order terms must also be retained, whether or not they are significant. For example,

- if the $A \times B$ interaction is retained in the factorial model A * B, then the main-effect terms A and B must also be retained

- if the nested term B-within-A is retained in the factorial model A/B, then the main-effect term A must also be retained

- if the term x^2 is retained in a polynomial model, then the term x must also be retained

- if the term $x_1 x_2$ is retained in a polynomial model, then the terms x_1 and x_2 must also be retained.

The marginality criterion applies recursively. Thus

- If the $A \times B \times C$ interaction is retained in the factorial model $A * B * C$, then the terms $A \times B$ and C must also be retained. Because $A \times B$ is retained, A and B must be retained. By symmetry, $B \times C$ and $A \times C$ must also be retained.

- If the term $x_1^2 x_2$ is retained in a polynomial model, the terms x_1^2 and x_2 must be retained. Because x_1^2 is retained, x_1 must be retained. Similarly, $x_1 x_2$ must be retained, which again leads to the conclusion that x_1 and x_2 must be retained.

A choice of model that may appear to breach the marginality criterion, though it does not really do so, is the replacement of the model

$$A * B = A + B + A.B$$

by the model

$$A/B = A + A.B.$$

When this change is made, the main-effect term B appears to be dropped while the interaction term A.B is retained. In reality, however, B is not deleted but assimilated: the variation and the degrees of freedom that it represents are *absorbed* into A.B, and the meaning of A.B changes from the crossed term '$A \times B$ interaction' to the nested term 'B-within-A'. For a more formal account of the marginality criterion, see McCullagh and Nelder (1989, Sections 3.5.1–3.5.4, pp. 63–70 and Section 3.9, p. 89).

In order to see whether 'Age' can be dropped from the model, we must first drop 'Agesq' and then re-fit the model. This is done by the following statements:

```
VCOMPONENTS [FIXED = Gender + Age + \
   Height + Weight + Htsq + HtxWt + Wtsq] \
   RANDOM = CentreNumber
REML [PRINT = model, components, Wald, deviance] CalL2_L4BMD
```

These produce the following output:

REML variance components analysis

Response variate: CalL2_L4BMD
Fixed model: Constant + Gender + Age + Height + Weight + Htsq +
        HtxWt + Wtsq
Random model: CentreNumber
Number of units: 2801 (117 units excluded due to zero weights or missing
        values)

Residual term has been added to model.

Sparse algorithm with AI optimization.
All covariates centred.
Analysis is subject to the restriction on CalL2_L4BMD.

Estimated variance components

Random term	Component	s.e.
CentreNumber	0.00066	0.00041

Residual variance model

Term	Model(order)	Parameter	Estimate	s.e.
Residual	Identity	Sigma2	0.0286	0.00077

Deviance: −2*Log-Likelihood

Deviance	d.f
−7013.78	2791

Note: deviance omits constants which depend on fixed model fitted.

Tests for fixed effects

Sequentially adding terms to fixed model

Fixed term	Wald statistic	n.d.f.	F statistic	d.d.f.	F pr.
Gender	67.34	1	67.34	2748.1	<0.001
Age	253.11	1	253.11	2790.3	<0.001
Height	77.59	1	77.59	2788.9	<0.001
Weight	116.22	1	116.22	2787.3	<0.001
Htsq	35.40	1	35.40	2788.5	<0.001
HtxWt	7.83	1	7.83	2787.5	0.005
Wtsq	10.80	1	10.80	2789.2	0.001

Dropping individual terms from full fixed model.

Fixed term	Wald statistic	n.d.f.	F statistic	d.d.f.	F pr.
Gender	2.93	1	2.93	2748.1	0.087
Age	178.06	1	178.06	2790.3	<0.001
Height	49.67	1	49.67	2788.9	<0.001
Weight	7.94	1	7.94	2787.3	0.005
Htsq	47.23	1	47.23	2788.5	<0.001
HtxWt	17.52	1	17.52	2787.5	<0.001
Wtsq	10.80	1	10.80	2789.2	0.001

Message: denominator degrees of freedom for approximate F tests are calculated using algebraic derivatives ignoring fixed/boundary/singular variance parameters.

The effect of 'Age' is now highly significant, even when all other terms are present in the model. The variance component for 'CentreNumber' is almost unchanged by dropping 'Agesq' from the model, and the diagnostic plots of residuals (not presented here) are also almost unchanged. The effect of 'Gender' is now nearly significant, but not quite: nevertheless, it may be prudent to retain this term in the model, and we will do so.

As the variance component for 'CentreNumber' is only slightly larger than its SE, it seems advisable to perform a more rigorous test of its significance. For this purpose, this term is dropped from the random-effect model (leaving only the residual term), and the deviance obtained by fitting this reduced model is noted. This is done by the following statements:

```
VCOMPONENTS [FIXED = Gender + Age + \
   Height + Weight + Htsq + HtxWt + Wtsq]
REML [PRINT = deviance] CalL2_L4BMD
```

The output of these statements is as follows:

Deviance: −2*Log-Likelihood

Deviance	d.f.
−6979.32	2792

Note: deviance omits constants which depend on fixed model fitted.

The difference between the deviances from the models with and without 'CentreNumber' is obtained, namely

$$\text{deviance}_{\text{reduced model}} - \text{deviance}_{\text{full model}} = -6979.32 - (-7013.78) = 34.46,$$

and

$$P(\text{deviance}_{\text{reduced model}} - \text{deviance}_{\text{full model}} > 34.46) = \frac{1}{2} \times P(\chi_1^2 > 34.46) = 2.18 \times 10^{-9},$$

as explained earlier (Section 3.12). Thus the variance due to 'CentreNumber' is highly significant: we can be very confident that there is real variation among the centres.

Having arrived at a satisfactory model, we can obtain estimates of its parameters using the following statements:

```
VCOMPONENTS [FIXED = Gender + Age + \
   Height + Weight + Htsq + HtxWt + Wtsq] \
   RANDOM = CentreNumber]
REML [PRINT = effects] CalL2_L4BMD
```

These produce the following output:

Table of effects for Constant

1.054   Standard error: 0.0113

Table of effects for Gender

Gender	1	2
	0.000000	0.016726

Standard error of differences: 0.009773.

Table of effects for Age

−0.002942   Standard error: 0.0002205

Table of effects for Height

0.07436   Standard error: 0.010550

Table of effects for Weight

−0.01397   Standard error: 0.004957

Table of effects for Htsq

−0.0002451   Standard error: 0.00003566

Table of effects for HtxWt

0.0001464   Standard error: 0.00003497

Table of effects for Wtsq

−0.00004883   Standard error: 0.000014857

These parameter estimates can be used to construct an arithmetical model giving predicted values of the response variable. However, in order to do this, we must take account of the GenStat's default parameterization of the model, which is centred on the mean values of the explanatory variates (cf. Section 1.8, where an option was set in the VCOMPONENTS statement to prevent this adjustment). These mean values are presented in Table 7.3.

Table 7.3 Mean values of the explanatory variables in the osteoporosis data.

Variable	Mean
Age	48.85
Height	168.1
Weight	70.42
Htsq	28353
HtxWt	11923
Wtsq	5159

Probands and BMD outliers are excluded from the calculation of these means.

The parameter estimates, together with the mean values of the explanatory variables, tell us that the best-fitting model (among those explored) is as follows:

$$
\begin{aligned}
\text{CalL2\_L4BMD} = {} & 1.054 + 0.016726 \times \text{Gender} - 0.002942 \times (\text{Age} - 48.85) \\
& + 0.07436 \times (\text{Height} - 168.1) - 0.01397 \times (\text{Weight} - 70.42) \\
& - 0.0002451 \times (\text{Height}^2 - 28\ 353) + 0.0001464 \times (\text{Height} \times \text{Weight} \\
& - 11923) - 0.00004883 \times (\text{Weight}^2 - 5159) + \Gamma + E
\end{aligned}
\tag{7.1}
$$

Table 7.4 Range of values of 'Height', 'Weight' and the higher-order terms derived from these terms, to be used for obtaining fitted values of BMD.

	A	B	C	D	E
	Htlev	Wtlev	Htsqlev	HtxWtlev	Wtsqlev
1					
2	133.0	35.000	17689.00	4655.000	1225.000
3	133.0	45.556	17689.00	6058.889	2075.309
4	133.0	56.111	17689.00	7462.778	3148.457
5	133.0	66.667	17689.00	8866.667	4444.444
6	133.0	77.222	17689.00	10270.556	5963.272
7	133.0	87.778	17689.00	11674.444	7704.938
8	133.0	98.333	17689.00	13078.333	9669.444
9	133.0	108.889	17689.00	14482.222	11856.790
10	133.0	119.444	17689.00	15886.111	14266.975
11	133.0	130.000	17689.00	17290.000	16900.000
12	166.5	35.000	27722.25	5827.500	1225.000
13	166.5	45.556	27722.25	7585.000	2075.309
14	166.5	56.111	27722.25	9342.500	3148.457
15	166.5	66.667	27722.25	11100.000	4444.444
16	166.5	77.222	27722.25	12857.500	5963.272
17	166.5	87.778	27722.25	14615.000	7704.938
18	166.5	98.333	27722.25	16372.500	9669.444
19	166.5	108.889	27722.25	18130.000	11856.790
20	166.5	119.444	27722.25	19887.500	14266.975
21	166.5	130.000	27722.25	21645.000	16900.000
22	200.0	35.000	40000.00	7000.000	1225.000
23	200.0	45.556	40000.00	9111.111	2075.309
24	200.0	56.111	40000.00	11222.222	3148.457
25	200.0	66.667	40000.00	13333.333	4444.444
26	200.0	77.222	40000.00	15444.444	5963.272
27	200.0	87.778	40000.00	17555.556	7704.938
28	200.0	98.333	40000.00	19666.667	9669.444
29	200.0	108.889	40000.00	21777.778	11856.790
30	200.0	119.444	40000.00	23888.889	14266.975
31	200.0	130.000	40000.00	26000.000	16900.000

Table 7.5 Range of values of 'Age' to be used for obtaining fitted values of BMD.

	A
1	Agelev
2	16
3	96

where
  Gender = 0 for a male, 1 for a female,
    Γ = the effect of the centre,
    E = the residual effect.

However, the individual parameters of a model as elaborate as this are not easy to interpret. It is more informative to inspect the fitted values given by the model over a range of realistic values of the explanatory variables. For this purpose, we first define variates to hold this set of values. Appropriate variates are shown in the spreadsheets in Tables 7.4 and 7.5. Note that in this it is necessary to specify every desired combination of 'Height' and 'Weight' explicitly, in order to specify the corresponding value of Height × Weight.

These variates are imported into GenStat. The predicted values given by these values of the explanatory variables are then obtained by the following statement:

```
VPREDICT [PREDICTIONS = BMDpredict; SE = BMDpredictse] \
    CLASSIFY = Age, Height, Weight, Htsq, HtxWt, Wtsq; \
    LEVELS = Agelev, Htlev, Wtlev, Htsqlev, HtxWtlev, Wtsqlev; \
    PARALLEL = Age, Height, Height, Height, Height, Height
```

The options PREDICTIONS and SE in this statement specify the names of tables to hold the predictions and their SEs, respectively. The parameter CLASSIFY indicates the model terms for which values are to be specified when making predictions, and the parameter LEVELS indicates the values of each term for which predictions are to be made. By default, the model terms specified in CLASSIFY are crossed: that is, all combinations of their specified levels are considered. However, this is not always appropriate: for example, each value of 'Height' should be combined only with the corresponding value of 'Htsq'. These two terms are to be considered *in parallel*, and this is specified by the parameter PARALLEL. In the present case, 'Height', 'Weight' and all the terms derived from them are considered in parallel with each other, whereas 'Age' is in parallel only with itself.

A representative sample of the output from this statement is as follows:

Predictions from REML analysis

Model terms included for prediction: Constant + Gender + Age + Height + Weight + Htsq + HtxWt + Wtsq
Model terms excluded for prediction: CentreNumber

Status of model variables in prediction:

Variable	Type	Status
Wtsq	variate	Classifies predictions
HtxWt	variate	Classifies predictions
Htsq	variate	Classifies predictions
Weight	variate	Classifies predictions
Height	variate	Classifies predictions
Age	variate	Classifies predictions
Gender	factor	Averaged over – equal weights
Constant	factor	Included in prediction
CentreNumber	factor	Ignored

Response variate: CalL2_L4BMD.

Predictions

Wtsq_HtxWt_Htsq_Weight_Height	1225, 4655, 17689, 35.00, 133.0
Age	
16.00	0.7704
96.00	0.5350

Wtsq_HtxWt_Htsq_Weight_Height	2075, 6059, 17689, 45.56, 133.0
Age	
16.00	0.7869
96.00	0.5515

.
.
.

Standard errors

Wtsq_HtxWt_Htsq_Weight_Height	1225, 4655, 17689, 35.00, 133.0
Age	
16.00	0.04208
96.00	0.04044

Wtsq_HtxWt_Htsq_Weight_Height	2075, 6059, 17689, 45.56, 133.0
Age	
16.00	0.04057
96.00	0.03831

.
.
.

Approximate average standard error of difference: 0.06207 (calculated on variance scale).

The way in which each model term is used when constructing the predictions is first specified. It is noted that 'Wtsq', 'HtxWt' and so on, classify the predictions. The constant is also used in forming the predictions, by default. The term *Gender* has not been mentioned in the VPREDICT statement, so the predictions are averaged over this term. When performing the averaging, equal weight is given to the males and the females, regardless of their relative numbers in the sample. The random-effect term 'CentreNumber', also not mentioned in the VPREDICT statement, is ignored when forming the predictions: that is, the effect of 'CentreNumber' is set to zero (though this term still contributes to the SEs of the predictions). The first of the predictions tells us that when Weight = 35.00 kg, Height = 133.0 cm and Age = 16.00 years, the predicted value of CalL2_L4BMD is 0.7704. When the value of Age is changed to 96.00 years, the predicted value is 0.5350, and so on. The table of predictions is followed by a table giving the SE of each prediction, following the same format. These predictions and their SEs are displayed graphically in Figure 7.3.

In every case, the curve for individuals aged 16 years predicts higher values of BMD than the corresponding curve for individuals aged 96 years, and this is consistent with the information from other sources. This figure shows that in most circumstances, a greater weight is associated with a higher value of BMD, but that this trend levels off at the greatest weights. At first glance the figure appears to show that in some circumstances, BMD is actually lower in heavy individuals, which is implausible. However, closer inspection shows that the lines that show this pattern strongly represent very short individuals, only 133 cm tall. The combination of this height with a weight of 100 or 120 kg lies well outside the range of the data: the heavy lines on the figure show that the greatest weight actually observed in such a short individual was below 50 kg. It is not to be expected that a model will give good predictions far outside the range of the data. This is particularly true of polynomial models, which are *curves of convenience*: they have no functional connection to the processes underlying the relationships in the data. The range of observed values of 'CalL2_L4BMD' in the data on which the predictions are based is from 0.5326 to 1.896. Thus the pattern of predictions presented here appears to account for

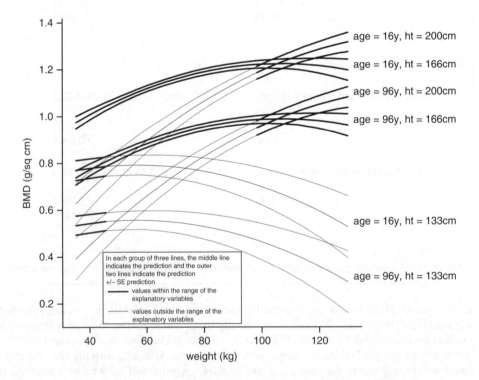

Figure 7.3   Predictions of BMD on the basis of weight, age and height, from the osteoporosis data.

individuals with low BMD fairly well, but fails to predict the occurrence of some individuals with very high values of BMD.

Other sets of predictions could be obtained from this model, and other curves could be drawn from them: in particular, the effect of 'Gender' could be represented. However, it is necessary to keep any display of predictions fairly simple in order to avoid a confusing mass of detail.

7.3   Use of R to fit the fixed-effects model with several variates and factors

The following commands import the data into R, convert Gender and CentreNumber to factors, calculate the derived variables to be used in the models (Agesq, Htsq, HtxWt and Wtsq) and fit Model 7.1 to the data from the non-proband individuals:

```
rm(list = ls())
osteoporosis.phenotypes <- read.table(
    "IMM edn 2\\Ch 7\\ost phenotypes.txt",
    header=TRUE)
attach(osteoporosis.phenotypes)
fGender <- factor(Gender)
fCentreNumber <- factor(CentreNumber)
Agesq = Age^2
Htsq = Height^2
HtxWt = Height * Weight
Wtsq = Weight^2
library(lme4)
ostpheno.model1lmer <- lmer(CalL2.L4BMD ~
    fGender + Age + Agesq + Height +
    Weight + Htsq + HtxWt + Wtsq + (1|fCentreNumber),
    data = osteoporosis.phenotypes, subset = Proband == 'No')
```

In the function `lmer()`, the argument 'subset' indicates that the probands are to be omitted from the modelling process.

The following commands extract the residual values and fitted values from this model, and produce diagnostic plots:

```
resostpheno <- residuals(ostpheno.model1lmer)
fitostpheno <- fitted(ostpheno.model1lmer)
windows()
hist(resostpheno)
windows()
par(pty = "s")
qqnorm(resostpheno)
qqline(resostpheno)
windows()
plot(fitostpheno, resostpheno)
```

The plots obtained are equivalent to those in Figure 7.1.

The following command fits the same model, this time omitting the individuals with outlying values of BMD as well as the probands:

```
ostpheno.model1lmer <- lmer(CalL2.L4BMD ~
    fGender + Age + Agesq + Height +
    Weight + Htsq + HtxWt + Wtsq + (1|fCentreNumber),
    data = osteoporosis.phenotypes,
    subset = Proband == 'No' & BMD_Outlier == 'No')
```

The diagnostic plots commands are then re-executed, giving plots equivalent to those in Figure 7.2. As the distribution of the residuals is now satisfactory, the results of the model are displayed by the following commands:

```
summary(ostpheno.model1lmer)
anova(ostpheno.model1lmer)
```

The output of these commands is as follows:

```
Linear mixed model fit by REML
Formula: CalL2.L4BMD ~ fGender + Age + Agesq + Height + Weight +
    Htsq + HtxWt + Wtsq + (1 | fCentreNumber)
    Data: osteoporosis.phenotypes
 Subset: Proband == "No" & BMD_Outlier == "No"
    AIC   BIC logLik deviance REMLdev
  -1838 -1773  930.1    -1998   -1860
Random effects:
 Groups        Name        Variance   Std.Dev.
 fCentreNumber (Intercept) 0.00065456 0.025584
 Residual                  0.02860951 0.169143
Number of obs: 2801, groups: fCentreNumber, 8

Fixed effects:
              Estimate Std. Error t value
(Intercept) -4.822e+00  8.137e-01  -5.926
fGender2     1.612e-02  9.811e-03   1.643
Age         -2.076e-03  1.248e-03  -1.664
Agesq       -8.873e-06  1.258e-05  -0.706
Height       7.346e-02  1.063e-02   6.911
Weight      -1.389e-02  4.959e-03  -2.801
Htsq        -2.422e-04  3.590e-05  -6.746
HtxWt        1.452e-04  3.502e-05   4.145
Wtsq        -4.818e-05  1.488e-05  -3.237

Correlation of Fixed Effects:
          (Intr) fGndr2 Age    Agesq Height Weight Htsq   HtxWt
fGender2 -0.063
Age       0.085 -0.040
Agesq    -0.100  0.087 -0.984
```

```
Height    -0.980  0.021 -0.112  0.121
Weight     0.387 -0.017  0.020 -0.022 -0.554
Htsq       0.927 -0.008  0.111 -0.114 -0.981  0.682
HtxWt     -0.358  0.078 -0.055  0.048  0.514 -0.923 -0.665
Wtsq       0.077 -0.147  0.077 -0.059 -0.120  0.226  0.233 -0.581

Analysis of Variance Table
         Df Sum Sq Mean Sq  F value
fGender   1 1.9265  1.9265  67.3380
Age       1 7.2397  7.2397 253.0516
Agesq     1 0.4094  0.4094  14.3089
Height    1 2.0675  2.0675  72.2648
Weight    1 3.1447  3.1447 109.9167
Htsq      1 0.9578  0.9578  33.4793
HtxWt     1 0.2215  0.2215   7.7424
Wtsq      1 0.2996  0.2996  10.4733
```

The F values given agree fairly closely with those given by GenStat when the terms are sequentially added to the fixed-effects model. R does not give the anova table obtained by dropping each term, but this is equivalent to the t values given by the summary() function. Thus for the term Agesq, $t = -0.705$ and $F = t^2 = 0.50$, which agrees with the value given by GenStat. As the effect of Agesq falls far short of significance, this term is dropped from the model. The resulting model is fitted, and its results displayed, by the following commands:

```
ostpheno.model2lmer <- lmer(CalL2.L4BMD ~
   fGender + Age + Height +
   Weight + Htsq + HtxWt + Wtsq + (1|fCentreNumber),
   data = osteoporosis.phenotypes,
   subset = Proband == 'No' & BMD_Outlier == 'No')
summary(ostpheno.model2lmer)
anova(ostpheno.model2lmer)
```

The output of these commands is as follows:

```
Linear mixed model fit by REML
Formula: CalL2.L4BMD ~ fGender + Age + Height + Weight + Htsq +
   HtxWt + Wtsq + (1 | fCentreNumber)
   Data: osteoporosis.phenotypes
 Subset: Proband == "No" & BMD_Outlier == "No"
  AIC   BIC logLik deviance REMLdev
 -1860 -1801  940.2    -1998   -1880
Random effects:
 Groups        Name        Variance   Std.Dev.
 fCentreNumber (Intercept) 0.00065773 0.025646
 Residual                  0.02860406 0.169127
Number of obs: 2801, groups: fCentreNumber, 8
```

```
Fixed effects:
               Estimate Std. Error t value
(Intercept) -4.879e+00  8.096e-01   -6.027
fGender2     1.672e-02  9.773e-03    1.711
Age         -2.942e-03  2.205e-04  -13.343
Height       7.437e-02  1.055e-02    7.049
Weight      -1.397e-02  4.957e-03   -2.818
Htsq        -2.451e-04  3.566e-05   -6.873
HtxWt        1.464e-04  3.497e-05    4.185
Wtsq        -4.879e-05  1.486e-05   -3.284

Correlation of Fixed Effects:
          (Intr) fGndr2 Age     Height Weight Htsq    HtxWt
fGender2 -0.055
Age      -0.076  0.257
Height   -0.980  0.011  0.040
Weight    0.387 -0.015 -0.012 -0.556
Htsq      0.926  0.002 -0.009 -0.981  0.684
HtxWt    -0.355  0.074 -0.039  0.513 -0.923 -0.665
Wtsq      0.071 -0.143  0.107 -0.114  0.225  0.228 -0.580

Analysis of Variance Table
          Df Sum Sq Mean Sq  F value
fGender    1 1.9264  1.9264   67.3482
Age        1 7.2396  7.2396  253.0978
Height     1 2.2172  2.2172   77.5123
Weight     1 3.3244  3.3244  116.2203
Htsq       1 1.0126  1.0126   35.4009
HtxWt      1 0.2241  0.2241    7.8339
Wtsq       1 0.3085  0.3085   10.7843
```

Once again, the F values given by R agree fairly closely with those given by GenStat, and the estimated coefficients of the fixed effects also agree, except for the intercept (constant), which differs due to the different parameterizations used by GenStat and R. That of GenStat is centred on the mean values of the explanatory variates by default (see Section 7.2), whereas in that of R the intercept represents the predicted value when all the explanatory variates have the value zero. Equation 7.1 can be adapted to give the best-fitting model in terms of R's parameterization, namely

$$CalL2\_L4BMD = -4.879 + 0.01672 \times Gender - 0.002942 \times Age$$
$$+ 0.07437 \times Height - 0.01397 \times Weight - 0.0002451 \times Height^2$$
$$+ 0.0001464 \times Height \times Weight - 0.00004879 \times Weight^2 + \Gamma + E$$
$$(7.2)$$

Suitable values of the explanatory variables can be substituted into this equation, in order to obtain predicted values of the response variable.

In order to compare the deviances from models with and without 'CentreNumber', the model omitting this term must be fitted. However, it cannot simply be dropped from the model, as this would give a model with no random-effect terms, which would produce an error. In order to overcome this problem, a random effect term that accounts for no variation is added to the model, as follows:

```
constant <- factor(rep(1, each = length(CalL2.L4BMD)))
ostpheno.model3lmer <- lmer(CalL2.L4BMD ~
   fGender + Age + Height +
   Weight + Htsq + HtxWt + Wtsq + (1|constant),
   data = osteoporosis.phenotypes,
   subset = Proband == 'No' & BMD_Outlier == 'No')
summary(ostpheno.model3lmer)
```

The function rep() repeats the value '1' as many times as there are values in the object 'CalL2.L4BMD', and the resulting list of 1s is then stored in a factor named 'constant'. This constant term is used as the random-effect term in the mixed model that follows. The relevant part of the output from the function summary() is as follows:

```
Linear mixed model fit by REML
Formula: CalL2.L4BMD ~ fGender + Age + Height + Weight + Htsq +
   HtxWt + Wtsq + (1 | constant)
   Data: osteoporosis.phenotypes
 Subset: Proband == "No" & BMD_Outlier == "No"
   AIC   BIC logLik deviance REMLdev
 -1826 -1767   923    -1964   -1846
```

Comparing these results with those obtained from the full model (including the term 'CentreNumber'), we obtain

$$\text{deviance}_{\text{reduced model}} - \text{deviance}_{\text{full model}} = -1846 - (-1880) = 34$$

which agrees with the value given by GenStat.

7.4 Use of SAS to fit the fixed-effects model with several variates and factors

The following SAS statements import the data, calculate the derived variables to be used in the models (Agesq, Htsq, HtxWt and Wtsq), eliminate proband individuals from the data set, fit Model 7.1 and produce diagnostic plots of the residuals:

```
PROC IMPORT OUT = ostphen DBMS = EXCELCS REPLACE
   DATAFILE = "&pathname.\IMM edn 2\Ch 7\ost phenotypes.xlsx";
   SHEET = "for SAS";
RUN;
```

```
DATA ostphen2; SET ostphen;
   Agesq = Age ** 2;
   Htsq = Height ** 2;
   HtxWt = Height * Weight;
   Wtsq = Weight ** 2;
   IF Proband = "No" THEN OUTPUT;
RUN;

ODS RTF;
ODS GRAPHICS ON;
PROC MIXED ASYCOV NOBOUND DATA = ostphen2;
   CLASS Gender CentreNumber;
   MODEL CalL2_L4BMD = Gender Age Agesq
      Height Weight Htsq HtxWt Wtsq /DDFM = KR HTYPE = 1 3 RESIDUAL;
   RANDOM CentreNumber;
RUN;
ODS GRAPHICS OFF;
ODS RTF CLOSE;
```

To fit the same model omitting the individuals with outlying values of BMD as well as the probands, the IF statement is changed to

```
IF Proband = "No" AND BMD_Outlier = "No" THEN OUTPUT;
```

In the MODEL statement in PROC MIXED, the option setting HTYPE indicates the types of null hypotheses to be tested: the two types chosen are explained below. Part of the tabular output from PROC MIXED is then as follows:

Convergence criteria met.

Covariance parameter estimates	
Cov Parm	**Estimate**
CentreNumber	0.000655
Residual	0.02861

Asymptotic covariance matrix of estimates			
Row	**Cov Parm**	**CovP1**	**CovP2**
1	CentreNumber	1.668E−7	−2.03E−9
2	Residual	−2.03E−9	5.877E−7

Fit statistics	
−2 Res Log Likelihood	−1860.4
AIC (smaller is better)	−1856.4
AICC (smaller is better)	−1856.4
BIC (smaller is better)	−1856.2

Null model likelihood ratio test		
DF	Chi-square	Pr > ChiSq
1	34.40	<0.0001

Type 1 tests of fixed effects				
Effect	Num DF	Den DF	F value	Pr > F
Gender	1	2788	67.34	<0.0001
Age	1	2790	247.93	<0.0001
Agesq	1	2790	16.09	<0.0001
Height	1	2633	73.53	<0.0001
Weight	1	2791	109.93	<0.0001
Htsq	1	2788	33.54	<0.0001
HtxWt	1	2786	7.72	0.0055
Wtsq	1	2788	10.49	0.0012

Type 3 tests of fixed effects				
Effect	Num DF	Den DF	F value	Pr > F
Gender	1	2748	2.68	0.1014
Age	1	2790	2.76	0.0967
Agesq	1	2789	0.50	0.4793
Height	1	2788	47.74	<0.0001
Weight	1	2786	7.85	0.0051
Htsq	1	2788	45.49	<0.0001
HtxWt	1	2787	17.19	<0.0001
Wtsq	1	2788	10.49	0.0012

The variance component estimates agree with those produced by GenStat, and the F values for the Type 1 tests of the fixed effects agree approximately with those obtained by sequentially adding terms to the fixed model in GenStat. The F values for the corresponding Type 3 tests agree approximately with those obtained in GenStat by dropping each term in turn after all terms have been added. Several sets of diagnostic plots are produced, among which those headed 'Conditional residuals for CalL2_L4BMD' are equivalent to those in Figure 7.2.

The model with the term 'Agesq' dropped is fitted, and the parameter estimates for the fixed-effect terms are included in the output, by the following statements:

```
ODS RTF;
PROC MIXED ASYCOV NOBOUND DATA = ostphen2;
   CLASS Gender CentreNumber;
   MODEL CalL2_L4BMD = Gender Age
      Height Weight Htsq HtxWt Wtsq /DDFM = KR HTYPE = 1 SOLUTION;
   RANDOM CentreNumber;
RUN;
ODS RTF CLOSE;
```

Part of the output from PROC MIXED is as follows:

| Convergence criteria met. |

Covariance parameter estimates	
Cov Parm	Estimate
CentreNumber	0.000658
Residual	0.02860

Asymptotic covariance matrix of estimates			
Row	Cov Parm	CovP1	CovP2
1	CentreNumber	1.683E−7	−2.05E−9
2	Residual	−2.05E−9	5.873E−7

Fit statistics	
−2 Res Log Likelihood	−1880.6
AIC (smaller is better)	−1876.6
AICC (smaller is better)	−1876.6
BIC (smaller is better)	−1876.4

Null model likelihood ratio test		
DF	Chi-square	Pr > ChiSq
1	34.46	<0.0001

Solution for fixed effects						
Effect	Gender	Estimate	Standard error	DF	t value	Pr > |t|
Intercept		−4.8619	0.8093	2790	−6.01	<0.0001
Gender	1	−0.01673	0.009798	2748	−1.71	0.0879
Gender	2	0	.	.	.	.
Age		−0.00294	0.000221	2790	−13.33	<0.0001
Height		0.07436	0.01055	2789	7.05	<0.0001
Weight		−0.01397	0.004957	2787	−2.82	0.0049
Htsq		−0.00025	0.000036	2789	−6.87	<0.0001
HtxWt		0.000146	0.000035	2788	4.19	<0.0001
Wtsq		−0.00005	0.000015	2789	−3.29	0.0010

Type 1 tests of fixed effects				
Effect	Num DF	Den DF	F value	Pr > F
Gender	1	2789	67.37	<0.0001
Age	1	2791	248.01	<0.0001
Height	1	2631	79.27	<0.0001
Weight	1	2792	116.28	<0.0001
Htsq	1	2788	35.46	<0.0001
HtxWt	1	2787	7.81	0.0052
Wtsq	1	2789	10.80	0.0010

Again the variance component estimates and F values agree approximately with those obtained from GenStat. Equation 7.1 can be adapted to give the best-fitting model in terms of SAS's parameterization, namely

$$\text{CalL2\_L4BMD} = -4.8619 + 0.01673 \times \text{Gender} - 0.00294 \times \text{Age}$$

$$+ 0.07436 \times \text{Height} - 0.01397 \times \text{Weight} - 0.00025 \times \text{Height}^2$$

$$+ 0.000146 \times \text{Height} \times \text{Weight} - 0.00005 \times \text{Weight}^2 + \Gamma + E \quad (7.3)$$

Suitable values of the explanatory variables can be substituted into this equation, in order to obtain predicted values of the response variable.

In order to compare the deviances from models with and without 'CentreNumber', the model omitting this term is fitted by omitting the statement

```
RANDOM CentreNumber;
```

from the PROC MIXED statements. Part of the output obtained is as follows:

Covariance parameter estimates	
Cov Parm	Estimate
Residual	0.02910

Fit statistics	
−2 Res Log Likelihood	−1846.1
AIC (smaller is better)	−1844.1
AICC (smaller is better)	−1844.1
BIC (smaller is better)	−1838.2

This output confirms that the residual term is now the only random-effect term in the model and gives the deviance from the reduced model. Comparing these results with those obtained from the full model (including the term 'CentreNumber') we obtain

$$\text{deviance}_{\text{reduced model}} - \text{deviance}_{\text{full model}} = -1846.1 - (-1880.6) = 34.5$$

which agrees with the value given by GenStat.

7.5   A random coefficient regression model

As noted earlier (Section 6.6), when repeated observations are made on each of a number of experimental units, it is often appropriate to fit a model comprising a set of regression lines, one for each unit, with random variation among their slopes and intercepts – a random coefficient model. An example of this situation is provided by a clinical trial conducted to test the efficacy of lithium as a treatment for ALS, a neurological disease (Aggarwal *et al.*, 2010). Eighty-four patients were randomized either to the lithium treatment or to a placebo, and their motor functions were assessed at the time of randomization (the baseline occasion), then regularly for a period of up to 34 weeks, on the revised Amyotrophic Lateral Sclerosis Functional Rating Scale (ALSFRSr), a standard clinical scale. The first and last few rows of the data are presented in Table 7.6. (Data reproduced by kind permission of the Northeast Amyotrophic Lateral Sclerosis consortium.) Graphical presentation of these results (Figure 7.4) shows that the patients' motor functions generally declined steadily from the baseline value, as expected for this incurable degenerative disease. The hypothesis to be investigated is that the lithium treatment reduces the rate of decline.

For each patient, a straight line can be fitted to the relationship between time and the change from the baseline value in the patient's ALSFRSr score. At time zero, the change from baseline is zero by definition, and the values at this time-point are therefore excluded from the analysis. The intercepts are obtained by extrapolating the lines obtained from the rest of the data. It is reasonable to suppose that there is random variation in the intercepts and slopes among the patients. These sources of variation can be represented by specifying 'id' and 'id.visit_day' as random-effect terms. The average slope is represented by the main effect of 'visit_day', and the difference in the average intercept between lithium-treated and placebo-treated patients is represented by the main effect of 'treatment', though this effect is expected to be small as the

Table 7.6   Results on the ALSFRSr scale from a clinical trial of lithium as a treatment for ALS. id = unique identifier of each patient in the study, visit_day = number of days since baseline occasion, visit_week = number of complete weeks since baseline occasion, baseline = patient's score on ALSFRSr scale at baseline, score = patient's score on the day of visit and change = change in score since baseline.

	A	B	C	D	E	F	G
	id!	visit_day	treatment!	visit_week!	baseline	score	change
1							
2	1	29	lithium	4	38	35	−3
3	1	65	lithium	9	38	32	−6
4	1	78	lithium	11	38	32	−6
5	1	93	lithium	13	38	32	−6
6	2	25	placebo	3	42	45	3
7	2	55	placebo	7	42	45	3
	.						.
	.						.
	.						.
442	84	222	lithium	31	45	41	−4
443	84	236	lithium	33	45	42	−3

Source: Data reproduced by kind permission of the Northeast Amyotrophic Lateral Sclerosis consortium.

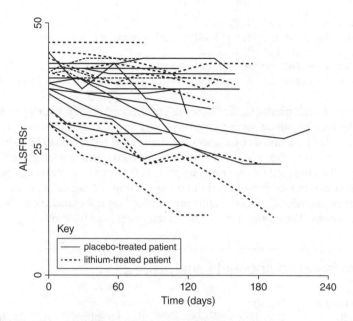

Figure 7.4 Scores on the ALSFRSr scale, over time, in ALS patients treated with lithium or a placebo. Results from a random subset of the patients are shown.

estimated intercept is expected to be close to zero for every patient. It is natural to specify 'visit_day' and 'treatment' as fixed-effect terms. The average difference in slope between lithium-treated and placebo-treated patients, the effect of primary interest, is represented by the 'treatment.visit' interaction term, which is also specified as a fixed-effect term. Although the baseline value has been subtracted from the score before model-fitting, it cannot be assumed that this fully accounts for any relationship between the baseline value and the subsequent scores. It may be that individuals who start with a higher baseline value decline faster or more slowly. 'baseline' is therefore also included in the model, specified as a fixed-effect term. Thus the model to be fitted is

Response variate: ALSFRSr
Fixed-effect model: baseline + treatment*visit_day
Random-effect model: id/visit_day.

Note that the terms 'treatment + visit_day + treatment.visit_day' have been abbreviated to 'treatment * visit_day', and the terms 'id + id.visit_day' to 'id / visit_day'. Although the intercepts are all expected to be close to zero, it cannot be assumed that the intercepts and slopes will vary independently among patients: for example, because of the extrapolation from which the intercepts are obtained, it may be that patients with steeper slopes will have higher intercepts. A correlation must therefore be specified between the intercepts and slopes, and its value must be estimated as part of the model-fitting process.

The following GenStat statements import the data and fit this model:

```
IMPORT 'IMM edn 2\\Ch 7\\ALS lithium.xlsx'
VCOMPONENTS [FIXED = baseline + treatment * visit_day; \
   CADJUST = none] \
```

```
    RANDOM = id / visit_day
VSTRUCTURE [id / visit_day; FORMATION = whole; \
CORRELATE = unrestricted]
REML [PRINT = model, components, Wald, effects; \
    PTERMS = 'constant' + baseline + treatment * visit_day] change
```

The VSTRUCTURE statement specifies how the covariance between the slopes and intercepts is to be modelled. The option setting 'FORMATION = whole' specifies that all values in the *covariance matrix* (the array of covariances between model terms) are to be individually specified: when a covariance matrix is specified for more complex models, more compact and restrictive specification methods are often helpful. This covariance matrix can be transformed to a correlation matrix (see below), and the option setting 'CORRELATE = unrestricted' indicates that no mathematical constraints are imposed on it, beyond those that apply to all correlation matrices. The output from the REML statement is as follows:

REML variance components analysis

Response variate:	change
Fixed model:	Constant + baseline + visit_day + treatment + visit_day.treatment
Random model:	id + id.visit_day
Number of units:	442

Residual term has been added to model.

Sparse algorithm with AI optimization.
Covariates not centred.

Covariance structures defined for random model

Correlated terms:

Set	Correlation across terms
1	Unstructured

Set	Terms	Covariance model within term
1	id	Identity
1	id.visit_day	Identity

Estimated parameters for covariance models

Random term(s)	Factor	Model(order)	Parameter	Estimate	s.e.
id + id.visit_day	Across terms	Unstructured	v_11	0.4535	0.2574
			v_21	0.008310	0.003063
			v_22	0.0003818	0.0000998
	Within terms	Identity	–	–	–

Note: the covariance matrix for each term is calculated as G or R where var(y) = Sigma2(ZGZ'+R), that is, relative to the residual variance, Sigma2.

Residual variance model

Term	Model(order)	Parameter	Estimate	s.e.
Residual	Identity	Sigma2	3.789	0.321

Tests for fixed effects

Sequentially adding terms to fixed model.

Fixed term	Wald statistic	n.d.f.	F statistic	d.d.f.	F pr.
baseline	0.00	1	0.00	74.2	0.966
visit_day	55.65	1	55.65	60.0	<0.001
treatment	0.02	1	0.02	71.1	0.895
visit_day.treatment	0.29	1	0.29	59.8	0.595

Dropping individual terms from full fixed model.

Fixed term	Wald statistic	n.d.f.	F statistic	d.d.f.	F pr.
baseline	0.03	1	0.03	74.2	0.870
visit_day.treatment	0.29	1	0.29	59.8	0.595

Message: denominator degrees of freedom for approximate F tests are calculated using algebraic derivatives ignoring fixed/boundary/singular variance parameters.

Table of effects for Constant

−1.013   Standard error: 1.9811

Table of effects for baseline

0.008357   Standard error: 0.0507502

Table of effects for visit_day

−0.03851   Standard error: 0.006876

Table of effects for treatment

Treatment	lithium	placebo
	0.00000	0.05588

Standard error of differences: 0.5250

Table of effects for visit_day.treatment

Treatment	lithium	placebo
	0.000000	0.005139

Standard error of differences: 0.009623

Among the estimated parameters for covariance models, v_11 and v_22 give the variance among the intercepts and slopes, respectively, expressed as multiples of the residual variance. The parameter v_12 gives the covariance between the intercepts and slopes, expressed in the same way, and this can be used together with the scaled variances to obtain a correlation coefficient,[1] $r = 0.008310/\sqrt{(0.4535 \times 0.0003818)} = 0.632$, indicating that patients with high intercepts tend to have less negative slopes. The F statistics indicate that the effect of visit_day is highly significant, as expected, and the corresponding table of effects confirms that this effect is negative, reflecting the decline in the patients' motor functions over time. However, there is no indication that a patient's baseline score had any effect on their subsequent values of 'change', nor that there was an effect of treatment on the value of 'change' at the time of randomization. Most importantly, from the point of view of research on ALS, the visit_day.treatment interaction is non-significant, giving no evidence that treatment with lithium changed the patients' rate of decline.

7.6   Use of R to fit the random coefficients model

The following R statements fit the random coefficients model to the ALS-lithium data:

```
rm(list = ls())
als.lithium <- read.table(
    "IMM edn 2\\Ch 7\\ALS lithium.txt",
    header=TRUE)
attach(als.lithium)
names(als.lithium)
fid <- factor(id)
ftreatment <- factor(treatment)
library(lme4)
als.lithium.model1 <- lmer(change ~ baseline + ftreatment *
    visit_day +
    ((1 + visit_day)|fid), data = als.lithium)
summary(als.lithium.model1)
anova(als.lithium.model1)
```

The model term '((1 + visit_day)|fid)' indicates that a separate intercept and a separate slope (the coefficient of 'visit_day') is to be fitted, as a random effect, for each level of 'fid', that is, for each patient. It is not necessary to specify that there may be a covariance between these intercepts and slopes, and that this covariance is to be estimated: in R, this model specification is the default. The summary() and anova() functions produce the following output:

```
Linear mixed model fit by REML
Formula: change ~ baseline + ftreatment * visit_day +
    ((1 + visit_day) | fid)
```

---

[1] The correlation coefficient r measures the association between two variables on a scale from $+1$ (perfect positive association) to -1 (perfect negative association). $r = 0$ indicates a perfect absence of association in the sample.

```
   Data: als.lithium
   AIC  BIC logLik deviance REMLdev
  2167 2204  -1075     2129     2149
Random effects:
  Groups    Name          Variance  Std.Dev. Corr
  fid       (Intercept)   1.7183164 1.310846
            visit_day     0.0014464 0.038032 0.632
  Residual                3.7888404 1.946494
Number of obs: 442, groups: fid, 83

Fixed effects:
                            Estimate Std. Error t value
(Intercept)                -1.013863   1.981110  -0.512
baseline                    0.008367   0.050751   0.165
ftreatmentplacebo           0.055900   0.525015   0.106
visit_day                  -0.038513   0.006876  -5.601
ftreatmentplacebo:visit_day 0.005139   0.009623   0.534

Correlation of Fixed Effects:
            (Intr) baseln ftrtmn vst_dy
baseline    -0.983
ftrtmntplcb -0.322  0.196
visit_day    0.000 -0.003  0.010
ftrtmntpl:_  0.019 -0.017 -0.050 -0.714

Analysis of Variance Table
                    Df  Sum Sq Mean Sq F value
baseline             1   0.006   0.006  0.0016
ftreatment           1   0.075   0.075  0.0198
visit_day            1 210.226 210.226 55.4856
ftreatment:visit_day 1   1.075   1.075  0.2837
```

The residual variance agrees with that produced by GenStat, as do the variances of the intercepts and slopes, taking into account GenStat's convention of presenting these relative to the residual variance: thus, for example, the variance of the intercepts given by GenStat is $0.4535 \times 3.789 = 1.718$. The correlation coefficient between the slopes and intercepts agrees with that derived from GenStat's scaled variances and covariance above. Most importantly, the fixed effect estimates and their SEs agree with those produced by GenStat.

7.7 Use of SAS to fit the random coefficients model

The following SAS statements fit the random coefficients model to the ALS-lithium data:

```
PROC IMPORT OUT = ALS_lithium DBMS = EXCELCS REPLACE
   DATAFILE = "&pathname.\IMM edn 2\Ch 7\ALS lithium.xlsx";
   SHEET = "lithium, for SAS";
RUN;
```

```
ODS RTF;
PROC MIXED ASYCOV NOBOUND;
   CLASS id treatment;
   MODEL change = baseline treatment visit_day treatment*visit_day
      / CHISQ DDFM = KR HTYPE = 3 SOLUTION;
   RANDOM intercept visit_day /SUBJECT = id TYPE = UN;
RUN;
ODS RTF CLOSE;
```

The RANDOM statement indicates that a separate intercept and a separate slope (the coefficient of 'visit_day') are to be fitted, as random effects, for each level of 'id', that is, for each patient. The option setting 'TYPE = UN' indicates that the correlation matrix between the slopes and intercepts is to be unrestricted.

Part of the output from PROC MIXED is as follows:

Convergence criteria met.

Covariance parameter estimates		
Cov Parm	**Subject**	**Estimate**
UN(1,1)	id	1.7183
UN(2,1)	id	0.03149
UN(2,2)	id	0.001446
Residual		3.7889

Asymptotic covariance matrix of estimates					
Row	**Cov Parm**	**CovP1**	**CovP2**	**CovP3**	**CovP4**
1	UN(1,1)	0.9383	−0.00038	−0.00008	−0.07383
2	UN(2,1)	−0.00038	0.000156	−5.25E−7	0.000851
3	UN(2,2)	−0.00008	−5.25E−7	1.722E−7	−0.00003
4	Residual	−0.07383	0.000851	−0.00003	0.1091

Fit statistics	
−2 Res Log Likelihood	2149.4
AIC (smaller is better)	2157.4
AICC (smaller is better)	2157.5
BIC (smaller is better)	2167.1

Null model likelihood ratio test		
DF	**Chi-square**	**Pr > ChiSq**
3	345.28	<0.0001

Solution for fixed effects						
Effect	Treatment	Estimate	Standard error	DF	t value	Pr > |t|
Intercept		−0.9576	1.9187	65.1	−0.50	0.6194
baseline		0.008357	0.05185	66.2	0.16	0.8725
treatment	lithium	−0.05588	0.5279	63	−0.11	0.9160
treatment	placebo	0				
visit_day		−0.03337	0.006760	42.9	−4.94	<0.0001
visit_day*treatment	lithium	−0.00514	0.009661	40.7	−0.53	0.5977
visit_day*treatment	placebo	0				

Type 3 tests of fixed effects						
Effect	Num DF	Den DF	Chi-Square	F value	Pr > ChiSq	Pr > F
baseline	1	66.2	0.03	0.03	0.8720	0.8725
treatment	1	63	0.01	0.01	0.9157	0.9160
visit_day	1	40.7	55.37	55.37	<0.0001	<0.0001
visit_day*treatment	1	40.7	0.28	0.28	0.5948	0.5977

The covariance parameter estimates UN(1,1), UN(2,1) and UN(2,2) relate to the variance of the intercepts, the covariance of the intercepts and slopes and the variance of the slopes, respectively. These estimates agree with those given by GenStat, taking into account that the GenStat values are presented relative to the residual variance: thus, for example, the variance of the intercepts given by GenStat is $0.4535 \times 3.789 = 1.718$. The F values for the fixed effects agree approximately with those given by GenStat. The discrepancy perhaps arises from the determination of the denominator d.f. by the Kenward–Roger method in SAS and a somewhat different method in GenStat. The table headed 'Solution for Fixed Effects' gives the parameter estimates for the fixed-effect model. Some of these differ from those given by GenStat because SAS uses 'placebo' as the reference level of 'treatment', whereas GenStat uses 'lithium'. Thus the Intercept and the effect of visit_day given by SAS relates to placebo-treated patients, whereas the Constant and the effect of visit_day given by GenStat relate to lithium-treated patients. Consequently, the effects of treatment given by SAS and GenStat, representing the difference between the intercepts, are equal and opposite (−0.05588 and 0.05588 respectively), as are the visit_day.treatment interaction effects, representing the difference between the slopes (−0.00514 and 0.005139, respectively).

7.8 A random-effects model with several factors

During a period when the area of canola (oilseed rape) grown in Western Australia was expanding rapidly, anecdotal evidence suggested that the variety that performed best at one location or season was not necessarily the best at another. A set of field trials was therefore undertaken with the aim of identifying a pattern in these genotype × environment interactions, in order to guide the choice of variety to be sown at different locations in future years. Eleven varieties, representative of those being grown in the region, were included in the trials, which were

Table 7.7 Data from a set of field trials of canola.

	A	B	C	D	E	F	G
1	location!	sowing_occasion!	sowing_date!	variety!	rainfall	mrainfall	oil
2	Mullewa	1	02-May	Karoo	79.6	80.73	43.82
3	Mullewa	1	02-May	Monty	93.6	80.73	46.99
4	Mullewa	1	02-May	Hyola 42	93.6	80.73	44.34
5	Mullewa	1	02-May	Oscar	62.4	80.73	42.89
6	Mullewa	1	02-May	Drum	51.6	80.73	41.12
7	Mullewa	1	02-May	Mustard	103.6	80.73	
8	Mullewa	2	19-May	Karoo	62.4	65.97	40.05
9	Mullewa	2	19-May	Monty	88.6	65.97	44.68
⋮							⋮
137	Beverley	6	14-Jul	Hyola 42	11.8	19.00	37.20
138	Beverley	6	14-Jul	Oscar	22.0	19.00	35.00
139	Beverley	6	14-Jul	Rainbow	22.0	19.00	37.00

Source: Data reproduced by kind permission of G. Walton, Department of Agriculture, Western Australia.

carried out at six representative locations. At each location, plots were sown at several dates, ranging from early to late in relation to the beginning of the growing season. The number of sown dates at each location varied from 3 to 6, and not all varieties were sown at every location.

The first and last few rows of the spreadsheet holding the data are shown in Table 7.7. Each row represents the mean of all the plots of a particular variety sown in a particular location on the same date. In addition to the actual sowing date, the data include a nominal value for the sowing occasion (an integer from 1 to 6). The values of rainfall represent the amount of rainfall (mm) encountered by the plant during the growing season. They vary from row to row even within a location.sowing occasion combination, because plots of different varieties generally matured and were harvested on different dates. The variate 'mrainfall' is the mean value of rainfall over the location.sowing date combination in question. The variate 'oil' indicates the percentage of oil in the harvested grain. (Data reproduced by kind permission of G. Walton, Department of Agriculture, Western Australia. An alternative analysis of these data has been presented by Si and Walton (2004).)

Because the locations and varieties are representative of those used in the region, it is reasonable to specify them as random-effect terms. It is reasonable to expect that there will be a trend over the sowings, from early to late, and it is therefore appropriate to specify sowing occasion as a fixed effect. The initial model (Model 7.4) is then as follows:

Response variate: oil
Fixed-effect model: sowing_occasion
Random-effect model: location + variety + sowing_occasion.location +
                           sowing_occasion.variety + location.variety +
                           sowing_occasion.location.variety.

The following GenStat statements import the data and fit this model:

```
IMPORT 'IMM edn 2\\Ch 7\\canola oil gxe.xlsx'
VCOMPONENTS [FIXED = sowing_occasion] \
   RANDOM = location + variety + \
   sowing_occasion.location + sowing_occasion.variety + \
   location.variety + sowing_occasion.location.variety
REML oil
```

The complete model fitted here is 'sowing_occasion*location*variety': however, in order to place 'sowing_occasion' in the fixed-effect model and the remaining terms in the random-effect model, it is necessary to spell out the terms individually. The output from the REML statement is as follows:

REML variance components analysis

Response variate:	oil
Fixed model:	Constant + sowing_occasion
Random model:	location + variety + location.sowing_occasion + variety.sowing_occasion + location.variety + location.variety.sowing_occasion
Number of units:	126 (12 units excluded due to zero weights or missing values)

location.variety.sowing_occasion used as residual term.

Sparse algorithm with AI optimization.

Estimated variance components

Random term	Component	s.e.
location	3.8538	2.6620
variety	2.1905	1.0549
location.sowing_occasion	1.1202	0.4467
variety.sowing_occasion	0.1183	0.0703
location.variety	0.1148	0.0713

Residual variance model

Term	Model(order)	Parameter	Estimate	s.e.
location.variety.sowing_occasion	Identity	Sigma2	0.293	0.0608

Tests for fixed effects

Sequentially adding terms to fixed model

Fixed term	Wald statistic	n.d.f.	F statistic	d.d.f.	F pr.
sowing_occasion	41.32	5	8.26	15.6	<0.001

Dropping individual terms from full fixed model.

Fixed term	Wald statistic	n.d.f.	F statistic	d.d.f.	F pr.
sowing_occasion	41.32	5	8.26	15.6	<0.001

Message: denominator degrees of freedom for approximate F tests are calculated using algebraic derivatives ignoring fixed/boundary/singular variance parameters.

GenStat detects that each location.variety.sowing_occasion combination specifies a unique observation, and therefore this is used as the residual term. As such, it need not be specified explicitly in future models. Each of the other random-effect terms is larger than its own SE, suggesting that they should all be retained in the model, and the F test shows that the main effect of 'sowing_occasion' is highly significant.

Because 'sowing_occasion' is a factor (as indicated by the exclamation mark (!) in its heading in the spreadsheet), the numbers used to specify its levels have been treated as arbitrary: the amount of variation among levels has been calculated, but no linear trend from Level 1 to Level 6 has been sought. In order to seek such a trend, it is necessary to create another data structure, 'vsowing_occasion', holding the same values but specified as a variate, not a factor, and to include this in the fixed-effects model. This is done by the following statements:

```
CALCULATE vsowing_occasion = sowing_occasion
VCOMPONENTS [FIXED = vsowing_occasion + sowing_occasion] \
   RANDOM = location + variety + sowing_occasion.location + \
   sowing_occasion.variety + location.variety
REML [PRINT = model, components, Wald, effects; \
   PTERMS = vsowing_occasion] oil
```

It might be thought that the actual sowing date, rather than the nominal sowing occasion, would provide a better estimate of any trend from early to late sowings. However, the effective difference between any two sowing dates in relation to plant development depends not only on the time elapsed, but also on the amount and distribution of rainfall in the interval. The sowing occasions were chosen to span the range from early to late as effectively as possible, and the numbers that designate them therefore constitute a reasonable explanatory variable. The output from these statements is as follows:

REML variance components analysis

Response variate:	oil
Fixed model:	Constant + vsowing_occasion + sowing_occasion
Random model:	location + variety + location.sowing_occasion + variety.sowing_occasion + location.variety
Number of units:	126 (12 units excluded due to zero weights or missing values)

Residual term has been added to model.

Sparse algorithm with AI optimization.
All covariates centred.

Estimated variance components

Random term	Component	s.e.
location	3.8538	2.6620
variety	2.1905	1.0549
location.sowing_occasion	1.1202	0.4467
variety.sowing_occasion	0.1183	0.0703
location.variety	0.1148	0.0713

Residual variance model

Term	Model(order)	Parameter	Estimate	s.e.
Residual	Identity	Sigma2	0.293	0.0608

Tests for fixed effects

Sequentially adding terms to fixed model

Fixed term	Wald statistic	n.d.f.	F statistic	d.d.f.	F pr.
vsowing_occasion	39.37	1	39.37	15.8	<0.001
sowing_occasion	1.95	4	0.49	15.5	0.745

Dropping individual terms from full fixed model

Fixed term	Wald statistic	n.d.f.	F statistic	d.d.f.	F pr.
sowing_occasion	1.95	4	0.49	15.5	0.745

Message: denominator degrees of freedom for approximate F tests are calculated using algebraic derivatives ignoring fixed/boundary/singular variance parameters.

Table of effects for vsowing_occasion

−1.890   Standard error: 0.6492

The variance component estimates for the random-effect terms are unchanged: the additional term affects only the fixed-effect part of the model. Moreover, the sum of the Wald statistics for sequentially adding 'vsowing_occasion' and 'sowing_occasion' to the model is equal to the Wald statistic for 'sowing_occasion' in the previous model, namely

$$39.37 + 1.95 = 41.32.$$

The term 'vsowing_occasion' does not account for any new variation: it accounts for the linear trend which is part of the variation due to 'sowing_occasion', and 'sowing_occasion' now accounts for the deviations from this trend. The F test shows that these deviations are non-significant. Note that we are not given an F test for dropping 'vsowing_occasion' from the model while retaining 'sowing_occasion', because the resulting model would not be legitimate: the linear trend is marginal to the deviations from the trend, and must be retained in the model if the term representing the deviations is retained (see the discussion of marginality

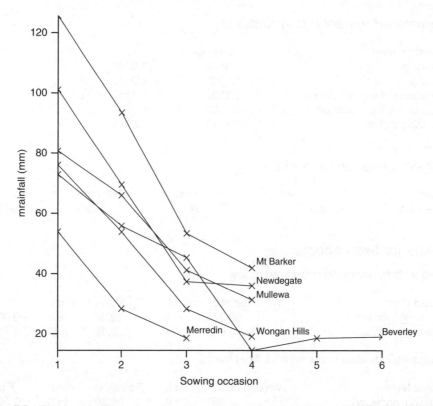

Figure 7.5   Relationship between sowing date and mean rainfall in field trials of canola at different locations.

in Section 7.2). The table of effects shows that on average, the oil content of the canola grain harvested falls by 1.890% from each sowing occasion to the next.

There is a strong relationship between the sowing date and the mean rainfall encountered by the plants at each location, as shown in Figure 7.5. An ordinary multiple regression analysis (not shown here) indicates that on average, the rainfall declines by 11.78 mm from each sowing occasion to the next. This may suggest that 'rainfall' could effectively replace 'vsowing_occasion' in the fixed-effect model. This change is made in the following statements:

```
VCOMPONENTS [FIXED = rainfall + sowing_occasion] \
   RANDOM = location + variety + \
   sowing_occasion.location + sowing_occasion.variety + \
   location.variety
REML [PRINT = model, components, Wald, effects; \
   PTERMS = rainfall] oil
```

This change to the model has little effect on the estimates of variance components (results not shown). However, the F tests and the estimate of the effect of rainfall from these statements are as follows:

Tests for fixed effects

Sequentially adding terms to fixed model

Fixed term	Wald statistic	n.d.f.	F statistic	d.d.f.	F pr.
rainfall	18.47	1	18.47	98.6	<0.001
sowing_occasion	24.49	5	4.89	17.2	0.006

Dropping individual terms from full fixed model

Fixed term	Wald statistic	n.d.f.	F statistic	d.d.f.	F pr.
rainfall	1.02	1	1.02	98.6	0.315
sowing_occasion	24.49	5	4.89	17.2	0.006

Message: denominator degrees of freedom for approximate F tests are calculated using algebraic derivatives ignoring fixed/boundary/singular variance parameters.

Table of effects for rainfall

0.006882   Standard error: 0.0068206

This outcome is perhaps surprising. The F tests indicate that the effect of rainfall is significant, as expected, but also that the effect of 'sowing_occasion' is once again highly significant. Moreover, the effect of rainfall on oil content is much weaker than might have been expected. The difference between successive sowings in the oil content of the canola grain and in the rainfall suggests that this effect should be about $(1.893\%)/(11.78\,\mathrm{mm}) = 0.1707\%$ per mm, whereas the present analysis gives a value of only 0.006882% per mm. The reason for this discrepancy is as follows.

Because the terms 'location', 'sowing_occasion' and 'location.sowing_occasion' are included in the model, and because the rainfall varies among varieties within each 'location.sowing_occasion' combination, the effect of rainfall accounts almost only for variation within each combination. Variation in oil content among 'location.sowing_occasion' combinations is accounted for almost entirely by the other terms: this gives the closest overall fit to the data. If we wish to discover the extent to which rainfall can account for the variation *among* 'location.sowing_occasion' combinations, we must specify a single rainfall value for each combination – that is, we must replace 'rainfall' by 'mrainfall' in the model. This is done in the following statements:

```
VCOMPONENTS [FIXED = mrainfall + sowing_occasion] \
    RANDOM = location + variety + \
    sowing_occasion.location + sowing_occasion.variety + \
    location.variety
REML [PRINT = model, components, Wald, effects; \
    PTERMS = mrainfall] oil
```

The F tests and the estimate of the effect of 'mrainfall' from these statements are now as follows:

Tests for fixed effects

Sequentially adding terms to fixed model

Fixed term	Wald statistic	n.d.f.	F statistic	d.d.f.	F pr.
mrainfall	34.83	1	34.83	13.9	<0.001
sowing_occasion	4.11	5	0.81	15.3	0.558

Dropping individual terms from full fixed model

Fixed term	Wald statistic	n.d.f.	F statistic	d.d.f.	F pr.
mrainfall	2.39	1	2.39	13.9	0.145
sowing_occasion	4.11	5	0.81	15.3	0.558

Message: denominator degrees of freedom for approximate F tests are calculated using algebraic derivatives ignoring fixed/boundary/singular variance parameters.

Table of effects for mrainfall

0.04173   Standard error: 0.027012

The *F* statistic for 'sowing_occasion' (sequentially added to the model) is once again reduced to non-significance, as it now accounts only for the small part of the variation among sowing occasions that is not related to rainfall. Conversely, the effect of rainfall is considerably increased, albeit not to the level that our preliminary calculation suggested.

Because the main effect of sowing_occasion is now non-significant, we may drop this term from the model. The variation that it accounted for will be absorbed by the terms 'location.sowing_occasion' and 'variety.sowing_occasion', and into the residual variation. (Note that this change does not breach the marginality criterion, for reasons explained in Section 7.2.) We should also consider the possibility that some varieties will respond more to rainfall than others, by adding the term 'variety.mrainfall' to the model. This term belongs in the random-effects model, because 'variety' is a random-effect term. Incorporating these changes, our model is specified by the following statements:

```
VCOMPONENTS [FIXED = mrainfall] RANDOM = location + \
   variety / mrainfall + sowing_occasion.location + \
   sowing_occasion.variety + location.variety
VSTRUCTURE [variety / mrainfall; FORMATION = whole; \
   CORRELATE = unrestricted]
REML [PRINT = model, components; PTERMS = mrainfall] oil
```

It is possible that there will be a correlation between the random main effects of 'variety' and the random 'variety.rainfall' interaction effects, and the VSTRUCTURE statement indicates how such a potential correlation is to be modelled (see Section 7.5). The output from these statements is as follows:

REML variance components analysis

Response variate: oil
Fixed model: Constant + mrainfall
Random model: location + variety + variety.mrainfall + location.sowing_occasion + variety.sowing_occasion + location.variety
Number of units: 126 (12 units excluded due to zero weights or missing values)

Residual term has been added to model.

Sparse algorithm with AI optimization.
All covariates centred.

Covariance structures defined for random model

Correlated terms:

Set	Correlation across terms
1	Unstructured

Set	Terms	Covariance model within term
1	variety	Identity
1	variety.mrainfall	Identity

Estimated variance components

Random term	Component	s.e.
location	1.6226	1.2940
location.sowing_occasion	1.3118	0.4585
variety.sowing_occasion	−0.0061	0.0330
location.variety	0.1487	0.0837

Estimated parameters for covariance models

Random term(s)	Factor	Model(order)	Parameter	Estimate	s.e.
variety + variety.mrainfall					
	Across terms	Unstructured	v_11	8.009	4.157
			v_21	0.03180	0.02842
			v_22	0.0005859	0.0004029
	Within terms	Identity	–	–	–

Note: the covariance matrix for each term is calculated as G or R where var(y) = Sigma2(ZGZ′+R), that is, relative to the residual variance, Sigma2.

Residual variance model

Term	Model(order)	Parameter	Estimate	s.e.
Residual	Identity	Sigma2	0.289	0.0579

Among the estimated parameters for covariance models, v_11 and v_22 give the variance for the terms 'variety' and 'variety.mrainfall', respectively, expressed as multiples of the residual variance. The parameter v_12 gives the covariance between these terms, expressed in the same way, giving the correlation coefficient $0.03180/\sqrt{(8.009 \times 0.0005859)} = 0.464$, indicating that varieties with high mean values at an average value of rainfall tend to have stronger responses to rainfall. The estimated variance components due to 'location', 'location.sowing_occasion', 'variety' and 'location.variety' are considerably larger than their respective SEs, indicating that these terms should be retained in the model. However, the estimated variance component due to 'variety.sowing_occasion' is now negative, and smaller than its SE, suggesting that the true value of this component is zero: this term can clearly be dropped from the model. The estimated variance component due to 'variety.mrainfall' appears very small, but it cannot be compared directly with the other components, as its magnitude depends on the units in which 'mrainfall' is measured. It requires further investigation, and the following statements obtain the deviance from models with and without this term:

```
VCOMPONENTS [FIXED = mrainfall] \
   RANDOM = location / sowing_occasion + variety / mrainfall + \
   location.variety
VSTRUCTURE [variety / mrainfall; FORMATION = whole; \
   CORRELATE = unrestricted]
REML [PRINT = deviance] oil
VCOMPONENTS [FIXED = mrainfall] \
   RANDOM = location / sowing_occasion + variety + \
   location.variety
REML [PRINT = deviance] oil
```

Note that the terms 'location' and 'location.sowing_occasion' are now expressed more succinctly as 'location / sowing_occasion'. The output of these statements is as follows:

Deviance: −2*Log-Likelihood

Deviance	d.f.
136.71	117

Note: deviance omits constants which depend on fixed model fitted.

Deviance: −2*Log-Likelihood

Deviance	d.f.
148.03	119

Note: deviance omits constants which depend on fixed model fitted.

The difference between the deviances from the models with and without 'variety.mrainfall' is obtained, namely

$$\text{deviance}_{\text{reduced model}} - \text{deviance}_{\text{full model}} = 148.03 - 136.71 = 11.32.$$

This difference is accounted for by two model sources of variation, namely the 'variety.mrainfall' variance and the covariance of this term with the main effect of variety.

Therefore the difference in deviance is related to a chi-square variable with two degrees of freedom, and

$$P(\chi_2^2 > 11.32) = 0.00348.$$

However, this significance test is conservative, so the true p-value is smaller than this, though not easily determined (see Section 3.12), and the variance component 'variety.mrainfall' should definitely be retained in the model.

The model-building process is now complete, and the full results from fitting our final model are obtained by the following statements:

```
VCOMPONENTS [FIXED = mrainfall] \
   RANDOM = location / sowing_occasion + variety / mrainfall + \
   location.variety
VSTRUCTURE [variety / mrainfall; FORMATION = whole; \
   CORRELATE = unrestricted]
REML [PRINT = model, components, Wald, effects; \
   PTERMS = mrainfall + variety.mrainfall; PSE = estimates] oil
REML [PRINT = means; PTERMS = 'constant'] oil
VPREDICT [PRINT = predictions, avesed] location
VPREDICT [PRINT = predictions, avesed] variety
```

Note that we fit the model twice, the first time to obtain the effects of 'mrainfall' and 'variety.mrainfall' (together with the rest of the usual output), the second time to obtain the estimated mean oil content at each location and in each canola variety. A GenStat REML statement is not able to obtain SEs of difference between means when the default option setting 'METHOD = AI' is used, and the alternative setting 'METHOD = Fisher', which enables these SEs to be obtained (see Section 5.1), is not available when the option setting 'CORRELATE = unrestricted' is specified in a VSTRUCTURE statement. Printing of the location and variety means is therefore not specified in a REML statement, but by the two VPREDICT statements. The output from these statements is as follows:

REML variance components analysis

Response variate:	oil
Fixed model:	Constant + mrainfall
Random model:	location + location.sowing_occasion + variety + variety.mrainfall + location.variety
Number of units:	126 (12 units excluded due to zero weights or missing values)

Residual term has been added to model.

Sparse algorithm with AI optimization.
All covariates centred.

Covariance structures defined for random model

Correlated terms:

Set	Correlation across terms
1	Unstructured

Set	Terms	Covariance model within term
1	variety	Identity
1	variety.mrainfall	Identity

Estimated variance components

Random term	Component	s.e.
location	1.6228	1.2944
location.sowing_occasion	1.3145	0.4592
location.variety	0.1481	0.0831

Estimated parameters for covariance models

Random term(s)	Factor	Model(order)	Parameter	Estimate	s.e.
variety + variety.mrainfall					
	Across terms	Unstructured	v_11	8.122	4.154
			v_21	0.03208	0.02867
			v_22	0.0005873	0.0004071
	Within terms	Identity	–	–	–

Note: the covariance matrix for each term is calculated as G or R where var(y) = Sigma2(ZGZ′ + R), that is, relative to the residual variance, Sigma2.

Residual variance model

Term	Model(order)	Parameter	Estimate	s.e.
Residual	Identity	Sigma2	0.284	0.0500

Tests for fixed effects

Sequentially adding terms to fixed model

Fixed term	Wald statistic	n.d.f.	F statistic	d.d.f.	F pr.
mrainfall	30.28	1	30.28	24.6	<0.001

Dropping individual terms from full fixed model

Fixed term	Wald statistic	n.d.f.	F statistic	d.d.f.	F pr.
mrainfall	30.28	1	30.28	24.6	<0.001

Message: denominator degrees of freedom for approximate F tests are calculated using algebraic derivatives ignoring fixed/boundary/singular variance parameters.

Table of effects for mrainfall

0.05666   Standard error: 0.010296

Table of effects for variety.mrainfall

variety	Drum	Dunkeld	Grouse	Hyola 42	Karoo
	-0.019345	0.002854	0.003919	0.007741	0.008694

variety	Monty	Mustard	Narendra	Oscar	Pinnacle
	0.012296	-0.022355	0.005795	0.000595	0.003535

variety	Rainbow
	-0.003727

Standard errors

Average:	0.007239
Maximum:	0.008770
Minimum:	0.005657

Table of predicted means for Constant

40.58   Standard error: 0.742

Predictions from REML analysis

Response variate: oil

Predictions

location	
Beverley	39.22
Merredin	40.76
Mt Barker	42.68
Mullewa	41.19
Newdegate	39.25
Wongan Hills	40.41

Approximate average standard error of difference: 0.5723 (calculated on variance scale).

Predictions from REML analysis

Response variate: oil

Predictions

variety	Drum	Dunkeld	Grouse	Hyola 42	Karoo
	37.90	42.55	42.26	41.44	38.91

variety	Monty	Mustard	Narendra	Oscar	Pinnacle
	42.31	39.74	40.15	40.08	40.13

variety	Rainbow
	40.94

Approximate average standard error of difference: 0.5336 (calculated on variance scale).

In order to compare the amounts of variation accounted for by the various terms, we must convert the variance component for 'variety.mrainfall' to the same scale as the other variance components. For this purpose we must obtain the following values:

$$n = \text{number of observations in the data set,}$$

$$\text{mean(mrainfall)} = \frac{\sum\limits_{i=1}^{n} \text{mrainfall}_i}{n}$$

where

$$\text{mrainfall}_i = \text{the } i\text{th value of mrainfall,}$$

and

$$\text{var(mrainfall)} = \frac{\sum\limits_{i=1}^{n} (\text{mrainfall}_i - \text{mean(mrainfall)})^2}{n}.$$

Note that var(mrainfall) is calculated on the basis of the individual values of 'mrainfall', using the number of values as the divisor: no attempt is made to determine the degrees of freedom of 'mrainfall', that is, the number of independent pieces of information to which the variation in 'mrainfall' is equivalent.

We can then define the transformed variable

$$\text{mrainfall}' = \frac{\text{mrainfall} - \text{mean(mrainfall)}}{\sqrt{\text{var(mrainfall)}}},$$

which has a variance of 1. If we fit our mixed model using 'mrainfall'' in place of 'mrainfall', we will obtain the transformed variance component

$$\sigma'^2_{\text{variety.mrainfall}} = \text{var(mrainfall)} \cdot \sigma^2_{\text{variety.mrainfall}},$$

which is directly comparable with the other variance components. Applying these formulae to the data we obtain

$$n = 138$$

$$\text{mean(mrainfall)} = \frac{80.733 + 80.733 + \cdots + 19.000}{138} = 52.429$$

$$\text{var(mrainfall)} = \frac{(80.733 - 52.429)^2 + (80.733 - 52.429)^2 + \cdots + (19.000 - 52.429)^2}{138}$$

$$= 811.142$$

and

$$\sigma'^2_{\text{variety.mrainfall}} = 811.142 \times 0.0005873 \times 0.284 = 0.1353.$$

The covariance between the terms variety.mrainfall and 'variety' must be similarly transformed: thus

$$\sigma'^2_{\text{variety, variety.mrainfall}} = \sqrt{\text{var(mrainfall)}} \cdot \sigma'_{\text{variety, variety.mrainfall}}.$$

Applying this formula to the data, we obtain

$$\sigma'^2_{\text{variety, variety.mrainfall}} = \sqrt{811.142} \times 0.03208 \times 0.284 = 0.2595.$$

The amount of variation accounted for by each term in the model can then be displayed in a fairly intuitive way, as shown in Table 7.8. Note that because there is a non-zero covariance between the effects of 'variety' and variety.mrainfall, separate variance components cannot be assigned unambiguously to these terms, which are therefore represented by a single combined component.

When interpreting this table, it must be remembered that

- $2 \times$ SD (standard deviation) does not cover the full range of values of a random variable and

- although variance components can be summed, the corresponding SDs cannot.

With these caveats, the table can be used to gain an impression of the relative magnitude of the different sources of variation. Of the total range in the oil content of the harvested grain – 13 percentage points, or 32% of its mean value of 40.62% – only 1.066 percentage points are due to residual variation: the terms in the model account for most of the variation observed. About half the variation is accounted for by the effect of rainfall: on average, each additional millimetre of rainfall results in an additional 0.05666 percentage points of oil in the harvested grain. About half is accounted for by the main effect of variety and the variety.mrainfall interaction, taken together. A lesser amount of variation is due to the remaining effects of location and sowing occasion. These values give an indication of the extent to which the decisions of farmers and researchers, in choosing varieties or developing new ones, can be expected to influence the outcome of a growing season.

A line of best fit describing the relationship between oil content and rainfall can be constructed for each variety, as follows:

 fitted oil content = variety mean oil content

$$+ \text{(rainfall effect} + \text{variety.rainfall effect)} \times \text{(rainfall} - \text{mean(mrainfall))}$$

For example, at the lowest value of rainfall encountered by the variety 'Drum', 11 mm, the fitted value of oil content of this variety is

$$37.90\% + (0.05666\%/\text{mm} - 0.019345\%/\text{mm}) \times (11\,\text{mm} - 52.43\,\text{mm}) = 36.35\%$$

and at the highest value, 86.4 mm, the fitted value is

$$37.90\% + (0.05666\%/\text{mm} - 0.019345\%/\text{mm}) \times (86.4\,\text{mm} - 52.43\,\text{mm}) = 39.17\%.$$

The fitted line for each variety, together with an indication of the range of variation due to terms not represented in the fitted lines, and the observed values for the contrasting varieties 'Drum' and 'Monty', are presented graphically in Figure 7.6.

This graph will give researchers and farmers a clear idea of the relative importance of rainfall, the choice of variety and other factors in determining the oil content of a harvested canola crop. It may indicate that particular varieties should be chosen for particular environments: Mustard (not strictly a canola variety) is among the low-oil varieties, but this disadvantage is

Table 7.8 Amount of variation in the oil content of canola accounted for by each term in the final model fitted.

Source of variation in oil content (%)	Amount of variation accounted for
Total range	$47.95 - 34.95 = 13.00$
$2 \times SD_{\text{Total of random effects}}$	$2 \times \sqrt{\sigma^2_{\text{location}} + \sigma^2_{\text{location.sowing occasion}} + \sigma^2_{\text{variety}} + 2\sigma'_{\text{variety, variety.mrainfall}} + \sigma'^2_{\text{variety.mrainfall}} + \sigma^2_{\text{location.variety}} + \sigma^2_{\text{Residual}}} =$ $2 \times \sqrt{1.6228 + 1.3145 + 8.122 \times 0.284 + 2 \times 0.2595 + 0.1353 + 0.1481 + 0.2840} = 5.032$
$2 \times SD_{\text{location}}$	$2 \times \sqrt{\sigma^2_{\text{location}}} = 2 \times \sqrt{1.6228} = 2.548$
$2 \times SD_{\text{location.sowing occasion}}$	$2 \times \sqrt{\sigma^2_{\text{location.sowing occasion}}} = 2 \times \sqrt{1.3145} = 2.293$
$2 \times SD_{\text{variety, variety.mrainfall}}$	$2 \times \sqrt{\sigma^2_{\text{variety}} + 2\sigma'_{\text{variety, variety.mrainfall}} + \sigma'^2_{\text{variety.mrainfall}}} = 2 \times \sqrt{8.122 \times 0.284 + 2 \times 0.2595 + 0.1353} = 3.441$
$2 \times SD_{\text{location.variety}}$	$2 \times \sqrt{\sigma^2_{\text{location.variety}}} = 2 \times \sqrt{0.1481} = 0.770$
$2 \times SD_{\text{Residual}}$	$2 \times \sqrt{\sigma^2_{\text{Residual}}} = 2 \times \sqrt{0.2840} = 1.066$
Range due to the effect of rainfall	Rainfall effect × rainfall range $= 0.05666 \times (125.6 - 14.45) = 6.299$

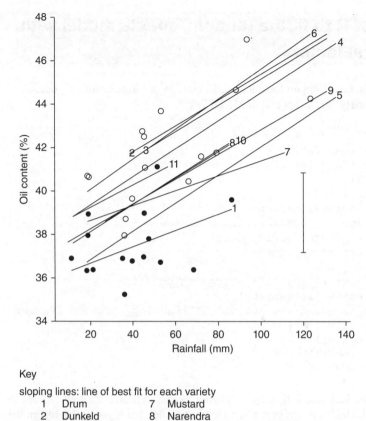

Figure 7.6   Predictions of oil content of canola on the basis of rainfall and variety, from the set of field trials.

less pronounced in low-rainfall environments and may be compensated by other advantages, making Mustard a more suitable choice for such environments. However, the graph shows that in general the variety.rainfall effects are small relative to the main effects of variety, so that such crossovers are fairly rare. The vertical bar, and the scatter of the observed values around their respective fitted lines, indicate the magnitude of the variation not accounted for by the lines. The variance component estimates show that most of this is not residual variation, but consistent effects of location and location.sowing date combination, which may be amenable to further investigation.

7.9 Use of R to fit the random-effects model with several factors

The following statements import the data into R, convert 'location', 'sowing_occasion', 'sowing_date' and 'variety' to factors and fit Model 7.4:

```
rm(list = ls())
canola.oil <- read.table(
    "IMM edn 2\\Ch 7\\canola oil gxe.txt",
    header=TRUE)
attach(canola.oil)
flocation <- factor(location)
fsowing_occasion <- factor(sowing_occasion)
fsowing_date <- factor(sowing_date)
fvariety <- factor(variety)
library(lme4)
canola.model1lmer <- lmer(oil ~ fsowing_occasion +
    (1|flocation) + (1|fvariety) +
    (1|fsowing_occasion:flocation) + (1|fsowing_occasion:fvariety) +
    (1|flocation:fvariety),
    data = canola.oil)
summary(canola.model1lmer)
anova(canola.model1lmer)
```

Note that the three-way interaction term 'sowing_occasion:location:variety' is not included explicitly in the model, as it was in the first GenStat model, but is recognized from the outset as the residual term. This is necessary in order to obtain a correct analysis from the function lmer().

The output of the summary() and anova() functions is as follows:

```
Linear mixed model fit by REML
Formula: oil ~ fsowing_occasion + (1 | flocation) +
    (1 | fvariety) + (1 | fsowing_occasion:flocation) +
    (1 | fsowing_occasion:fvariety) + (1 | flocation:fvariety)
   Data: canola.oil
   AIC   BIC logLik deviance REMLdev
 378.1 412.1 -177.1    362.6   354.1

Random effects:
 Groups                     Name        Variance Std.Dev.
 fsowing_occasion:fvariety  (Intercept) 0.11829  0.34393
 flocation:fvariety         (Intercept) 0.11481  0.33883
 fsowing_occasion:flocation (Intercept) 1.12021  1.05840
 fvariety                   (Intercept) 2.19054  1.48005
 flocation                  (Intercept) 3.85380  1.96311
 Residual                               0.29342  0.54168
```

```
Number of obs: 126, groups: fsowing_occasion:fvariety, 51;
   flocation:fvariety, 31; fsowing_occasion:flocation, 25;
   fvariety, 11; flocation, 6

Fixed effects:
                  Estimate Std. Error t value
(Intercept)        42.5770    1.0303     41.32
fsowing_occasion2  -1.8902    0.6492     -2.91
fsowing_occasion3  -2.4664    0.6492     -3.80
fsowing_occasion4  -3.6535    0.6884     -5.31
fsowing_occasion5  -5.0774    1.2931     -3.93
fsowing_occasion6  -5.2024    1.2931     -4.02

Correlation of Fixed Effects:
            (Intr) fswn_2 fswn_3 fswn_4 fswn_5
fswng_ccsn2 -0.315
fswng_ccsn3 -0.315  0.500
fswng_ccsn4 -0.297  0.472  0.472
fswng_ccsn5 -0.158  0.251  0.251  0.250
fswng_ccsn6 -0.158  0.251  0.251  0.250  0.269

Analysis of Variance Table
                 Df Sum Sq Mean Sq F value
fsowing_occasion  5  12.16  2.4319  8.2884
```

The variance component estimates agree with those given by GenStat and so approximately does the *F* value for 'fsowing_occasion'.

In an attempt to add 'sowing_occasion' to the model, to represent the linear trend over sowing occasions, the following command is run:

```
canola.model2lmer <- lmer(oil ~
   sowing_occasion + fsowing_occasion +
   (1|flocation) + (1|fvariety) +
   (1|fsowing_occasion:flocation) +
   (1|fsowing_occasion:fvariety) + (1|flocation:fvariety),
   data = canola.oil)
```

However, this produces the following message:

```
Error in mer_finalize(ans)  : Downdated X'X is not positive
   definite, 7.
```

Because of the close relationship between 'sowing_occasion' and 'fsowing_occasion' (technically known as *partial aliasing*), these terms cannot both be included in the fixed effect model.

However, if 'fsowing_occasion' is moved to the random-effects model, the resulting model can be successfully fitted, as follows:

```
canola.model3lmer <- lmer(oil ~ sowing_occasion +
    (1|fsowing_occasion) + (1|flocation) + (1|fvariety) +
    (1|fsowing_occasion:flocation) + (1|fsowing_occasion:fvariety) +
    (1|flocation:fvariety),
    data = canola.oil)
summary(canola.model3lmer)
anova(canola.model3lmer)
```

The output of these commands is as follows:

```
Linear mixed model fit by REML
Formula: oil ~ sowing_occasion + (1 | fsowing_occasion) +
    (1 | flocation) + (1 | fvariety) +
    (1 | fsowing_occasion:flocation) +
    (1 | fsowing_occasion:fvariety) + (1 | flocation:fvariety)
    Data: canola.oil
    AIC BIC logLik deviance REMLdev
 382.4 408 -182.2    364.4    364.4
Random effects:
 Groups                     Name         Variance     Std.Dev.
 fsowing_occasion:fvariety  (Intercept)  1.1436e-01   3.3817e-01
 flocation:fvariety         (Intercept)  1.1430e-01   3.3808e-01
 fsowing_occasion:flocation (Intercept)  9.9284e-01   9.9641e-01
 fvariety                   (Intercept)  2.1928e+00   1.4808e+00
 flocation                  (Intercept)  3.8423e+00   1.9602e+00
 fsowing_occasion           (Intercept)  1.5142e-12   1.2305e-06
 Residual                                2.9520e-01   5.4332e-01
Number of obs: 126, groups: fsowing_occasion:fvariety, 51;
    flocation:fvariety, 31; fsowing_occasion:flocation, 25;
    fvariety, 11; flocation, 6; fsowing_occasion, 6

Fixed effects:
                Estimate Std. Error t value
(Intercept)      43.3608     1.0420   41.61
sowing_occasion  -1.1102     0.1676   -6.62

Correlation of Fixed Effects:
            (Intr)
sowing_ccsn -0.418

Analysis of Variance Table
                Df Sum Sq Mean Sq F value
sowing_occasion  1 12.953  12.953  43.877
```

The output does not specify the degrees of freedom of the denominator of the F statistic for 'sowing_occasion', so we will convert it to the corresponding Wald statistics using Equations 1.5 and 1.6 (Section 1.8), as follows:

$$\text{Wald statistic} = F \times \text{DF}_{\text{Numerator of F}} = 43.877 \times 1 = 43.877$$

$$\text{DF}_{\text{Wald statistic}} = \text{DF}_{\text{Numerator of } F} = 1.$$

Interpreting the Wald statistic as an χ^2 statistic, its significance is given by

$$P(\chi_1^2 > 43.87) < 0.001.$$

The linear trend over sowing occasions is highly significant and should be retained in the model unless 'mrainfall' is substituted for it.

This substitution is made in the following commands:

```
canola.model4lmer <- lmer(oil ~ mrainfall + fsowing_occasion +
    (1|flocation) + (1|fvariety) +
    (1|fsowing_occasion:flocation) + (1|fsowing_occasion:fvariety) +
    (1|flocation:fvariety),
    data = canola.oil)
summary(canola.model4lmer)
anova(canola.model4lmer)
```

Note that it is now possible to return 'fsowing_occasion' to the fixed-effects part of the model, to obtain the same model as was fitted using GenStat at the corresponding stage of the modelling process. The output of these commands is as follows:

```
Linear mixed model fit by REML
Formula: oil ~ mrainfall + fsowing_occasion + (1 | flocation) +
    (1 | fvariety) + (1 | fsowing_occasion:flocation) +
    (1 | fsowing_occasion:fvariety) + (1 | flocation:fvariety)
    Data: canola.oil
 AIC    DIC logLik deviance REMLdev
 384 420.9   -179    361.3     358
Random effects:
 Groups                      Name        Variance Std.Dev.
 fsowing_occasion:fvariety   (Intercept) 0.12077  0.34752
 flocation:fvariety          (Intercept) 0.11474  0.33873
 fsowing_occasion:flocation  (Intercept) 1.30155  1.14086
 fvariety                    (Intercept) 2.19657  1.48208
 flocation                   (Intercept) 2.14492  1.46456
 Residual                                0.29218  0.54054
Number of obs: 126, groups: fsowing_occasion:fvariety, 51;
    flocation:fvariety, 31; fsowing_occasion:flocation, 25;
    fvariety, 11; flocation, 6
```

```
Fixed effects:
                   Estimate Std. Error t value
(Intercept)        39.02428    2.47241  15.784
mrainfall           0.04172    0.02701   1.545
fsowing_occasion2  -0.88965    0.94963  -0.937
fsowing_occasion3  -0.46514    1.46867  -0.317
fsowing_occasion4  -1.07479    1.80995  -0.594
fsowing_occasion5  -2.55058    2.17933  -1.170
fsowing_occasion6  -2.69435    2.16993  -1.242


Correlation of Fixed Effects:
            (Intr) mrnfll fswn_2 fswn_3 fswn_4 fswn_5
mrainfall   -0.932
fswng_ccsn2 -0.738  0.682
fswng_ccsn3 -0.887  0.881  0.774
fswng_ccsn4 -0.905  0.914  0.763  0.896
fswng_ccsn5 -0.766  0.774  0.645  0.758  0.772
fswng_ccsn6 -0.764  0.772  0.644  0.756  0.770  0.703


Analysis of Variance Table
                 Df Sum Sq Mean Sq F value
mrainfall         1 10.067  10.067 34.4553
fsowing_occasion  5  1.250   0.250  0.8557
```

The F values for 'mrainfall' and 'fsowing_occasion' are similar, though not identical, to those given by GenStat.

In order to drop the non-significant main effect of 'fsowing_occasion' from the model, and to add to the model possible variation in the effect of rainfall between the varieties, the following command is next run:

```
canola.model5lmer <- lmer(oil ~ mrainfall +
   (1|flocation) + (1 + mrainfall|fvariety) +
   (1|fsowing_occasion:flocation) + (1|fsowing_occasion:fvariety) +
   (1|flocation:fvariety),
   data = canola.oil)
summary(canola.model5lmer)
anova(canola.model5lmer)
```

The main effect of variety (i.e. the variation in the intercept among varieties) and the rainfall × variety interaction, that is, the variation in the effect of rainfall among varieties, are represented by the composite model term '(1 + mrainfall|fvariety)'. As in the random coefficients model (Section 7.6), it is not necessary to specify that there may be a covariance between these intercepts and slopes, and that this covariance is to be estimated: in R, this model specification is the default.

The output of these commands is as follows:

```
Linear mixed model fit by REML
Formula: oil ~ mrainfall + (1 | flocation) +
   (1 + mrainfall | fvariety) + (1 | fsowing_occasion:flocation) +
   (1 | fsowing_occasion:fvariety) + (1 | flocation:fvariety)
   Data: canola.oil
   AIC BIC logLik deviance REMLdev
 384.6 413 -182.3   358.5   364.6
Random effects:
 Groups                     Name        Variance    Std.Dev.    Corr
 fsowing_occasion:fvariety  (Intercept) 1.6612e-11  4.0758e-06
 flocation:fvariety         (Intercept) 1.4805e-01  3.8477e-01
 fsowing_occasion:flocation (Intercept) 1.3146e+00  1.1466e+00
 fvariety                   (Intercept) 1.8093e+00  1.3451e+00
                            mrainfall   1.6681e-04  1.2916e-02  0.014
 flocation                  (Intercept) 1.6227e+00  1.2739e+00
 Residual                               2.8403e-01  5.3295e-01
Number of obs: 126, groups: fsowing_occasion:fvariety, 51;
   flocation:fvariety, 31; fsowing_occasion:flocation, 25; fvariety, 11;
   flocation, 6

Fixed effects:
            Estimate Std. Error t value
(Intercept) 37.57064    0.86639   43.36
mrainfall    0.05666    0.01030    5.50

Correlation of Fixed Effects:
          (Intr)
mrainfall -0.527

Analysis of Variance Table
          Df Sum Sq Mean Sq F value
mrainfall  1 8.5338  8.5338  30.045
```

The variance component estimate for 'fsowing_occasion:fvariety' is effectively zero, whereas the corresponding estimate given by GenStat was negative: the function lmer() does not permit negative estimates of variance components. This constraint has an influence on the estimates of the other variance components: nevertheless, those for 'flocation:fvariety', 'fsowing_occasion:flocation', 'flocation' and the residual term agree fairly well with those given by GenStat. For the variance components related to the main effect of 'fvariety' and the 'fvariety:mrainfall' interaction, the corresponding values from GenStat, scaled to take into account their relationship to the residual variance, are as follows:

- For 'fvariety (Intercept)', that is, the main effect of 'fvariety', the component $v\_11 = 8.009 \times 0.289 = 2.3146$. This is not the same as the value given by R, 1.8093, due to the different parameterizations used by the two software systems: the variance component from GenStat is related to the mean value for each variety at an average value of rainfall, whereas that from R is related to the intercept for each variety, that is, its estimated value at zero rainfall.

- For 'fvariety mrainfall', that is, the 'fvariety:mrainfall' interaction, the component $v\_22 = 0.0005859 \times 0.289 = 0.00016933$. This agrees fairly well with the value given by R, 0.00016681.

The correlation between these two terms, 0.014, indicates that the response of a variety to rainfall is not much related to its intercept. This correlation coefficient is not the same as that obtained from GenStat, 0.464, again due to the different parameterizations used by the two software systems.

The following commands fit the model from which the non-significant term 'variety.sowing_occasion' is dropped, and the model from which 'variety.mrainfall' is also dropped, and obtain the deviance from each model:

```
canola.model6lmer <- lmer(oil ~ mrainfall +
    (1|flocation) + (1 + mrainfall|fvariety) +
    (1|fsowing_occasion:flocation) +
    (1|flocation:fvariety),
    data = canola.oil)
deviance(canola.model6lmer)

canola.model7lmer <- lmer(oil ~ mrainfall +
    (1|flocation) + (1|fvariety) +
    (1|fsowing_occasion:flocation) +
    (1|flocation:fvariety),
    data = canola.oil)
deviance(canola.model7lmer)
```

Note that the term '(1 + mrainfall|fvariety)' is replaced by '(1|fvariety)', so that the main effect of variety is retained in the model.

The output from these commands is as follows:

```
364.6084
375.9291
```

Comparison of these deviances gives

$$\text{deviance}_{\text{reduced model}} - \text{deviance}_{\text{full model}} = 375.9291 - 364.6084 = 11.3207,$$

the same value as was obtained from GenStat, and on this basis the term 'variety.mrainfall' is retained. Full output from the final model is produced by the following commands:

```
summary(canola.model6lmer)
anova(canola.model6lmer)
coef(canola.model6lmer)
```

The output obtained from the functions summary() and anova(), and part of that from coef(), is as follows:

```
Linear mixed model fit by REML
Formula: oil ~ mrainfall + (1 | flocation) +
    (1 + mrainfall | fvariety) + (1 | fsowing_occasion:flocation) +
    (1 | flocation:fvariety)
    Data: canola.oil
    AIC   BIC logLik deviance REMLdev
  382.6 408.1 -182.3    358.5    364.6
Random effects:
 Groups                     Name         Variance   Std.Dev. Corr
 flocation:fvariety         (Intercept)  0.14804524 0.384766
 fsowing_occasion:flocation (Intercept)  1.31459320 1.146557
 fvariety                   (Intercept)  1.80932148 1.345110
                            mrainfall    0.00016681 0.012916 0.014
 flocation                  (Intercept)  1.62274682 1.273871
 Residual                                0.28403195 0.532946
Number of obs: 126, groups: flocation:fvariety, 31;
    fsowing_occasion:flocation, 25; fvariety, 11; flocation, 6

Fixed effects:
            Estimate Std. Error t value
(Intercept) 37.57064    0.86639   43.36
mrainfall    0.05666    0.01030    5.50

Correlation of Fixed Effects:
          (Intr)
mrainfall -0.527

Analysis of Variance Table
          Df Sum Sq Mean Sq F value
mrainfall  1 8.5267  8.5267   30.02

$fvariety
          (Intercept)  mrainfall
Drum         35.93635 0.03731213
Dunkeld      39.35590 0.05951190
Grouse       39.02273 0.06057730
Hyola 42     38.01332 0.06439837
Karoo        35.46521 0.06535054
Monty        38.63066 0.06895221
Mustard      37.90970 0.03430254
Narendra     36.83633 0.06245203
Oscar        37.04542 0.05725297
Pinnacle     36.94009 0.06019239
Rainbow      38.12136 0.05293071
```

The variance component estimates agree with those given by GenStat, with exceptions in those relating to 'mrainfall' and the 'fvariety:mrainfall' interaction, as follows:

- For 'fvariety (Intercept)', the GenStat component $v\_11 = 8.122 \times 0.284 = 2.3066$.

- For 'fvariety mrainfall', the component $v\_22 = 0.0005873 \times 0.284 = 0.00016679$.

- For the correlation between these terms, the value based on covariance $v\_21 = 0.03208/\sqrt{(8.122 \times 0.0005873)} = 0.464$.

The second of these values agrees with that obtained from R, but the others do not, again due to the different parameterizations used by the two software systems. The correlation between 'mrainfall' and 'variety.mrainfall', 0.014, confirms that the response of a variety to rainfall is not much related to its intercept. The F value for 'mrainfall' agrees approximately with that given by GenStat. In terms of the parameter estimates given by R, the line of best fit describing the relationship between oil content and rainfall is constructed for each variety as follows:

fitted oil content = variety-specific intercept + variety-specific rainfall effect × rainfall

For example, the coefficients under the heading '`$fvariety`' show that at the lowest value of rainfall encountered by the variety 'Drum', 11 mm, the fitted value of oil content of this variety is

$$35.93635\% + 0.03731213 \times 11\,\text{mm} = 36.35\%$$

and at the highest value, 86.4 mm, the fitted value is

$$35.93635\% + 0.03731213 \times 84\,\text{mm} = 39.07\% \,.$$

These fitted values agree with those obtained from GenStat.

7.10 Use of SAS to fit the random-effects model with several factors

The following statements import the data into SAS and fit Model 7.4:

```
PROC IMPORT OUT = canola DBMS = EXCELCS REPLACE
   DATAFILE = "&pathname.\IMM edn 2\Ch 7\canola oil gxe.xlsx";
   SHEET = "for SAS";
RUN;

ODS RTF;
PROC MIXED ASYCOV DATA = canola;
   CLASS sowing_occasion location variety;
   MODEL oil = sowing_occasion /DDFM = KR HTYPE = 1;
   RANDOM location variety sowing_occasion*location
sowing_occasion*variety
      location*variety;
RUN;
ODS RTF CLOSE;
```

Note that the three-way interaction term 'sowing_occasion*location*variety' is not included explicitly in the model, as it was in the first GenStat model, but is recognized from the outset as the residual term: this avoids slight clumsiness in the output from PROC MIXED. Note also that in the PROC MIXED statement the NOBOUND option has not been set, and negative estimates of variance components will therefore not be permitted. This is because when it is set, these statements produce the warning message 'Stopped because of infinite likelihood'. Part of the output from PROC MIXED is as follows:

Convergence criteria met.

Covariance parameter estimates	
Cov Parm	**Estimate**
location	3.8538
variety	2.1905
sowing_occa*location	1.1202
sowing_occas*variety	0.1183
location*variety	0.1148
Residual	0.2934

Asymptotic covariance matrix of estimates							
Row	**Cov Parm**	**CovP1**	**CovP2**	**CovP3**	**CovP4**	**CovP5**	**CovP6**
1	location	7.0891	−0.06082	−0.05049	−0.00035	0.001603	0.000246
2	variety	−0.06082	1.1072	−0.00055	−0.00212	−0.00208	0.000851
3	sowing_occa*location	−0.05049	−0.00055	0.1989	0.001580	0.000382	−0.00140
4	sowing_occas*variety	−0.00035	−0.00212	0.001580	0.005086	0.000474	−0.00177
5	location*variety	0.001603	−0.00208	0.000382	0.000474	0.005002	−0.00099
6	Residual	0.000246	0.000851	−0.00140	−0.00177	−0.00099	0.003783

Fit statistics	
−2 Res Log Likelihood	354.1
AIC (smaller is better)	366.1
AICC (smaller is better)	366.8
BIC (smaller is better)	364.9

Type 1 tests of fixed effects				
Effect	**Num DF**	**Den DF**	**F value**	**Pr > F**
sowing_occasion	5	15.6	8.18	0.0006

The variance component estimates agree with those produced by GenStat. The F value is similar but not identical, and its denominator d.f. are the same.

The following statements add a variate named 'vsowing_occasion', representing a linear trend over sowing occasions, to the model:

```
DATA canola2; SET canola;
   vsowing_occasion = sowing_occasion;
RUN;

ODS RTF;
PROC MIXED ASYCOV DATA = canola2;
   CLASS sowing_occasion location variety;
   MODEL oil = vsowing_occasion sowing_occasion /DDFM = KR HTYPE = 1
SOLUTION;
   RANDOM location variety sowing_occasion*location
sowing_occasion*variety
      location*variety;
RUN;
ODS RTF CLOSE;
```

Note that whereas 'sowing_occasion' is specified as a CLASS variable, 'vsowing_occasion' is not.

Part of the output from PROC MIXED is as follows:

Convergence criteria met.

Covariance parameter estimates	
Cov Parm	**Estimate**
location	3.8538
variety	2.1905
sowing_occa*location	1.1202
sowing_occas*variety	0.1183
location*variety	0.1148
Residual	0.2934

Asymptotic covariance matrix of estimates							
Row	**Cov Parm**	**CovP1**	**CovP2**	**CovP3**	**CovP4**	**CovP5**	**CovP6**
1	location	7.0891	−0.06082	−0.05049	−0.00035	0.001603	0.000246
2	variety	−0.06082	1.1072	−0.00055	−0.00212	−0.00208	0.000851
3	sowing_occa*location	−0.05049	−0.00055	0.1989	0.001580	0.000382	−0.00140
4	sowing_occas*variety	−0.00035	−0.00212	0.001580	0.005086	0.000474	−0.00177
5	location*variety	0.001603	−0.00208	0.000382	0.000474	0.005002	−0.00099
6	Residual	0.000246	0.000851	−0.00140	−0.00177	−0.00099	0.003783

Type 1 tests of fixed effects				
Effect	**Num DF**	**Den DF**	**_F_ value**	**Pr > _F_**
vsowing_occasion	1	15.8	38.99	<0.0001
sowing_occasion	4	15.6	0.49	0.7463

The variance component estimates agree with those given by GenStat. The F values, and denominator d.f. values, produced by SAS for 'vsowing_occasion' and 'sowing_occasion' are similar, but not identical, to those given by GenStat.

The following statements replace 'vsowing_occasion' by 'mrainfall':

```
ODS RTF;
PROC MIXED ASYCOV DATA = canola2;
   CLASS sowing_occasion location variety;
   MODEL oil = mrainfall sowing_occasion /DDFM = KR HTYPE = 1
SOLUTION;
   RANDOM location variety sowing_occasion*location
sowing_occasion*variety
       location*variety;
RUN;
ODS RTF CLOSE;
```

Part of the output from PROC MIXED is as follows:

Type 1 tests of fixed effects				
Effect	Num DF	Den DF	F value	Pr > F
mrainfall	1	18.2	17.17	0.0006
sowing_occasion	5	14.7	0.76	0.5894

These F values, and their denominator d.f., do not agree with those produced by GenStat: the reason for this discrepancy is not known to the author at the time of writing.

The following statements drop the non-significant term 'sowing_occasion' and add the term 'variety.mrainfall' to the model:

```
ODS RTF;
PROC MIXED ASYCOV DATA = canola2;
   CLASS sowing_occasion location variety;
   MODEL oil = mrainfall /DDFM = KR HTYPE = 1 SOLUTION;
   RANDOM location variety sowing_occasion*location
sowing_occasion*variety
       location*variety;
   RANDOM intercept mrainfall /SUBJECT = variety TYPE = UN SOLUTION;
RUN;
ODS RTF CLOSE;
```

The terms 'variety' and 'variety.mrainfall' are effectively specified in the second RANDOM statement, which indicates that a separate intercept and a separate coefficient of 'mrainfall' are to be fitted, as random effects, for each level of 'variety'. Part of the output from PROC MIXED is as follows:

> Convergence criteria met but final hessian is not positive definite.

Covariance parameter estimates		
Cov Parm	**Subject**	**Estimate**
location		1.6228
variety		1.0918
sowing_occa*location		1.3145
sowing_occas*variety		0
location*variety		0.1481
UN(1,1)	variety	0.7175
UN(2,1)	variety	0.000239
UN(2,2)	variety	0.000167
Residual		0.2840

Asymptotic covariance matrix of estimates									
Row	**Cov Parm**	**CovP1**	**CovP2**	**CovP3**	**CovP4**	**CovP5**	**CovP6**	**CovP7**	**CovP8**	**CovP9**
1	Location	1.6827	−0.03100	−0.05761		−0.00021		−0.00002	−2.85E−6	0.000722
2	Variety	−0.03100	0.9710	−0.00380		−0.00332		−0.00180	0.000011	−0.00300
3	sowing_occa*location	−0.05761	−0.00380	0.2116		0.000314		−0.00003	1.904E−6	−0.00083
4	sowing_occas*variety									
5	location*variety	−0.00021	−0.00332	0.000314		0.006877		−8.56E−6	9.076E−7	−0.00081
6	UN(1,1)									
7	UN(2,1)	−0.00002	−0.00180	−0.00003		−8.56E−6		0.000056	−2.98E−7	0.000059
8	UN(2,2)	−2.85E−6	0.000011	1.904E−6		9.076E−7		−2.98E−7	1.354E−8	−1.36E−6
9	Residual	0.000722	−0.00300	−0.00083		−0.00081		0.000059	−1.36E−6	0.002634

The variance component estimates for 'location' and 'location.sowing_occasion' agree with those given by GenStat, and the Residual component is similar though not identical. That for 'variety.sowing_occasion' is zero, which agrees with the negative estimate given by Gen-Stat, taking into account the fact that the NOBOUND option was not set in SAS's PROC MIXED statement. For the variance components related to the main effect of 'variety' and the 'variety.mrainfall' interaction, the corresponding values from GenStat, scaled to take into account their relationship to the residual variance, are as follows:

- For UN(1, 1), that is, the main effect of 'variety', the component $v\_11 = 8.009 \times 0.289 = 2.315$. This is not the same as the value given by SAS, 0.7175. Agreement is not expected here, due to the different parameterizations used by the two software systems: the variance component from GenStat is related to the mean value for each variety at an average value of rainfall, whereas that from SAS is related to the intercept for each variety, that is, its estimated value at zero rainfall. However, the software R also relates this variance component to the intercept, and the value given by R, 1.8093, also disagrees with that from SAS. The reason for this discrepancy is not known to the author.

- For UN(2, 1), that is, the covariance between the 'variety.mrainfall' interaction and the main effect of 'variety', the covariance $v\_21 = 0.03180 \times 0.289 = 0.00919$. This is not the same as the value given by SAS, 0.000239, again due to the different parameterizations.

- For UN(2, 2), that is, the 'variety.mrainfall' interaction, the component $v\_22 = 0.0005859 \times 0.289 = 0.000169$. This agrees fairly well with the value given by SAS, 0.000167.

The covariance UN(2, 1) corresponds to a correlation coefficient of $0.000239/\sqrt{(0.7175 \times 0.000167)} = 0.0218$, indicating that the response of a variety to rainfall is not much related to its intercept.

The following statements fit the model from which the non-significant term 'sowing_occasion.variety' has been dropped:

```
ODS RTF;
PROC MIXED ASYCOV DATA = canola2;
   CLASS sowing_occasion location variety;
   MODEL oil = mrainfall /DDFM = KR HTYPE = 1;
   RANDOM location variety sowing_occasion*location
      location*variety;
   RANDOM intercept mrainfall /SUBJECT = variety TYPE = UN SOLUTION;
RUN;
ODS RTF CLOSE;
```

Part of the output from PROC MIXED is as follows:

Fit statistics	
−2 Res Log Likelihood	364.6
AIC (smaller is better)	380.6
AICC (smaller is better)	381.9
BIC (smaller is better)	378.9

The following statement fits the model from which 'variety.mrainfall' has been dropped, in order to determine whether this term is significant:

```
ODS RTF;
PROC MIXED ASYCOV DATA = canola2;
   CLASS sowing_occasion location variety;
   MODEL oil = mrainfall /DDFM = KR HTYPE = 1;
   RANDOM location variety sowing_occasion*location
      location*variety;
RUN;
ODS RTF CLOSE;
```

Part of the output from PROC MIXED is as follows:

Fit statistics	
−2 Res Log Likelihood	375.9
AIC (smaller is better)	385.9
AICC (smaller is better)	386.4
BIC (smaller is better)	384.9

The difference between the deviances from the models with and without 'variety.mrainfall' is

$$\text{deviance}_{\text{reduced model}} - \text{deviance}_{\text{full model}} = 375.9 - 364.6 = 11.3,$$

which agrees with the value given by GenStat, and on this basis the term 'variety.mrainfall' is retained.

The following statements restore the term 'variety.mrainfall' to the model, and obtain estimates of parameters and variety means for the final model:

```
ODS RTF;
PROC MIXED ASYCOV DATA = canola2;
    CLASS sowing_occasion location variety;
    MODEL oil = mrainfall /DDFM = KR HTYPE = 1 SOLUTION;
    RANDOM location sowing_occasion*location
        location*variety /SOLUTION;
    RANDOM intercept mrainfall /SUBJECT = variety TYPE = UN SOLUTION;
RUN;
ODS RTF CLOSE;
```

Part of the output from PROC MIXED is as follows:

Convergence criteria met.

Covariance parameter estimates		
Cov Parm	**Subject**	**Estimate**
location		1.6228
sowing_occa*location		1.3145
location*variety		0.1481
UN(1,1)	variety	1.8091
UN(2,1)	variety	0.000241
UN(2,2)	variety	0.000167
Residual		0.2840

		Asymptotic covariance matrix of estimates						
Row	**Cov Parm**	**CovP1**	**CovP2**	**CovP3**	**CovP4**	**CovP5**	**CovP6**	**CovP7**
1	location	1.6827	−0.05761	−0.00021	−0.03099	−0.00002	−2.85E−6	0.000722
2	sowing_occa*location	−0.05761	0.2116	0.000314	−0.00380	−0.00003	1.904E−6	−0.00083
3	location*variety	−0.00021	0.000314	0.006878	−0.00332	−8.55E−6	9.076E−7	−0.00081
4	UN(1,1)	−0.03099	−0.00380	−0.00332	0.9706	−0.00180	0.000011	−0.00299
5	UN(2,1)	−0.00002	−0.00003	−8.55E−6	−0.00180	0.000056	−2.97E−7	0.000059
6	UN(2,2)	−2.85E−6	1.904E−6	9.076E−7	0.000011	−2.97E−7	1.353E−8	−1.36E−6
7	Residual	0.000722	−0.00083	−0.00081	−0.00299	0.000059	−1.36E−6	0.002634

Solution for fixed effects					
Effect	**Estimate**	**Standard error**	**DF**	**_t_ value**	**Pr > |_t_|**
Intercept	37.5706	0.8771	17.9	42.83	<0.0001
mrainfall	0.05666	0.01058	23.9	5.36	<0.0001

Solution for random effects								
Effect	**location**	**variety**	**sowing_occasion**	**Estimate**	**Std Err Pred**	**DF**	**_t_ value**	**Pr > |_t_|**
Intercept		Drum		−1.6343	0.5913	16.8	−2.76	0.0134
mrainfall		Drum		−0.01935	0.007630	15.8	−2.54	0.0222
Intercept		Dunkeld		1.7852	0.8358	18.8	2.14	0.0461
mrainfall		Dunkeld		0.002855	0.008557	13.3	0.33	0.7439
Intercept		Grouse		1.4520	0.8358	18.8	1.74	0.0987
mrainfall		Grouse		0.003920	0.008557	13.3	0.46	0.6543
Intercept		Hyola 42		0.4427	0.5231	14.1	0.85	0.4116
mrainfall		Hyola 42		0.007740	0.006017	13.9	1.29	0.2194
Intercept		Karoo		−2.1054	0.5540	15.8	−3.80	0.0016
mrainfall		Karoo		0.008693	0.006446	14.6	1.35	0.1981
Intercept		Monty		1.0600	0.5913	16.8	1.79	0.0911
mrainfall		Monty		0.01230	0.007630	15.8	1.61	0.1268
Intercept		Mustard		0.3390	0.8849	14.7	0.38	0.7071
mrainfall		Mustard		−0.02235	0.01032	9.5	−2.17	0.0570
Intercept		Narendra		−0.7343	0.7285	17.8	−1.01	0.3270
mrainfall		Narendra		0.005794	0.01026	11.9	0.56	0.5827
Intercept		Oscar		−0.5252	0.5231	14.1	−1.00	0.3324
mrainfall		Oscar		0.000595	0.006017	13.9	0.10	0.9227
Intercept		Pinnacle		−0.6305	0.8358	18.8	−0.75	0.4600
mrainfall		Pinnacle		0.003534	0.008557	13.3	0.41	0.6862
Intercept		Rainbow		0.5507	0.7285	17.8	0.76	0.4596
mrainfall		Rainbow		−0.00373	0.01026	11.9	−0.36	0.7228

Type 1 tests of fixed effects				
Effect	**Num DF**	**Den DF**	**_F_ value**	**Pr > _F_**
mrainfall	1	23.9	28.69	<0.0001

The variance component estimates for 'location', 'sowing_occasion.location' and 'location.variety' and the Residual component agree with those given by GenStat. For the variance components related to the main effect of 'variety' and the variety.mrainfall interaction, the corresponding values from GenStat, scaled to take into account their relationship to the residual variance, are as follows:

- For UN(1, 1), the component v_11 = 8.122 × 0.284 = 2.3066.

- For UN(2, 1), the covariance v_21 = 0.03208 × 0.284 = 0.00911.

- For UN(2, 2), the component v_22 = 0.0005873 × 0.284 = 0.000167.

The value for UN(2, 2) agrees with that obtained from GenStat, but those for UN(1, 1) and UN(2, 1) do not, due to the difference in parameterization between the two software systems, as described above. The covariance UN(2, 1) corresponds to a correlation coefficient of $0.000241/\sqrt{(1.8091 \times 0.000167)} = 0.014$, confirming that the response of a variety to rainfall is not much related to its intercept. The F value for 'mrainfall' and its denominator d.f. do not agree closely with those given by GenStat, but like those from GenStat indicate that the effect of rainfall is highly significant.

In terms of the parameter estimates given by SAS, the line of best fit describing the relationship between oil content and rainfall is constructed for each variety as follows:

$$\text{fitted oil content} = \text{intercept} + \text{variety effect}$$

$$+ (\text{rainfall effect} + \text{variety.rainfall effect}) \times \text{rainfall}$$

For example, at the lowest value of rainfall encountered by the variety 'Drum', 11 mm, the fitted value of oil content of this variety is

$$37.5704\% - 1.6343\% + (0.05666\%/\text{mm} - 0.01935\%/\text{mm}) \times 11\,\text{mm} = 36.35\%$$

and at the highest value, 86.4 mm, the fitted value is

$$37.5704\% - 1.6343\% + (0.05666\%/\text{mm} - 0.01935\%/\text{mm}) \times 86.4\,\text{mm} = 39.16\%$$

These fitted values agree with those obtained from GenStat.

7.11 Summary

The concepts introduced in earlier chapters are applied to three more elaborate data sets, concerning

- the causes of variation in BMD among human patients,

- the potential value of lithium as a treatment for ALS and

- the causes of variation in oil content in a grain crop.

Some widely applicable new concepts and additional features of the mixed-modelling process are also introduced. These are summarized in Section 7.1.

7.12 Exercises

7.1 Return to the data presented in Exercise 1.2.

(a) Fit the following model to these data:

Response variate: available chlorine
Fixed-effect model: linear, quadratic and cubic effects of time
Random-effect model: deviation of mean available chlorine at each time from
                                    the value predicted by the fixed-effects model.

(b) Investigate whether any terms can reasonably be dropped from this model.

(c) Obtain diagnostic plots of the residuals and consider whether the assumptions on which the analysis is based are reasonable.

(d) Plot the estimate of each random effect against the corresponding time and consider whether there is any evidence of a trend over time that is not accounted for by the fixed-effect model.

N.B. As the level of available chlorine declines, the rate of decline becomes lower, and this exponential decay process is more naturally modelled by a non-linear function than by the polynomial function used for convenience here. The model fitted by Draper and Smith (1998, Section 24.3, pp. 518–519) is

$$Y = \alpha + (0.49 - \alpha)e^{-\beta(X-8)} + \epsilon$$

where

$Y =$ available chlorine,

$X =$ time and

α and β are parameters to be estimated.

7.2 In an experiment to determine the effect of four experimental diets on the growth of pigs of both sexes, 32 pigs were allocated to four randomized blocks, with eight pigs of each sex in each block. Within each block, one pig of each sex, chosen at random, was allocated to each diet. The weight (kg) of each animal was recorded on nine occasions at weekly intervals. The first and last few rows of the spreadsheet holding the data are presented in Table 7.9: the complete data set is held in the file 'PigGrowth.xlsx' on this book's website (see Preface).

(a) Produce a plot of the values of Weight against those of Time, connecting the points representing each pig with a line. Note any trends common to all pigs and any patterns of variation among the pigs.

(b) Consider whether the weight of each pig at Time 1 should be used as a baseline value in the analysis of these data. Give reasons for your judgement.

For the rest of this exercise, do not use a baseline value.

(c) Specify an appropriate mixed model for these data. Your model should include the specification of a covariance structure between random-effect terms, where appropriate. Fit your model to the data.

From the results of your mixed-model analysis, answer the following questions:

(d) What is the correlation between the estimated weight of each pig when Time $= 0$ and its rate of growth over time?

(e) Is there a significant main effect of Sex or of Diet, or a significant Sex $\times$ Diet interaction? What interpretation should be placed on significant effects among these terms?

(f) How should the main effect of Time be interpreted? Do the sexes differ significantly in their growth rate over time? Do the diets have significantly different effects on

Table 7.9 Weight of pigs of both sexes on experimental diets, recorded on nine occasions.

	A	B	C	D	E	F
1	Pig!	Block!	Diet!	Sex!	Time	Weight
2	1	1	1	1	1	25
3	2	1	1	2	1	23
4	3	1	2	1	1	24
5	4	1	2	2	1	22
⋮						⋮
287	30	4	3	2	9	64
288	31	4	4	1	9	70
289	32	4	4	2	9	80

Sex 1 = male; 2 = female.

growth rate? Is there a significant interaction between the effects of Sex and Diet on growth rate?

(g) Do the results of your mixed-model analysis indicate that any of the model terms are unimportant with regard to their influence on weight? If so, drop these terms from the model and fit the resulting model to the data. From your new model, obtain the predicted values of weight at each time for each combination of Sex and Diet. Make a graphical display of the relationship between these variables. Comment on the effects seen in this plot.

7.3 An experiment was performed to determine the effect of the enzyme lactase (which hydrolyses the sugar lactose) on the composition of the milk in a sheep's udder. In each side of the udder in each of eight lactating sheep, the levels of fat and of lactose in the milk (%) were measured during a pre-experimental period. This was followed by a treatment period, at the beginning of which lactase was injected into one side of the udder, chosen at random. It was assumed that this injection did not affect the composition of the milk in the other side, which therefore served as a control. A day later, the levels of fat and lactose were measured again. This was followed a week later by a second treatment period, at the beginning of which lactase was injected into the other side of the udder. Again, a day later the fat and lactose levels were measured. The results are presented in Table 7.10 (Data reproduced by kind permission of Roberta Bencini, Faculty of Natural and Agricultural Sciences, The University of Western Australia).

The final levels of fat and lactose are the response variables in this experiment.

(a) Identify the block and treatment terms in this experiment. ('Period' might be placed in either category: specify it as a treatment term.) Specify the block-structure and treatment-structure models.

(b) Determine the effects of the block and treatment terms on each response variable, by analysis of variance and by mixed modelling. For each significant relationship observed, state the nature of the effect – for example, does lactase raise or lower the level of lactose?

Table 7.10 Fat and lactose content of sheep's milk, with and without injection of lactase into the udder.

Sheep	Side	Period	Treatment	Fat (%)		Post-treatment	Lactose (%)		Post-treatment
				Initial			Initial		
				Individual value	Sheep mean		Individual value	Sheep mean	
Surprise	Right	1	lactase	6.06	5.990	4.81	3.17	3.600	3.57
Surprise	Left	1	control	5.92	5.990	5.70	4.03	3.600	4.06
159	Right	1	lactase	5.92	6.095	3.58	3.93	3.930	2.00
159	Left	1	control	6.27	6.095	5.06	3.93	3.930	2.98
338	Right	1	control	4.35	5.130	3.46	3.06	3.705	3.19
338	Left	1	lactase	5.91	5.130	5.82	4.35	3.705	3.22
356	Right	1	lactase	6.39	6.410	4.99	4.50	4.540	3.54
356	Left	1	control	6.43	6.410	4.86	4.58	4.540	4.01
369	Right	1	lactase	6.27	6.485	4.73	4.28	4.325	2.16
369	Left	1	control	6.70	6.485	5.21	4.37	4.325	4.03
389	Right	1	lactase	5.80	6.045	5.04	4.44	4.540	3.53
389	Left	1	control	6.29	6.045	5.11	4.64	4.540	4.56
477	Right	1	control	6.15	6.085	5.52	3.89	4.085	4.44
477	Left	1	lactase	6.02	6.085	5.43	4.28	4.085	3.62
486	Right	1	control	4.48	4.480	3.50	3.66	3.660	3.26
486	Left	1	lactase	4.48	4.480	2.95	3.66	3.660	0.37
Surprise	Right	2	control	6.06	5.990	5.79	3.17	3.600	3.93
Surprise	Left	2	lactase	5.92	5.990	7.00	4.03	3.600	0.66
159	Right	2	control	5.92	6.095	5.46	3.93	3.930	3.73
159	Left	2	lactase	6.27	6.095	5.48	3.93	3.930	4.60
338	Right	2	lactase	4.35	5.130	3.88	3.06	3.705	2.35
338	Left	2	control	5.91	5.130	4.37	4.35	3.705	3.73
356	Right	2	control	6.39	6.410	4.76	4.50	4.540	3.97
356	Left	2	lactase	6.43	6.410	6.12	4.58	4.540	3.71
369	Right	2	control	6.27	6.485	4.75	4.28	4.325	3.75
369	Left	2	lactase	6.70	6.485	5.52	4.37	4.325	3.79
389	Right	2	control	5.80	6.045	3.41	4.44	4.540	1.13
389	Left	2	lactase	6.29	6.045	4.11	4.64	4.540	4.39
477	Right	2	lactase	6.15	6.085	3.67	3.89	4.085	0.59
477	Left	2	control	6.02	6.085	3.66	4.28	4.085	3.73
486	Right	2	lactase	4.48	4.480	4.02	3.66	3.660	0.68
486	Left	2	control	4.48	4.480	4.01	3.66	3.660	3.76

Source: Data reproduced by kind permission of Roberta Bencini, Faculty of Natural and Agricultural Sciences, The University of Western Australia.

In each mixed-model analysis, the mean initial value of the response variable for each sheep can be added to the model. The difference between this mean and the corresponding individual initial value on each side of the udder in each sheep can also be added.

(c) Should these terms be placed in the fixed-effect or the random-effect model?

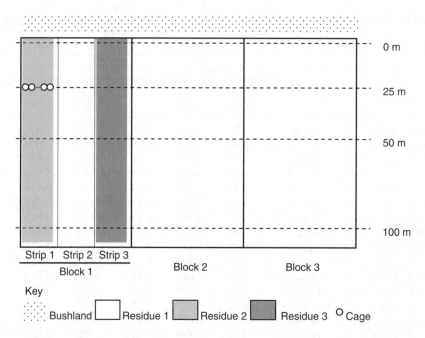

Figure 7.7 Layout of an experiment to investigate influences on predation of seeds. Crop residues are shown only in Block 1. Cages are shown only at 25 m from the bush in Strip 1 of Block 1.

    (d) Make this change to each mixed model and repeat the analysis. Is the final value of each response variate related to its initial value? Does the adjustment for the initial value give any improvement in the precision with which the effects of the treatments are estimated?

7.4 An experiment was performed to investigate the factors that influence the predation of seeds lying on the ground in an area on which a crop has been grown. Seeds of three species were placed on the ground in cages of four designs that excluded different types of predators, in areas on which three types of crop residue were present, and at four distances from an area of bushland. The levels of these treatment factors are given in Table 7.11. This experimental area was divided into three blocks, and each block was further divided into three strips. Each strip had a boundary with the bushland at one end. Each strip received a crop residue treatment, each treatment being applied to one randomly chosen strip in each block. In each strip, one cage of each design was placed at each distance from the edge of the bushland. At each distance, the four cages in each strip were positioned at random. Each cage contained seeds of all three species. After 15 days, the number of seeds of each species that remained in each cage was recorded, and the percent predation was calculated. The layout of the experiment is illustrated in Figure 7.7.

    The first and last few rows of the spreadsheet holding the data are presented in Table 7.12: the full data set is held in the file 'seed predation.xlsx' on this book's website (see Preface). (Data reproduced by kind permission of Robert Gallagher.)

Distance, residue, cage_type and species are to be classified as treatment factors.

(a) Identify any treatment factors for which randomization cannot be performed.

Table 7.11 Factors that were varied in an experiment to investigate influences on predation of seeds.

		Factor				
species		cage_type		distance (m)	residue	
level	label	level no.	level		level	level no.
wild oat (20 seeds/cage)	wo	1	no exclusion	0	no residue	1
wild radish (20 seeds/cage)	wr	2	exclusion by 2 cm mesh	25	standing wheat stubble	2
annual ryegrass (90 seeds/cage)	arg	3	exclusion by 1 cm mesh	50	standing wheat stubble + cut wheat straw (i.e. cages obscured by straw)	3
		4	exclusion by 'tac' gel around sides, but top remained open	100		

Table 7.12   Percentage of the predation of seeds of three species at different levels of three experimental variables.

	A	B	C	D	E	F
1	distance	block	residue	cage_type	species	%pred
2	0	1	2	1	wo	100.0
3	0	1	2	2	wo	95.0
4	0	1	2	3	wo	30.0
5	0	1	2	4	wo	10.0
6	0	1	1	1	wo	95.0
7	0	1	1	2	wo	100.0
8	0	1	1	3	wo	100.0
9	0	1	1	4	wo	10.0
10	0	1	3	1	wo	100.0
11	0	1	3	2	wo	30.0
12	0	1	3	3	wo	100.0
13	0	1	3	4	wo	10.0
14	0	2	3	1	wo	10.0
15	0	2	3	2	wo	0.0
16	0	2	3	3	wo	5.0
17	0	2	3	4	wo	0.0
⋮						⋮
430	100	3	3	1	arg	100.0
431	100	3	3	2	arg	12.2
432	100	3	3	3	arg	23.3
433	100	3	3	4	arg	11.1

Source: Data reproduced by kind permission of Robert Gallagher.

(b) Determine the block and treatment structures of this experiment.

(c) Analyse the data by analysis of variance.

(d) Perform an equivalent analysis on the data by mixed modelling and obtain diagnostic plots of the residuals. Note any indications that the assumptions underlying the analysis may not be fulfilled and consider why this may be the case.

(e) Create a variate holding the same values as the factor 'distance', and add this to the mixed model. Consider whether the factor 'distance' should now be specified as a fixed-effect term or a random-effect term. Consider also the consequences of your decision with regard to the interaction of 'distance' with other terms. Fit your new model to the data.

(f) Assuming that any departures from the assumptions underlying analysis of variance and mixed modelling are not so serious as to render these analyses invalid, interpret the results of the analysis performed in Section (e). In particular, consider the following points:

(i) Is there a linear trend in the level of predation over distance? If so, does the level increase or decrease with increasing distance from the bushland?

(ii) Does this trend vary among the species? If so, what is the nature of this variation?

(iii) Does this trend vary according to the crop residue? If so, what is the nature of this variation?

(iv) Does this trend vary according to the cage type? If so, what is the nature of this variation?

(v) Is there evidence of variation among the levels of distance that cannot be accounted for by the linear trend?

(vi) Is there evidence of three-way or four-way interactions among the treatment factors?

(g) Obtain predicted values of predation at representative distances from the bushland for each species, crop residue and cage type. Make a graphical display of the relationship between these variables.

A refinement of this analysis, taking account of the problems identified in the diagnostic plots, is introduced in Exercise 10.2. These data have also been analysed and interpreted by Spafford-Jacob *et al.* (2005).

7.5 An experiment was conducted to determine the effect of anoxia (lack of oxygen) on the porosity of roots in nine genotypes of wheat. At an early stage of development, control plants of each genotype were grown in a well-aerated solution, while treated plants of each genotype were grown in stagnant agar which deprived the roots of oxygen. Following this treatment, the plants were grown in pots on a glasshouse bench, in a randomized complete block design with three replications. At the end of the experiment, seminal roots (i.e. those that develop directly from the seed) and nodal roots (i.e. those that originate from a node of the plant at a later stage) were harvested from each plant, and their porosity (% air space per root volume) was measured. On three additional plants of each genotype, the porosity of the seminal root was measured at an early stage of development, prior to the stagnant agar/aerated solution treatment. One of these three plants was assigned to each of the three blocks. The first and last few rows of the spreadsheet holding the data are presented in Table 7.13: the full data set is held in the file 'root porosity.xlsx' on this book's website (see Preface). (Data reproduced by kind permission of Michael McDonald.)

Genotype, treatment and root type are to be classified as treatment factors.

(a) Determine the block and treatment structures of this experiment.

(b) Analyse the response variable 'porosity_final' by analysis of variance and by mixed modelling.

The term 'porosity_initial' can be added to the mixed model.

(c) Should this term be placed in the fixed-effect or the random-effect model?

(d) Make this change to the mixed model and repeat the analysis. Is the final value of porosity related to its initial value? Does the adjustment for the initial value give any improvement in the precision with which the effects of the treatments are estimated?

(e) Interpret the results of your analysis. In particular, consider the following points:

Table 7.13   Porosity of roots of a range of wheat genotypes with and without exposure to anoxia.

	A	B	C	D	E	F	G
1	block	plant	genotype	treatment	root_type	porosity_initial	porosity_final
2	1	1	1	anoxic	seminal	3.572	4.630
3	1	1	1	anoxic	nodal	3.572	17.145
4	1	2	7	anoxic	seminal	2.722	2.589
5	1	2	7	anoxic	nodal	2.722	19.791
6	1	3	9	anoxic	seminal	3.477	7.048
7	1	3	9	anoxic	nodal	3.477	17.819
8	1	4	2	anoxic	seminal	3.341	5.110
9	1	4	2	anoxic	nodal	3.341	20.098
10	1	5	5	anoxic	seminal	0.904	0.481
11	1	5	5	anoxic	nodal	0.904	20.040
12	1	6	3	anoxic	seminal	2.701	6.204
13	1	6	3	anoxic	nodal	2.701	19.670
14	1	7	1	anoxic	seminal	3.572	1.146
15	1	7	1	anoxic	nodal	3.572	3.109
⋮	⋮					⋮	
106	3	17	5	anoxic	seminal	2.562	2.469
107	3	17	5	anoxic	nodal	2.562	16.963
108	3	18	5	control	seminal	2.562	1.754
109	3	18	5	control	nodal	2.562	4.435

Source: Data reproduced by kind permission of Michael McDonald.

(i) Is the porosity of roots affected by anoxia, and does it vary among genotypes and root types? If so, what is the direction of the effects of anoxia and root type? Which genotype has the most porous roots, and which the least?

(ii) Is there evidence of two-way interactions between anoxia, root type and genotype? If so, what is the nature of these interactions?

(iii) Is there evidence of a three-way interaction among these factors?

These data have also been analysed and interpreted by McDonald et al. (2001).

7.6 An experiment was performed to determine the effect of nutritionally inadequate diets on the quality of sheep's wool. The experiment consisted of a 21-day pre-experimental period (ending on 22 October 1997), a 12-week restricted-intake period (ending on 14 January 1998) and a 6-week recovery period (ending on 25 February). Three experimental treatments were defined, namely

– Treatment 1. A control dietary regime.
– Treatment 2. An energy-deficient/protein-deficient nutritional regime.
– Treatment 3. An energy-deficient/protein-adequate nutritional regime.

Table 7.14 Live weight and fibre diameter of sheep on a range of nutritional regimes.

Tag	Treatment	Live weight (kg), 20 October 1997	Fibre diameter (µm)			
			22 October 1997	7 January 1998	28 January 1998	18 February 1998
100	3	36.5	20.44	13.92	15.43	19.90
170	2	33.5	16.62	16.38	17.09	16.68
1519	1	34.0	18.12	16.30	16.74	19.22
162	1	35.0	18.55	15.74	15.25	18.02
1582	3	32.0	19.33	15.77	16.41	19.22
175	2	35.0	18.26	16.85	17.47	18.08
124	1	35.0	20.35	19.39	18.94	21.68
95	1	33.0	18.27	16.30	16.95	19.70
1497	2	35.0	17.95	18.55	19.78	20.01
1337	3	34.0	19.12	14.78	16.52	19.27
62	1	31.5	20.88	17.58	18.52	21.22
1420	3	33.5	17.05	13.18	14.35	17.12
171	2	33.0	18.21	17.77	18.44	19.14
1556	2	33.5	18.05	16.59	17.55	17.93
167	1	35.5	18.16	14.67	17.61	20.87
1048	1	33.0	20.63	14.93	15.58	19.40
1083	3	35.0	17.97	13.64	14.86	18.66
1422	2	35.5	17.70	16.69	17.74	18.64
1042	2	31.5	18.41	16.90	17.96	19.18
1065	3	32.5	18.57	14.64	15.86	19.09
142	3	34.0	20.08	16.20	18.48	21.71

Source: Data reproduced by kind permission of Rachel Kirby.

Animals in the control group were fed at 1.3 times the level required for maintenance (1.3 M) throughout the experiment. The energy-deficient/protein-deficient group were fed at 0.75 M during the restricted-intake period, then *ad libitum* up to 2 M during the recovery period. The energy-deficient/protein-adequate group received the same diet as the energy-deficient/protein-deficient group except that they received a supplement of 3.3 g/day of active methionine throughout the experimental period.

Twenty-one Merino wether hoggets, each identified by a tag, were used in the experiment. Following the 21-day adjustment period, the animals were divided into seven groups on the basis of live weight, and one randomly chosen animal from each group was allocated to each treatment. Measurements of the live weight and the fibre diameter of the wool of each animal were taken at the end of the pre-experimental period, and measurements of fibre diameter were also taken during the recovery period. The results are presented in Table 7.14. (Data reproduced by kind permission of Rachel Kirby.)

The experimental design includes a block factor which is not shown explicitly in Table 7.14.

(a) Assign an appropriate block number to each row of the data.

(b) Analysing measurements made on each occasion separately, and using the block factor and the treatment factor as model terms, perform analyses of variance on the measurements of fibre diameter made during the restricted-intake period and the recovery period. Interpret the results.

(c) Stack the measurements made on these three occasions into a single variate and perform a new analysis by mixed modelling. Include the following variables in your initial model:

(i) The time at which the measurement was made.

(ii) The fibre diameter from the same sheep during the pre-experimental period.

(iii) The live weight of the sheep during the pre-experimental period.

Determine whether any terms can reasonably be dropped from the model. If so, make the appropriate changes and re-fit the model.

(d) Interpret the results of your analysis. In particular, consider the following points:

(i) Is the fibre diameter influenced by the sheep's pre-experiment live weight?

(ii) Is the fibre diameter during the experiment related to the fibre diameter in the pre-experimental period?

(iii) Is the fibre diameter influenced by the treatment, the date (i.e. the time elapsed since the restricted-intake period) and the treatment × date interaction? If so, what is the direction or nature of these effects?

(e) Obtain predicted values of fibre diameter for each treatment at representative times and make a graphical display of the relationship between these variables.

7.7 The data on the yield of F_3 families of wheat presented in Exercise 3.3 in Chapter 3 are a subset from a larger investigation, in which two crosses were studied, and the F_3 families were grown with and without competition from ryegrass. The experiment had a split-split plot design, with the following relationship between the block and treatment structures:

- 'cross' was the treatment factor that varied only among main plots
- 'family' was the treatment factor that varied among sub-plots within each main plot, but not among sub-sub-plots within the same sub-plot
- presence or absence of ryegrass was the treatment factor that varied among sub-sub-plots within the same sub-plot.

The mean grain yield per plant was determined in each sub-sub-plot. (When the subset of the data is considered in isolation, it has a randomized block design, as described in the earlier exercise.) The first and last few rows of the spreadsheet holding the data are presented in Table 7.15: the full data set is held in the file 'wheat with ryegrass.xlsx' on this book's website (see Preface). (Data reproduced by kind permission of Soheila Mokhtari.)

(a) Analyse the data according to the experimental design, both by analysis of variance and by mixed modelling.

(b) Modify your mixed model so that family-within-cross is specified as a random-effect term.

(c) Interpret the results of your analysis. In particular, consider the following points:

(i) Does the presence of ryegrass affect the yield of the wheat plants? If so, in which direction?

(ii) Is there a difference in mean yield between the crosses?

(iii) Do the families within each cross vary in yield?

(iv) Do the effects of crosses and families interact with the effect of ryegrass?

Table 7.15 Yield per plant (g) of F_3 families from two crosses between inbred lines of wheat, grown with and without competition from ryegrass in a split-split plot design. 0 = ryegrass absent; 1 = ryegrass present.

	A	B	C	D	E	F	G	H
1	rep	mainplot	subplot	subsubplot	cross	family	ryegrass	yield
2	1	1	1	1	1	29	0	15.483
3	1	1	1	2	1	29	1	5.333
4	1	1	2	1	1	26	0	13.4
5	1	1	2	2	1	26	1	3.483
6	1	1	3	1	1	40	0	11.817
7	1	1	3	2	1	40	1	3.583
⋮								⋮
382	2	2	47	1	2	4	0	7.55
383	2	2	47	2	2	4	1	4.02
384	2	2	48	1	2	15	0	6.867
385	2	2	48	2	2	15	1	3.183

(d) Compare the results with those that you obtained from the subset of the data in Chapter 3.

7.8 Return to the data set concerning the vernalization of F_3 chickpea families, introduced in Exercise 5.2 in Chapter 5.

(a) Use analysis of variance methods to obtain the estimated mean number of days to the flowering of vernalized plants of each F_3 family.

(b) Create a new variable, with one value for each control plant, giving the corresponding unshrunk family mean for vernalized plants.

(c) Fit a mixed model in which the response variable is the number of days to the flowering of the control plants, and the model terms are

  – the corresponding family-mean value for vernalized plants
  – family
  – plant group within family and
  – plant within group.

Interpret the results.

(d) Obtain estimates of the constant and of the effect of days to flowering in vernalized plants.

(e) Display the results of your analysis graphically, showing the linear trend relating the number of days to flowering in the control plants to that in the vernalized plants and the mean values of these variables for each family.

(f) Obtain estimates of the amount of variation in the number of days to the flowering of the control plants that is accounted for by each term in the mixed model.

(g) How can the analysis be interpreted to give an estimate of the effect of omitting the low-temperature stimulus on the subsequent development of plants in each family?

References

Aggarwal, S.P., Zinman, L., Simpson, E. *et al.* (2010) Safety and efficacy of lithium in combination with riluzole for treatment of amyotrophic lateral sclerosis: a randomised, double-blind, placebo-controlled trial. *Lancet Neurology*, **9**, 481–488.

Draper, N.R. and Smith, H. (1998) *Applied Regression Analysis*, 3rd edn, John Wiley and Sons, Inc., New York, 706 pp.

McCullagh, P. and Nelder, J.A. (1989) *Generalized Linear Models*, 2nd edn, Chapman & Hall, London, 511 pp.

McDonald, M.P., Galwey, N.W., Ellneskog-Staam, P. and Colmer, T.D. (2001) Evaluation of *Lophopyrum elongatum* as a source of genetic diversity to increase the waterlogging tolerance of hexaploid wheat (*Triticum aestivum*). *New Phytologist*, **151**, 369–380.

Ralston, S.H., Galwey, N., MacKay, I. *et al.* (2005) Loci for regulation of bone mineral density in men and women identified by genome wide linkage scan: the FAMOS study. *Human Molecular Genetics*, **14**, 943–951.

Si, P. and Walton, G.H. (2004) Determinants of oil concentration and seed yield in canola and Indian mustard in the lower rainfall areas of Western Australia. *Australian Journal of Agricultural Research*, **55**, 367–377.

Spafford-Jacob, H., Minkey, D., Gallagher, R. and Borger, C. (2005) Variation in post-dispersal weed seed predation in a cropping field. *Weed Science*, **54**, 148–155.

8

Meta-analysis and the multiple testing problem

8.1 Meta-analysis: Combined analysis of a set of studies

It has been noted that 'Evaluation of health-care interventions rarely concerns a single summary statistic. "Multiplicity" is everywhere' (Spiegelhalter, Abrams and Myles, 2004, Section 3.17, p. 91). The same is true of most other areas of research. This chapter explores two types of multiplicity, namely

- the synthesis in a meta-analysis of multiple studies that estimate the same response variable and

- the combined interpretation of multiple response variables measured in a single study,

and shows how mixed models can assist the interpretation of the results that they produce.

When several studies have been performed addressing the same question, it is natural to seek to assess the strength of the combined evidence. This situation commonly occurs in medical research, when a number of clinical trials have been conducted to assess the same treatment. None of the trials may provide sufficient evidence on its own to conclude that the treatment is effective, but the combined evidence may nevertheless permit a confident conclusion to be reached. There may be a similar case for the combined analysis of a set of agricultural field experiments, or a set of studies in any other discipline. Such a combined analysis is known as a *meta-analysis*. Meta-analyses are becoming increasingly common, particularly in medical research, motivated by the need to obtain

- the fullest possible value from clinical trials, often conducted at great expense,

- reliable estimates of the effects of marginal improvements in treatment, which are a much commoner result than spectacular, easily evaluated breakthroughs and

Introduction to Mixed Modelling: Beyond Regression and Analysis of Variance, Second Edition. N. W. Galwey.
© 2014 John Wiley & Sons, Ltd. Published 2014 by John Wiley & Sons, Ltd.
Companion website: http://www.wiley.com/go/beyond_regression

- evidence concerning a wider range of outcomes, including unanticipated side-effects or benefits, which may not be clearly detectable in an individual trial.

The studies included in a meta-analysis will usually be assumed to be representative of a larger population of studies that were, or might have been, conducted, which suggests that a mixed-modelling approach may be appropriate. However, it is first necessary to consider whether this assumption is reasonable. For example, if only the results of studies that show a significant effect in the expected direction are published, and if only published studies are included in the meta-analysis, then the combined estimate of the effect size will be biased upwards. The detection and avoidance of bias in the choice of studies for meta-analysis is the subject of a considerable literature, for example the pioneering study of Simes (1986) and the more recent review by Dwan *et al.* (2008), but will not be pursued further here.

If we decide to proceed with the analysis, we should consider the possibility that the true effect of the treatment varies among the studies in the sample, in which case we can take account of this source of variation by specifying the study × treatment interaction as a random-effect term. The main effect of treatment is usually specified as fixed, so it follows from the guidelines given in Section 6.3 that the main effect of study should also be specified as random. However, on closer investigation this decision turns out to be less straightforward than might be expected, as discussed below (Section 8.4), and we will therefore first consider the simpler situation in which study × treatment interaction is assumed to be absent. A good account of the various statistical methods that are appropriate for meta-analysis in different situations is given by Whitehead (2002): the connections between her account and the methods presented here will be noted.

8.2 Fixed-effect meta-analysis with estimation only of the main effect of treatment

We will examine the concepts of meta-analysis in the context of part of a set of clinical trials of antidepressant medications that were jointly analysed by Kirsch *et al.* (2008). In each trial, some patients, chosen at random, were allocated to an active antidepressant treatment (a selective serotonin reuptake inhibitor (SSRI) medication), and the remainder were allocated to a placebo treatment. The severity of each patient's depression was assessed on a numerical scale (the Hamilton Rating Scale of Depression, HRSD) at the time of randomization (the baseline assessment), and after several weeks of treatment, and the change in HRSD score from the baseline value was noted for each patient. A positive value of change indicates improvement, that is, a reduction in the degree of depression. The studies analysed by Kirsch *et al.* covered four antidepressant medications: those that investigated the compound fluoxetine are summarized in Table 8.1. (Results reproduced by permission under an open-access agreement.)

Ideally, a meta-analysis should be based on the individual-subject data from each of the contributing studies. It has not proved possible to obtain a data set of this type for presentation here: such data relating to human subjects are highly sensitive, due to the requirements of patient confidentiality. However, a set of simulated individual-patient data has been produced which gives the summary-statistic values presented in Table 8.1, and these simulated data will

Table 8.1 Summary data from a meta-analysis of clinical trials of the antidepressant fluoxetine. The meanings of the column headings are as follows: study = unique identifier of study; change1 = mean change from baseline in score on HRSD scale in patients in active-treatment arm; n1 = number of patients in active arm; sd1 = standard deviation of the sample of values used to calculate change1; change2, n2, sd2 = corresponding variables for patients in the placebo-treated arm. The meanings of diff, s2, se_diff and weight are explained in the text in Section 8.7.

	A	B	C	D	E	F	G	H	I	J	K
1	study	change1	n1	sd1	change2	n2	sd2	diff	s2	se_diff	weight
2	19	12.50	22	8.6806	5.50	24	8.7302	7.00	75.8035	2.5698	0.1514
3	25	7.20	18	8.6747	8.80	24	8.5437	−1.60	73.9533	2.6814	0.1391
4	27	11.00	181	9.5652	8.40	163	9.5455	2.60	91.3145	1.0318	0.9392
5	62 (mild)	5.89	299	5.7745	5.82	56	5.5429	0.07	32.9365	0.8356	1.4320
6	62 (moderate)	8.82	297	7.8053	5.69	48	7.9028	3.13	61.1327	1.2163	0.6759

Source: Results reproduced by permission under an open-access agreement.

Table 8.2 Simulated individual-patient data from the clinical studies of fluoxetine summarized in Table 8.1.

Treatment 0 = placebo; 1 = active. The meanings of the column headings b and w are explained in the text in Section 8.3.

	A	B	C	D	E	F
		study!	patient	treatment!	b	w	change
1	study!	patient	treatment!	b	w	change
2	19	1	0	−0.02174	−0.47826	16.75
3	19	2	1	−0.02174	0.52174	31.74
4	19	3	1	−0.02174	0.52174	19.94
5	19	4	0	−0.02174	−0.47826	−1.80
	.					.
	.					.
	.					.
1131	62 (moderate)	343	1	0.360869565	0.139130435	17.37
1132	62 (moderate)	344	1	0.360869565	0.139130435	17.54
1133	62 (moderate)	345	1	0.360869565	0.139130435	4.23

be used to illustrate the process of individual-subject meta-analysis. The first and last few lines of this data set are presented in Table 8.2.

The simplest way to analyse these data is with the following model (Model 8.1):

Response variate:	change
Fixed-effect model:	study + treatment
Random-effect model:	none

This is the fixed-effects meta-analysis model presented by Whitehead (2002, Sections 5.2.1–5.2.3, pp. 100–103). It is fitted using the following GenStat statements:

```
IMPORT [FORDER = unsorted] \
   'IMM edn 2\\Ch 8\\ssri indiv patient.xlsx'; \
   SHEET = 'fluoxetine'
VCOMPONENTS [FIXED = study + treatment]
REML [PRINT = model, components, Wald, means, effects, deviance; \
   PTERMS = treatment] change
```

The option setting 'FORDER = unsorted' in the IMPORT statement ensures that the levels of the factors 'study' and 'treatment' are retained in the order in which they occur in the data, rather than being sorted into alphanumeric order. The output of the REML statement is as follows:

REML variance components analysis

Response variate: change
Fixed model: Constant + study + treatment
Number of units: 1132

Residual term has been added to model

Sparse algorithm with AI (average information) optimization

Residual variance model

Term	Model(order)	Parameter	Estimate	s.e.
Residual	Identity	Sigma2	62.87	2.65

Deviance: −2*Log-Likelihood

Deviance	d.f.
5819.28	1125

Note: deviance omits constants which depend on fixed model fitted.

Tests for fixed effects

Sequentially adding terms to fixed model

Fixed term	Wald statistic	n.d.f.	F statistic	d.d.f.	F pr.
study	43.83	4	10.96	1126.0	<0.001
treatment	14.28	1	14.28	1126.0	<0.001

Dropping individual terms from full fixed model

Fixed term	Wald statistic	n.d.f.	F statistic	d.d.f.	F pr.
study	54.81	4	13.70	1126.0	<0.001
treatment	14.28	1	14.28	1126.0	<0.001

Message: denominator degrees of freedom for approximate F-tests are calculated using algebraic derivatives ignoring fixed/boundary/singular variance parameters.

Table of effects for treatment

treatment	0	1
	0.000	2.140

Standard error of differences: 0.5663

Table of predicted means for treatment

treatment	0	1
	6.856	8.997

Standard error of differences: 0.5663

The F statistics indicate that there is a highly significant effect of antidepressant treatment on the patients' change from baseline on the HRSD scale. The table of predicted means gives the mean change in score during the trial for patients on each treatment arm (0 = placebo; 1 = active), and the table of effects for treatment shows that the mean difference between the effect of the placebo and active treatments is 2.140 units on the HRSD scale: that is, patients on the active-treatment study arm show a greater increase in their score, on average, than those on the placebo-treated arm. (It is a common finding in clinical trials that even patients on the placebo-treated arm show some improvement.)

However, the F statistics also indicate that there is a highly significant variation among the studies in the magnitude of the change. This leads us to ask whether the effect of treatment varies between the studies, and we can obtain an initial answer to this question by adding study.treatment interaction as a fixed-effect term to the model, as follows (Model 8.2):

Response variate:	change
Fixed-effect model:	study + treatment + study.treatment
Random-effect model:	none

The following statements fit the resulting model:

```
VCOMPONENTS [FIXED = study * treatment]
REML [PRINT = model, components, Wald, means; \
   PTERMS = study.treatment] change
```

The output of the REML statement is as follows:

REML variance components analysis

Response variate:	change
Fixed model:	Constant + study + treatment + study.treatment
Number of units:	1132

Residual term has been added to model

Sparse algorithm with AI optimization

Residual variance model

Term	Model(order)	Parameter	Estimate	s.e.
Residual	Identity	Sigma2	62.49	2.64

Tests for fixed effects

Sequentially adding terms to fixed model

Fixed term	Wald statistic	n.d.f.	F statistic	d.d.f.	F pr.
study	44.10	4	11.02	1122.0	<0.001
treatment	14.37	1	14.37	1122.0	<0.001
study.treatment	10.81	4	2.70	1122.0	0.029

Dropping individual terms from full fixed model

Fixed term	Wald statistic	n.d.f.	F statistic	d.d.f.	F pr.
study.treatment	10.81	4	2.70	1122.0	0.029

Message: denominator degrees of freedom for approximate F-tests are calculated using algebraic derivatives ignoring fixed/boundary/singular variance parameters.

Table of predicted means for study.treatment

treatment	0	1
study		
19	5.501	12.499
25	8.800	7.200
27	8.400	11.000
62 (mild)	5.820	5.890
62 (moderate)	5.690	8.820

Standard errors of differences

Average:	1.663
Maximum:	2.512
Minimum:	0.6476

Average variance of differences: 3.017

Standard error of differences for the same level of factor:

	study	treatment
Average:	1.607	1.667
Maximum:	2.465	2.512
Minimum:	0.8536	0.6476

Average variance of differences:

3.017	3.017

The F statistics indicate that the study.treatment interaction effect is just significant. Inspection of the means shows that the observed treatment effect varies from $12.499 - 5.501 = 6.998$ in Study 19 to $7.2 - 8.8 = -1.6$ in Study 25 and suggests that the possibility of interaction is worth investigating further.

8.3   Random-effects meta-analysis with estimation of study × treatment interaction effects

If the studies are regarded as a representative sample from an effectively infinite population of possible studies, it is natural to specify 'study' as a random-effect term, and following the guidelines given in Section 6.3, study.treatment interaction is then specified as a random-effect term. This gives the following model (Model 8.3):

Response variate: change
Fixed-effect model: treatment
Random-effect model: study + study.treatment

The appropriate VCOMPONENTS statement is then:

```
VCOMPONENTS [FIXED = treatment] RANDOM = study + study.treatment
```

The output of the REML statement is then as follows:

REML variance components analysis

Response variate: change
Fixed model: Constant + treatment
Random model: study + study.treatment
Number of units: 1132

Residual term has been added to model

Sparse algorithm with AI optimization

Estimated variance components

Random term	Component	s.e.
study	1.24	2.43
study.treatment	2.20	2.60

Residual variance model

Term	Model(order)	Parameter	Estimate	s.e.
Residual	Identity	Sigma2	62.51	2.64

Tests for fixed effects

Sequentially adding terms to fixed model

Fixed term	Wald statistic	n.d.f.	F statistic	d.d.f.	F pr.
treatment	3.15	1	3.15	2.9	0.177

Dropping individual terms from full fixed model

Fixed term	Wald statistic	n.d.f.	F statistic	d.d.f.	F pr.
treatment	3.15	1	3.15	2.9	0.177

Message: denominator degrees of freedom for approximate F-tests are calculated using algebraic derivatives ignoring fixed/boundary/singular variance parameters.

Table of effects for treatment

treatment	0	1
	0.000	2.063

Standard error of differences: 1.163

Table of predicted means for treatment

treatment	0	1
	6.881	8.944

Standard error of differences: 1.163

The main consequence of these changes in the model is a substantial increase in the SE (standard error) of the main effect of 'treatment', from 0.5663 to 1.163. This is because the component of variance due to study.treatment interaction now makes its proper contribution to this SE, just as in Chapter 1 it was necessary to specify 'town' as a random-effect term in order for this term to make its proper contribution to the SE of the effect of 'latitude' (Sections 1.8 and 4.2). But there is also a slight change in the estimate of the main effect of 'treatment', from 2.140 to 2.063. This is perhaps unexpected: the role of the studies in a meta-analysis is very similar to the role of the blocks in a randomized complete block design, and in the latter design it makes no difference to the estimate of the treatment effects whether block is specified as a fixed- or a random-effect term, and whether or not a block.treatment term is specified in the model. There are two causes of this change. Firstly, the specification of a random-effect study.treatment interaction term changes the weight given to each study in the calculation of the estimate, giving relatively more weight to smaller studies. When 'study' is specified as a fixed-effect term, the weight given to each study depends solely on its number of observations, whereas the study.treatment interaction variance component contributes equally to the weight of every study. Secondly, whereas the variation among the block means in a randomized complete block design gives no information on the treatment effects, each treatment occurring the same number of times in each block, in the present case the variation among the study means may be partly due to the effect of the treatment, as the active treatment accounts for a much greater proportion of the observations in studies 62 (mild) and 62 (moderate) than in the other studies. However, the mean values of 'change' in these studies are not sufficiently distinctive for this phenomenon to have a strong influence on the estimate of the main effect of 'treatment'. So in the present example, the change in the estimate due to the inclusion of the random-effect terms 'study' and 'study.treatment' is slight, but this will not always be the case. This effect has no influence when 'study' is specified as a fixed-effect term: the effect of treatment is then estimated *conditional on* study, that is, the difference between the means of different treatments *in observations in the same study* is estimated.

8.4   A random-effect interaction between two fixed-effect terms

There is a consensus that in a meta-analysis of randomized trials, like the present one, the pooled estimate of the treatment effect should be based on within-study effects only. Within each study, care has been taken to eliminate systematic bias by randomizing the allocation of patients to treatments. But between studies there is no such defence against bias. It is possible that the rather low mean values of change in study '62 (mild)', in which a large proportion of patients $(299/(299 + 56) = 0.84)$ is allocated to the active treatment, are no accident: perhaps clinical trials conducted on patients who have little capacity for response to treatment generally have a large proportion of patients allocated to the active treatment. The allocation of patients to treatments would then be *partially confounded* with another variable, and the estimate of the treatment effect would be biased. In unrandomized studies, such as many epidemiological studies of the effect of exposure to a specified risk, the case for basing estimation on within-study effects is much less clear. Much of the available information may come from comparison of exposed individuals in one study with unexposed individuals in another, and the assumption that confounding is unimportant may be worth making so that this information will not be wasted.

It is possible to obtain an estimate of the treatment effect based entirely on comparisons between observations in the same study, while still allowing for the possibility that a random study × treatment interaction is present. The most explicit way of doing this is to partition the specification of each patient's treatment between variables representing between- and within-study components: the variables 'b' and 'w' in Table 8.2. This approach is illustrated in Exercise 8.1 at the end of this chapter. However, it is a clumsy approach and becomes much more so if there are more than two treatments. But there is a simpler way to obtain the required effect estimates, namely to specify 'study' as a fixed-effect term, giving the following model (Model 8.4):

Response variate: change
Fixed-effect model: study + treatment
Random-effect model: study.treatment

We then have a model in which study.treatment is specified as a random-effect term although both the corresponding main-effect terms are specified as fixed, which cannot be justified on the basis of the guidelines given in Section 6.3. If the studies comprise a fixed set of 35, and the treatments a fixed set of 2, then the possible study.treatment combinations are limited to 70 – not an infinite population. Nevertheless, the model gives the required estimate of the main effect of treatment, and it can be justified on the grounds that when a variable is specified as a fixed-effect term, the coefficient of any other fixed-effect term represents the effect of the latter variable when the former is held constant. By specifying 'study' as a fixed-effect term, we wish only to indicate that the effect of 'treatment' should be estimated with 'study' held constant – that is, we are interested in the difference between the means of the active and placebo treatments *in individuals in the same study*. The studies may nevertheless be a random sample from a large population and the within-study treatment effect may vary between studies, in which case we want such variation to contribute to the SE of this effect, and study.treatment must be specified as random in order to achieve this. Alternatively, we can justify the specification of a limited set of study.treatment combinations as a random-effect term by the 'empirical-Bayesian' approach (Spiegelhalter, Abrams and Myles, 2004, Sections 3.17 and 3.20), using the concept of exchangeability (Sections 1.6 and 5.6).

Model 8.4 is the random-effects model for normally distributed data presented by Whitehead (2002, Section 5.8, pp. 131–136). The appropriate mixed modelling statements are then as follows:

```
VCOMPONENTS [FIXED = study, treatment] RANDOM = study.treatment
REML [PRINT = model, components, Wald, means, effects, deviance; \
    PTERMS = treatment] change
```

The output of the REML statement is as follows:

REML variance components analysis

Response variate:	change
Fixed model:	Constant + study + treatment
Random model:	study.treatment
Number of units:	1132

Residual term has been added to model

Sparse algorithm with AI optimization

Estimated variance components

Random term	Component	s.e.
study.treatment	1.83	2.35

Residual variance model

Term	Model(order)	Parameter	Estimate	s.e.
Residual	Identity	Sigma2	62.56	2.64

Deviance: −2*Log-Likelihood

Deviance	d.f.
5817.63	1124

Note: deviance omits constants which depend on fixed model fitted.

Tests for fixed effects

Sequentially adding terms to fixed model

Fixed term	Wald statistic	n.d.f.	F statistic	d.d.f.	F pr.
study	6.95	4	1.58	3.1	0.364
treatment	3.95	1	3.95	2.7	0.150

Dropping individual terms from full fixed model

Fixed term	Wald statistic	n.d.f.	F statistic	d.d.f.	F pr.
study	7.78	4	1.77	3.1	0.329
treatment	3.95	1	3.95	2.7	0.150

Message: denominator degrees of freedom for approximate F-tests are calculated using algebraic derivatives ignoring fixed/boundary/singular variance parameters.

Table of effects for treatment

treatment	0	1
	0.000	2.177

Standard error of differences: 1.096

Table of predicted means for treatment

treatment	0	1
	6.877	9.054

Standard error of differences: 1.096

The estimate of the main effect of 'treatment', 2.177, is slightly larger than the value of 2.063 obtained when variation among the study means was allowed to contribute, and the estimate of the study.treatment variance component is also changed, from 2.20 to 1.83.

The study.treatment term is now the only difference between our current model and the original fixed-effects model presented in Section 8.2. Hence its significance can be tested using the test for the addition of a single random-effect term presented in Section 3.12, as an alternative to the F test of its significance as a fixed-effect term presented in Section 8.2. Comparing the output with and without study.treatment included as a random-effect term we obtain

$$\text{deviance}_{\text{reduced model}} - \text{deviance}_{\text{full model}} = 5819.28 - 5817.63 = 1.65,$$

and

$$P(\chi_1^2 > 1.65) = 0.20.$$

Hence

$$P(\text{deviance}_{\text{reduced model}} - \text{deviance}_{\text{full model}} > 1.65) = \frac{1}{2} \times 0.20 = 0.10.$$

On this basis the study.treatment variance component is not quite significant, but the F statistic should be preferred (see Section 3.13).

We can change the settings of the last REML statement so as to obtain the study.treatment means produced by this analysis, as follows:

```
REML [PRINT = means; PTERMS = treatment + study.treatment] change
```

The output of this statement is as follows:

Table of predicted means for treatment

treatment   0   1
                  6.877   9.054

Standard error of differences: 1.096

Table of predicted means for study.treatment

treatment	0	1
study		
19	6.879	10.995
25	7.790	8.546
27	8.437	10.967
62 (mild)	5.349	5.978
62 (moderate)	5.930	8.781

Standard errors of differences of means cannot be formed using random terms when METHOD = sparse.

The treatment effect within each study calculated from these shrunk means can be compared with that calculated from the unshrunk means presented in Section 8.2. For example, for Study 27, the unshrunk means give

$$\text{mean(active treatment)} - \text{mean(placebo)} = 11.000 - 8.400 = 2.600$$

whereas the shrunk means give

$$10.967 - 8.437 = 2.530.$$

The estimates of treatment effect obtained from shrunk and unshrunk means are compared with each other, and with the pooled estimate obtained from the most recent model, in Figure 8.1. As might be expected, the treatment-effect estimates obtained from shrunk means are themselves shrunk towards the pooled estimate, and those from small studies (e.g. Study 25) are more strongly shrunk than those from large studies (e.g. Study 62 (mild)). Our belief concerning the true effect in Study 19 is revised downward when we consider it in the context of the other studies, whereas our belief concerning the true effect in Study 25 is revised upward.

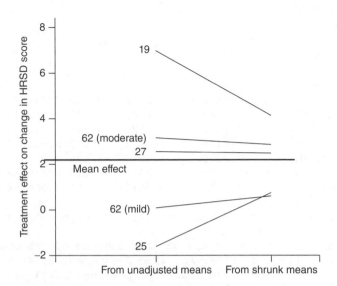

Figure 8.1   Estimates of the effect of treatment with antidepressant medication within each clinical study, obtained from unadjusted means, compared with estimates obtained from shrunk means.

8.5   Meta-analysis of individual-subject data using R

The following commands import the data, convert 'study' and 'treatment' to factors, and perform the fixed-effects meta-analysis with estimation only of the main effect of treatment (Model 8.1):

```
rm(list = ls())
ssri <- read.table(
    "IMM edn 2\\Ch 8\\ssri indiv patient, fluoxetine.txt",
    header=TRUE, sep = "\t")
attach(ssri)
fstudy <- factor(study)
ftreatment <- factor(treatment)
ssri.model1 <- lm(change ~ fstudy + ftreatment)
summary(ssri.model1)
anova(ssri.model1)
```

The output of the summary() and anova() functions is as follows:

```
Call:
lm(formula = change ~ fstudy + ftreatment)
```

```
Residuals:
      Min        1Q    Median        3Q       Max
 -28.2518   -5.3366    0.1429    5.4092   28.0281

Coefficients:
                       Estimate Std. Error t value Pr(>|t|)
(Intercept)              7.8243     1.2001   6.520 1.06e-10 ***
fstudy25                -0.6272     1.6925  -0.371 0.711023
fstudy27                 0.8175     1.2451   0.657 0.511603
fstudy62 (mild)         -3.7479     1.2595  -2.976 0.002986 **
fstudy62 (moderate)     -1.2819     1.2633  -1.015 0.310458
ftreatment1              2.1402     0.5663   3.779 0.000166 ***
---
Signif. codes:  0 '***' 0.001 '**' 0.01 '*' 0.05 '.' 0.1 ' ' 1

Residual standard error: 7.929 on 1126 degrees of freedom
Multiple R-squared: 0.04908, Adjusted R-squared: 0.04486
F-statistic: 11.62 on 5 and 1126 DF,  p-value: 5.723e-11

Analysis of Variance Table

Response: change
             Df Sum Sq Mean Sq F value     Pr(>F)
fstudy        4   2756  688.99  10.959 1.015e-08 ***
ftreatment    1    898  897.80  14.280 0.0001658 ***
Residuals  1126  70794   62.87
---
Signif. codes:  0 '***' 0.001 '**' 0.01 '*' 0.05 '.' 0.1 ' ' 1
```

The residual mean square agrees with that given by GenStat, as do the F statistics for 'fstudy' and 'ftreatment' when these terms are sequentially added to the fixed model, and, most importantly, the estimate of the treatment effect (reported as the coefficient of ftreatment1, 2.1402) and its SE (=0.5663).

The following commands fit the model in which the main effects of study and treatment and the study × treatment interaction are all specified as fixed-effect terms (Model 8.2):

```
ssri.model2 <- lm(change ~ fstudy * ftreatment)
summary(ssri.model2)
anova(ssri.model2)
```

The output of the summary() and anova() functions is as follows:

```
Call:
lm(formula = change ~ fstudy * ftreatment)
```

```
Residuals:
     Min      1Q   Median      3Q      Max
 -28.010   -5.130   0.050   5.187   27.810


Coefficients:
                                Estimate Std. Error t value Pr(>|t|)
(Intercept)                       5.5008     1.6137   3.409 0.000675 ***
fstudy25                          3.2992     2.2821   1.446 0.148543
fstudy27                          2.8989     1.7284   1.677 0.093774 .
fstudy62 (mild)                   0.3192     1.9287   0.165 0.868594
fstudy62 (moderate)               0.1894     1.9763   0.096 0.923679
ftreatment1                       6.9983     2.3334   2.999 0.002766 **
fstudy25:ftreatment1             -8.5983     3.3942  -2.533 0.011436 *
fstudy27:ftreatment1             -4.3982     2.4846  -1.770 0.076969 .
fstudy62 (mild):ftreatment1      -6.9283     2.6018  -2.663 0.007859 **
fstudy62 (moderate):ftreatment1  -3.8682     2.6376  -1.467 0.142773
---
Signif. codes:  0 '***' 0.001 '**' 0.01 '*' 0.05 '.' 0.1 ' ' 1


Residual standard error: 7.905 on 1122 degrees of freedom
Multiple R-squared: 0.05815, Adjusted R-squared: 0.0506
F-statistic: 7.697 on 9 and 1122 DF,  p-value: 4.849e-11


Analysis of Variance Table

Response: change
                  Df Sum Sq Mean Sq F value    Pr(>F)
fstudy             4   2756  688.99 11.0248 8.995e-09 ***
ftreatment         1    898  897.80 14.3663 0.0001585 ***
fstudy:ftreatment  4    676  168.89  2.7025 0.0293034 *
Residuals       1122  70118   62.49
---
Signif. codes:  0 '***' 0.001 '**' 0.01 '*' 0.05 '.' 0.1 ' ' 1
```

The F statistic obtained for the study × treatment interaction (the term fstudy:ftreatment) agrees with that given by GenStat.

The following commands fit the model in which both the main effect of study and the study × treatment interaction are specified as random-effect terms (Model 8.3):

```
library(lme4)
ssri.model3 <- lmer(change ~ ftreatment + (1|fstudy/ftreatment),
 data = ssri)
summary(ssri.model3)
```

The output from the `summary()` function is as follows:

```
Linear mixed model fit by REML
Formula: change ~ ftreatment + (1 | fstudy/ftreatment)
```

```
   Data: ssri
  AIC  BIC logLik deviance REMLdev
  7914 7939  -3952     7908     7904
Random effects:
 Groups              Name           Variance Std.Dev.
 ftreatment:fstudy (Intercept)  2.2015  1.4838
 fstudy            (Intercept)   1.2426  1.1147
 Residual                        62.5074 7.9062
Number of obs: 1132, groups: ftreatment:fstudy, 10; fstudy, 5


Fixed effects:
             Estimate Std. Error t value
(Intercept)   6.8810     0.9879   6.965
ftreatment1   2.0624     1.1626   1.774


Correlation of Fixed Effects:
            (Intr)
ftreatment1 -0.625
```

The estimate of the treatment effect (reported as the effect of ftreatment1, 2.0624) and its SE (=1.1626) agree with those given by GenStat, as do the estimated variance components for the study, study × treatment and residual terms.

The following commands fit the model in which the study × treatment interaction is specified as a random-effect term although the main effects of both study and treatment are specified as fixed-effect terms (Model 8.4):

```
ssri.model4 <- lmer(change ~ fstudy + ftreatment +
  (1|fstudy:ftreatment), data = ssri)
summary(ssri.model4)
anova(ssri.model4)
coef(ssri.model4)
```

The output of the summary(), anova() and coef() functions is as follows:

```
Linear mixed model fit by REML
Formula: change ~ fstudy + ftreatment + (1 | fstudy:ftreatment)
   Data: ssri
  AIC  BIC logLik deviance REMLdev
  7903 7943  -3944     7900     7887
Random effects:
 Groups             Name           Variance Std.Dev.
 fstudy:ftreatment (Intercept)  1.8331  1.3539
 Residual                        62.5564 7.9093
Number of obs: 1132, groups: fstudy:ftreatment, 10
```

```
Fixed effects:
                      Estimate Std. Error t value
(Intercept)             7.8490     1.6001   4.905
fstudy25               -0.7688     2.1663  -0.355
fstudy27                0.7644     1.8367   0.416
fstudy62 (mild)        -3.2740     1.8692  -1.752
fstudy62 (moderate)    -1.5813     1.8784  -0.842
ftreatment1             2.1765     1.0949   1.988


Correlation of Fixed Effects:
             (Intr) fstd25 fstd27 fstdy62(ml) fstdy62(md)
fstudy25     -0.662
fstudy27     -0.771  0.572
fstdy62(ml)  -0.741  0.561  0.664
fstdy62(md)  -0.734  0.558  0.660  0.652
ftreatment1  -0.333  0.016 -0.010 -0.061       -0.069


Analysis of Variance Table
            Df Sum Sq Mean Sq F value
fstudy       4 434.99  108.75  1.7384
ftreatment   1 240.75  240.75  3.8485


$'fstudy:ftreatment'
                 (Intercept)    fstudy25    fstudy27 fstudy62 (mild) fstudy62 (moderate)
19:0                6.879616 -0.7688386 0.7643863        -3.27403            -1.58133
19:1                8.818430 -0.7688386 0.7643863        -3.27403            -1.58133
25:0                8.559017 -0.7688386 0.7643863        -3.27403            -1.58133
25:1                7.139029 -0.7688386 0.7643863        -3.27403            -1.58133
27:0                7.672250 -0.7688386 0.7643863        -3.27403            -1.58133
27:1                8.025797 -0.7688386 0.7643863        -3.27403            -1.58133
62 (mild):0         8.622783 -0.7688386 0.7643863        -3.27403            -1.58133
62 (mild):1         7.075263 -0.7688386 0.7643863        -3.27403            -1.58133
62 (moderate):0     7.511450 -0.7688386 0.7643863        -3.27403            -1.58133
62 (moderate):1     8.186597 -0.7688386 0.7643863        -3.27403            -1.58133
                 ftreatment1
19:0                2.176453
19:1                2.176453
25:0                2.176453
25:1                2.176453
27:0                2.176453
27:1                2.176453
62 (mild):0         2.176453
62 (mild):1         2.176453
62 (moderate):0     2.176453
62 (moderate):1     2.176453
```

The F statistics for 'study' and 'treatment' agree fairly closely with those given by GenStat when these terms are sequentially added to the fixed model, and the estimate of the effect of treatment (the estimate for the fixed-effect ftreatment1) and its SE agree with those given by GenStat. The method used to fit model ssri.model1 does not produce a value of deviance in the output, so we cannot determine the effect on the deviance of including study.treatment as a random-effect term by comparing the results of ssri.model1 and ssri.model4. However, the F test for the study:treatment interaction in Model 8.2 tests the same null hypothesis, and is preferable to the test based on the comparison of deviances.

The results from model ssri.model2 can be used to obtain the difference between unshrunk treatment means for each study. For example, the difference in Study 27 is obtained as follows:

unshrunk mean (active treatment) − unshrunk mean (placebo)

$$= \text{estimate (ftreatment1)} + \text{estimate (fstudy27 : ftreatment1)}$$

$$= 6.9983 + (-4.3982) = 2.6001.$$

Similarly, the results from model `ssri.model4` can be used to obtain the difference between unshrunk treatment means for each study. For example, the difference in Study 27 is obtained as follows:

shrunk mean (active treatment) − shrunk mean (placebo)

$$= \text{estimate (ftreatment1)} + \text{coef (27 : 1 (intercept))} - \text{coef (27 : 0 (intercept))}$$

$$= 2.1765 + 8.025797 - 7.672250 = 2.530047.$$

These values agree with corresponding differences between the study.treatment means produced by GenStat.

8.6   Meta-analysis of individual-subject data using SAS

The following SAS statements import the data and perform the fixed-effects meta-analysis with estimation only of the main effect of treatment (Model 8.1):

```
PROC IMPORT OUT= ssri DBMS = EXCELCS REPLACE
    DATAFILE= "&pathname.\IMM edn 2\Ch 8\ssri indiv patient.xlsx";
    SHEET="fluoxetine for SAS";
RUN;


ODS RTF;
PROC MIXED ASYCOV NOBOUND;
    CLASS study treatment;
    MODEL change = study treatment
        / CHISQ DDFM = KR HTYPE = 3;
    LSMEANS treatment;
    ESTIMATE 'trt effect' treatment -1 1;
RUN;
ODS RTF CLOSE;
```

Part of the output produced by `PROC MIXED` is as follows:

Covariance parameter estimates	
Cov Parm	**Estimate**
Residual	62.8720

Fit statistics	
−2 Res Log Likelihood	7888.7
AIC (smaller is better)	7890.7
AICC (smaller is better)	7890.7
BIC (smaller is better)	7895.8

Type 3 tests of fixed effects					
Effect	Num DF	Den DF	Chi-square	F value	Pr > ChiSq	Pr > F
study	4	1126	54.81	13.70	<0.0001	<0.0001
treatment	1	1126	14.28	14.28	0.0002	0.0002

Estimates					
Label	Estimate	Standard error	DF	t value	Pr > |t|
trt effect	2.1402	0.5663	1126	3.78	0.0002

Least squares means						
Effect	treatment	Estimate	Standard error	DF	t value	Pr > |t|
treatment	0	6.8564	0.5123	1126	13.38	<0.0001
treatment	1	8.9965	0.4252	1126	21.16	<0.0001

The residual covariance parameter estimate (residual mean square) agrees with that given by GenStat, as do the F statistics for 'study' and 'treatment' when each of these terms is individually dropped from the full fixed model, the treatment means and, most importantly, the estimate of the effect of treatment and its SE.

The following commands fit the model in which the main effects of 'study' and 'treatment' and the study × treatment interaction are all specified as fixed-effect terms (Model 8.2):

```
ODS RTF;
PROC MIXED ASYCOV NOBOUND;
    CLASS study treatment;
    MODEL change = study treatment study*treatment
        / CHISQ DDFM = KR HTYPE = 3;
    LSMEANS treatment study*treatment;
    ESTIMATE 'trt effect' treatment -1 1;
RUN;
ODS RTF CLOSE;
```

Part of the output produced by PROC MIXED is as follows:

Covariance parameter estimates	
Cov Parm	Estimate
Residual	62.4940

Fit statistics	
−2 Res Log Likelihood	7865.6
AIC (smaller is better)	7867.6
AICC (smaller is better)	7867.6
BIC (smaller is better)	7872.6

Type 3 tests of fixed effects						
Effect	**Num DF**	**Den DF**	**Chi-square**	**F value**	**Pr > ChiSq**	**Pr > F**
study	4	1122	31.74	7.94	<0.0001	<0.0001
treatment	1	1122	8.31	8.31	0.0039	0.0040
study*treatment	4	1122	10.81	2.70	0.0288	0.0293

Estimates					
Label	**Estimate**	**Standard error**	**DF**	**t value**	**Pr > |t|**
trt effect	2.2397	0.7768	1122	2.88	0.0040

Least squares means							
Effect	**study**	**treatment**	**Estimate**	**Standard error**	**DF**	**t value**	**Pr > |t|**
treatment		0	6.8422	0.5660	1122	12.09	<0.0001
treatment		1	9.0818	0.5321	1122	17.07	<0.0001
study*treatment	19	0	5.5008	1.6137	1122	3.41	0.0007
study*treatment	19	1	12.4991	1.6854	1122	7.42	<0.0001
study*treatment	25	0	8.8000	1.6137	1122	5.45	<0.0001
study*treatment	25	1	7.2000	1.8633	1122	3.86	0.0001
study*treatment	27	0	8.3998	0.6192	1122	13.57	<0.0001
study*treatment	27	1	10.9998	0.5876	1122	18.72	<0.0001
study*treatment	62 (mild)	0	5.8200	1.0564	1122	5.51	<0.0001
study*treatment	62 (mild)	1	5.8900	0.4572	1122	12.88	<0.0001
study*treatment	62 (moderate)	0	5.6902	1.1410	1122	4.99	<0.0001
study*treatment	62 (moderate)	1	8.8202	0.4587	1122	19.23	<0.0001

The value $F = 2.70$ obtained for the study $\times$ treatment interaction agrees with that given by GenStat.

The following statements fit the model in which both the main effect of study and the study $\times$ treatment interaction are specified as random-effect terms (Model 8.3):

```
ODS RTF;
PROC MIXED ASYCOV NOBOUND;
   CLASS study treatment;
   MODEL change = treatment
      / CHISQ DDFM = KR HTYPE = 3;
   RANDOM study study*treatment;
   LSMEANS treatment;
   ESTIMATE 'trt effect' treatment -1 1;
RUN;
ODS RTF CLOSE;
```

Part of the output produced by PROC MIXED is as follows:

> Convergence criteria met.

Covariance parameter estimates	
Cov Parm	Estimate
study	1.2409
study*treatment	2.2015
Residual	62.5075

Asymptotic covariance matrix of estimates				
Row	Cov Parm	CovP1	CovP2	CovP3
1	study	6.3253	−4.8381	0.1610
2	study*treatment	−4.8381	8.6607	−0.2770
3	Residual	0.1610	−0.2770	6.9666

Fit statistics	
−2 Res Log Likelihood	7904.1
AIC (smaller is better)	7910.1
AICC (smaller is better)	7910.1
BIC (smaller is better)	7908.9

Null model likelihood ratio test		
DF	Chi-square	Pr > ChiSq
2	44.34	<0.0001

Type 3 tests of fixed effects						
Effect	Num DF	Den DF	Chi-square	F value	Pr > ChiSq	Pr > F
treatment	1	2.28	2.86	2.86	0.0911	0.2176

Estimates					
Label	Estimate	Standard error	DF	t value	Pr > |t|
trt effect	2.0624	1.2206	2.28	1.69	0.2176

Least squares means						
Effect	treatment	Estimate	Standard error	DF	t value	Pr > |t|
treatment	0	6.8810	1.0068	8.4	6.83	0.0001
treatment	1	8.9435	0.9800	6.54	9.13	<0.0001

The treatment means and the estimate of the effect of treatment agree with those produced by GenStat, but for the SE of the treatment effect the agreement is not good (SAS: SE = 1.2206; GenStat: SE = 1.163). The covariance parameter (variance component) estimates for the study, study.treatment and residual terms agree with those produced by GenStat. The F statistic for the effect of treatment does not agree well with that produced by GenStat (SAS: $F = 2.86$, denominator d.f. = 2.28; GenStat: $F = 3.15$, denominator d.f. = 2.9). This discrepancy probably arises from the determination of the denominator d.f. by the Kenward–Roger method in

SAS and a somewhat different method in GenStat, and this probably also accounts for the discrepancy in the SE of the treatment effect.

The following statements fit the model in which study.treatment interaction is specified as a random-effect term although the main effects of both 'study' and 'treatment' are specified as fixed-effect terms (Model 8.4):

```
ODS RTF;
PROC MIXED ASYCOV NOBOUND;
    CLASS study treatment;
    MODEL change = study treatment
        / CHISQ DDFM = KR HTYPE = 3;
    RANDOM study*treatment;
    LSMEANS treatment;
    ESTIMATE 'trt effect' treatment -1 1;
    %MACRO loop();
    %DO i=1 %TO 5;
    ESTIMATE  "trt mean, placebo, in SAS-order study &i"
        intercept 1 treatment 1 0 |
        study*treatment%DO j=1 %TO &i-1; 0 0 %END; 1 0;
    ESTIMATE  "trt mean, active, in SAS-order study &i"
        intercept 1 treatment 0 1 |
        study*treatment%DO j=1 %TO &i-1; 0 0 %END; 0 1;
    %END;
    %MEND loop;
    OPTIONS MPRINT;
    %loop;
RUN;
ODS RTF CLOSE;
```

The ESTIMATE statement with the label 'trt effect in SAS-order study &i' indicates that the treatment effect within each study is to be estimated by obtaining the main effect of 'treatment' and adding to it the study × treatment interaction effect for each study in turn. A somewhat fuller explanation of such statements, and the 'macro' loop around them, is given in Section 5.5. Part of the output produced by PROC MIXED is as follows:

Convergence criteria met.

Covariance parameter estimates	
Cov Parm	**Estimate**
study*treatment	1.8321
Residual	62.5565

Asymptotic covariance matrix of estimates			
Row	**Cov Parm**	**CovP1**	**CovP2**
1	study*treatment	7.5446	−0.3432
2	Residual	−0.3432	6.9905

Fit statistics	
−2 Res Log Likelihood	7887.1
AIC (smaller is better)	7891.1
AICC (smaller is better)	7891.1
BIC (smaller is better)	7891.7

Null model likelihood ratio test		
DF	Chi-square	Pr > ChiSq
1	1.65	0.1986

Type 3 tests of fixed effects						
Effect	Num DF	Den DF	Chi-square	F value	Pr > ChiSq	Pr > F
study	4	3.38	7.57	0.35	0.1086	0.8328
treatment	1	2	3.49	3.49	0.0616	0.2025

Estimates					
Label	Estimate	Standard error	DF	t value	Pr > |t|
trt effect	2.1765	1.1643	2	1.87	0.2025
trt mean, placebo, in SAS-order study 1	5.9078	1.4648	2.01	4.03	0.0560
trt mean, active, in SAS-order study 1	10.0228	1.4839	1.82	6.75	0.0269
trt mean, placebo, in SAS-order study 2	7.5870	1.4704	1.96	5.16	0.0370
trt mean, active, in SAS-order study 2	8.3436	1.5195	1.79	5.49	0.0401
trt mean, placebo, in SAS-order study 3	6.7004	1.0454	1.76	6.41	0.0320
trt mean, active, in SAS-order study 3	9.2302	1.0578	1.65	8.73	0.0224
trt mean, placebo, in SAS-order study 4	7.6504	1.1875	2.12	6.44	0.0202
trt mean, active, in SAS-order study 4	8.2802	1.0968	1.7	7.55	0.0263
trt mean, placebo, in SAS-order study 5	6.5397	1.2229	2.17	5.35	0.0277
trt mean, active, in SAS-order study 5	9.3909	1.1138	1.72	8.43	0.0211

Least squares means					
Effect	treatment	Estimate	Standard error	DF	t value	Pr > |t|
treatment	0	6.8770	0.8357	2.74	8.23	0.0052
treatment	1	9.0535	0.8073	2.06	11.21	0.0071

The treatment means and the estimate of the effect of treatment agree with those produced by GenStat, but for the SE of the treatment effect the agreement is not good (SAS: SE = 1.1643; GenStat: SE = 1.096). The covariance parameter (variance component) estimates for the study.treatment and residual terms agree with those produced by GenStat. The F statistic for the effect of treatment, when this term is individually dropped from the full fixed model, does not agree well with that produced by GenStat (SAS: $F = 3.49$, denominator d.f. = 2; GenStat: $F = 3.95$, denominator d.f. = 2.7). There is a similar discrepancy for the effect of study. As in the case of Model 8.3, these discrepancies are probably due to the different methods for the determination of the denominator d.f. used by the two software systems, and this probably also accounts for the discrepancy in the SE of the treatment effect.

Comparison of the results from this model with those from the model omitting the interaction term gives

$$\text{deviance}_{\text{reduced model}} - \text{deviance}_{\text{full model}} = 7888.7 - 7887.1 = 1.6,$$

approximately agreeing with the value given by GenStat.

The results from Model 8.2 can be used to obtain the difference between unshrunk treatment means for each study. For example, the difference in Study 27 is obtained as follows:

unshrunk mean (active treatment) − unshrunk mean (placebo)

= Least Squares mean (treatment1) − Least Squares mean (treatment0)

= 10.9998 − 8.3998 = 2.6.

Similarly, the results from Model 8.4 can be used to obtain the difference between unshrunk treatment means for each study. For example, the difference in Study 27 – which is the third study in the ordering used by SAS – is obtained as follows:

shrunk mean (active treatment) − shrunk mean (placebo) = 9.2302 − 6.7004.

These values agree with those produced by GenStat.

8.7 Meta-analysis when only summary data are available

In the preceding sections, meta-analysis has been performed on the basis of raw data from each of the clinical trials, and in this respect has been similar to most other statistical analysis. However, the purpose of meta-analysis is often to combine studies from which only published summary information is available, or from which confidential individual-patient data cannot be shared. The methods used in this situation can be illustrated using the results in Table 8.1.

The first steps in a meta-analysis of summary data are the following preliminary calculations:

$$d_i = \bar{x}_{2i} - \bar{x}_{1i} \tag{8.5}$$

$$s_i^2 = \frac{(n_{1i} - 1) \cdot \text{SD}_{1i}^2 + (n_{2i} - 1) \cdot \text{SD}_{2i}^2}{n_{1i} + n_{2i} - 2} \tag{8.6}$$

$$\text{SE}_{d_i} = \sqrt{\left(\frac{1}{n_{1i}} + \frac{1}{n_{2i}}\right) s_i^2} \tag{8.7}$$

$$w_i = \frac{1}{\text{SE}_{d_i}^2} \tag{8.8}$$

where

d_i = estimated difference between the mean responses to the treatments in the ith study,

$\bar{x}_{ki}$ = mean value of the response in the kth treatment in the ith study,

s_i^2 = residual mean square within the ith study,

n_{ki} = number of patients allocated to the kth treatment in the ith study,

SD_{ki} = between-patient standard deviation of the response in the kth treatment in the ith study,

SE_{d_i} = SE of d_i,

w_i = weight to be given to the ith study.

The results of Equations 8.5–8.8 are given in the columns headed diff, s2, se_diff and weight, respectively, in Table 8.1.

Fixed- and random-effect meta-analyses of these summary data, corresponding to Models 8.1 and 8.4, respectively, in the analyses of individual-subject data, are then performed by the following GenStat statements:

```
IMPORT 'IMM edn 2\\Ch 8\\ssri.xlsx'; SHEET = 'fluoxetine'
META LABELS = study; ESTIMATES = diff; SEESTIMATES = se_diff
```

The output of the META statement is as follows:

Meta analysis

Random estimate formed using residual maximum likelihood, with confidence intervals formed using profile likelihood.

Test for heterogeneity

Q	d.f.	pr.
11.706	4	0.020

Estimates

	θ_i	se(θ_i)	95% CI
19	7.000	2.570	(1.963, 12.037)
25	−1.600	2.681	(−6.855, 3.655)
27	2.600	1.032	(0.578, 4.622)
62 (mild)	0.070	0.836	(−1.568, 1.708)
62 (moderate)	3.130	1.216	(0.746, 5.514)
Fixed	1.646	0.547	(0.574, 2.719)
Random	2.060	1.063	(−0.849, 5.237)

Random effect variance: 3.302

Overall test probabilities

Fixed: 0.001
Random: 0.026

(under null hypothesis that estimates are not greater than zero)

A guide to the interpretation of such meta-analyses of summary data is given by Whitehead (2002, Chapter 4, pp. 57–98). The statistic Q provides a significance test of the null hypothesis that there is no variation in the true treatment effect (no heterogeneity) between the studies. On this hypothesis it is distributed as χ^2 with $k-1$ degrees of freedom, where k is the number of studies. In the present case, there is significant evidence of heterogeneity. A measure of the percentage of variation among studies that is due to heterogeneity rather than to variation among individual patients is given by

$$I^2 = 100\% \times \frac{Q - (k-1)}{Q} \tag{8.9}$$

(Higgins and Thompson, 2002). In the present case,

$$I^2 = 100\% \times \frac{11.71 - (5-1)}{11.71} = 65.8.$$

I^2 is approximately equivalent to a typical value of the shrinkage factor applied to the estimate of the treatment effect in each study in Figure 8.1, expressed as a percentage,

$$100 \times \frac{\hat{\sigma}^2_{\text{study.treatment}}}{\hat{\sigma}^2_{\text{study.treatment}} + \frac{\hat{\sigma}^2_{\text{Resid}}}{\bar{n}}} = 100 \times \frac{1.83}{1.83 + \frac{62.56}{66.16}} = 65.93$$

where

$\bar{n} = \exp\left(\frac{\sum_{i=1}^{k} \sum_{j=1}^{2} \log_e (n_{ij})}{2k} \right)$ = the harmonic mean of n_{ij}, the number of patients in the jth

treatment arm in the ith study,

$\hat{\sigma}^2_{\text{study.treatment}}$ = the estimate of the study.treatment variance component from Model 8.4.

In the special case where n_{ij} is constant over both treatment arms and over all studies, and where $\hat{\sigma}^2_{\text{Resid}}$ is the same in all studies, this equivalence is exact.

The estimate of the treatment effect from each study is presented next, then the pooled fixed- and random-effect estimates. Each estimate is accompanied by its SE and a 95% confidence interval. There is a considerable difference between the fixed- and random-effect pooled estimates, and as expected, the SE of the random-effect estimate is larger. The fixed-effect estimate is approximately equivalent to the estimate obtained from the fitting of Model 8.1 to individual-patient data (Section 8.2), but does not make the assumption that the true variance among patients is constant over the trials: if the standard deviation is the same in every arm of every trial, the estimates and their SEs are identical. Similarly, the random-effect estimate is approximately equivalent to that obtained when Model 8.4 is fitted to individual-patient data

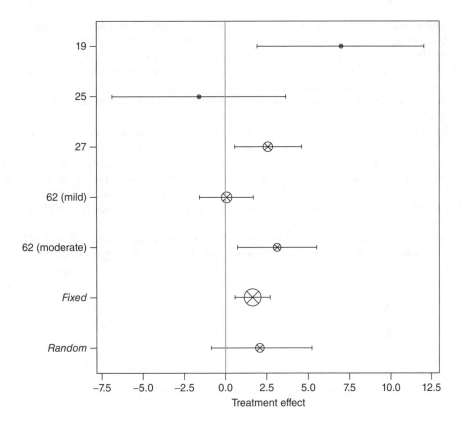

Figure 8.2 Forest plot of the meta-analysis of summary data from clinical trials of the antidepressant fluoxetine.
The size of the symbol representing each trial indicates the number of subjects in the trial.

(Section 8.4). These point and interval estimates of the treatment effect are also presented in a *forest plot* (Figure 8.2). The features of this method of display are reviewed by Whitehead (2002, Section 7.3.1, pp. 183–186) and by Anzures-Cabrera and Higgins (2010), but the origin of the term is uncertain – perhaps the series of lines was thought to resemble trees scattered in a forest. An estimate of the variance component for this heterogeneity is next presented. In the special case mentioned above,

$$\text{Random effect variance} = 2\hat{\sigma}^2_{\text{study.treatment}}. \tag{8.10}$$

The factor 2 in this expression reflects the fact that the random-effect variance relates to the difference between the treatment means in each study, not the individual means. Finally, one-sided significance tests are presented using the fixed- and random-effect estimates, with the null hypothesis that the true treatment effect is zero and the alternative hypothesis that it is greater than zero.

These methods for the analysis of summary data can be applied directly to studies with a binary outcome (e.g. responded to treatment/did not respond or survived/died), as well as to continuous outcomes like the change in the HRSD scale considered here. The estimates of treatment effects from such studies, for example the change in the proportion of responders, are usually presented with an approximate SE (or a confidence interval from which a SE can be derived), providing the values required for these analyses. In contrast, the methods presented for meta-analysis based on individual observations must be adapted for application to binary responses, by using generalized-linear mixed models instead of ordinary mixed models (see Exercise 10.7).

Model-fitting using the REML criterion is not the only available method for performing a random-effects meta-analysis on summary data. An alternative is the method of DerSimonian and Laird (1986), which partitions the total variance among the study means into within-study and between-study components using simple formulae that can be implemented with spreadsheet software or a pocket calculator (though this method is also offered by the GenStat META statement). However, the model-fitting approach described here is preferable to this non-iterative method when statistical software is available.

Meta-analysis is conveniently performed in R using the package 'metafor', available from the R website.[1] The fixed-effect analysis is performed by the following R commands:

```
rm(list = ls())
ssri <- read.table(
    "IMM edn 2\\Ch 8\\ssri, fluoxetine.txt",
    header=TRUE, sep = "\t")
attach(ssri)
library(metafor)
s2_diff <- se_diff^2
fma.ssri <- rma(diff, s2_diff, method = "FE")
print(fma.ssri)
```

The output from the `print()` function is as follows:

```
Fixed-Effects Model (k = 5)

Test for Heterogeneity:
Q(df = 4) = 11.7069, p-val = 0.0197

Model Results:

estimate       se      zval      pval      ci.lb      ci.ub
   1.6464   0.5473    3.0080    0.0026     0.5736     2.7192        **

---
Signif. codes:   0 '***' 0.001 '**' 0.01 '*' 0.05 '.' 0.1 ' ' 1
```

These results agree with those produced by GenStat. A forest plot similar to that produced by GenStat is given by the commands

---

[1] For information on the installation of packages from the R website, see the footnote in Section 3.16.

```
windows()
forest(fma.ssri)
```

The random-effect analysis is performed by the following commands:

```
rma.ssri <- rma(diff, s2_diff, method = "REML")
print(rma.ssri)
```

The output from the `print()` function is as follows:

```
Random-Effects Model (k = 5; tau^2 estimator: REML)

tau^2 (estimated amount of total heterogeneity): 3.3024 (SE = 3.9079)
tau (square root of estimated tau^2 value):       1.8173
I^2 (total heterogeneity / total variability):    65.60%
H^2 (total variability / sampling variability):   2.91


Test for Heterogeneity:
Q(df = 4) = 11.7069, p-val = 0.0197


Model Results:

estimate      se     zval      pval     ci.lb     ci.ub
  2.0598  1.0634   1.9370    0.0527   -0.0244    4.1440            .


---
Signif. codes:  0 '***' 0.001 '**' 0.01 '*' 0.05 '.' 0.1 ' ' 1
```

These results agree with those produced by GenStat. In the R output, the sources of variation are related by the equation

$$\text{total variability} = \text{total heterogeneity} + \text{sampling variability},$$

from which it follows that

$$H^2 = \frac{1}{(1 - I^2)} = \frac{1}{(1 - 0.656)} = 2.91, \tag{8.11}$$

as stated. A forest plot showing the random-effect estimate is obtained by replacing the object 'fma.ssri' by 'rma.ssri' in the `forest()` function. A fuller account of the facilities for meta-analysis available in the package 'metafor' is given by Viechtbauer (2010).

The following statements perform the fixed-effects analysis in SAS:

```
PROC IMPORT OUT = ssri DBMS = EXCELCS REPLACE
   DATAFILE = "&pathname.\IMM edn 2\Ch 8\ssri.xlsx";
   SHEET = "fluoxetine SAS";
RUN;


ODS RTF;
```

```
PROC MIXED DATA = ssri;
   MODEL diff = /SOLUTION;
   WEIGHT weight;
   PARMS 1 /HOLD = 1;
RUN;
ODS RTF CLOSE;
```

The SOLUTION option in the MODEL statement indicates that fixed-effect parameter estimates are to be presented – in the present case just for the intercept, which is the treatment effect of interest. The PARMS statement indicates that the residual variance (the only random-effect term) is to be initialized to 1, and the option 'HOLD = 1' in this statement indicates that the first random-effect term (i.e. the residual variance) is to be held at the value to which it is initialized.

Part of the output produced by PROC MIXED is as follows:

Convergence criteria met.

Covariance parameter estimates	
Cov Parm	**Estimate**
Residual	1.0000

Fit statistics	
−2 Res Log Likelihood	24.2
AIC (smaller is better)	24.2
AICC (smaller is better)	24.2
BIC (smaller is better)	24.2

PARMS model likelihood ratio test		
DF	**Chi-Square**	**Pr > ChiSq**
0	0.00	1.0000

Solution for fixed effects					
Effect	**Estimate**	**Standard error**	**DF**	**t value**	**Pr > $|t|$**
Intercept	1.6464	0.5474	4	3.01	0.0396

The table of covariance parameters (Cov Parm) confirms that the residual variance has been fixed at 1. The intercept and its SE agree with the values produced by GenStat.

The following statements perform the random-effects analysis in SAS:

```
ODS RTF;
PROC MIXED DATA = ssri ASYCOV;
   CLASS study;
   MODEL diff = /SOLUTION DDFM = KR;
   WEIGHT weight;
   RANDOM study;
```

```
PARMS 1 1/HOLD = 2;
RUN;
ODS RTF CLOSE;
```

The RANDOM statement indicates that 'study' is now specified as a random-effect term. The PARMS statement now indicates that both random-effect terms ('study' and the residual variance) are to be initialized to 1, and the option 'HOLD = 2' in this statement indicates that the second random-effect term (i.e. the residual variance) is to be held at the value to which it is initialized.

Part of the output produced by PROC MIXED is as follows:

Convergence criteria met.

Covariance parameter estimates	
Cov Parm	**Estimate**
study	3.3020
Residual	1.0000

Asymptotic covariance matrix of estimates			
Row	**Cov Parm**	**CovP1**	**CovP2**
1	study	21.7952	
2	Residual		

Fit statistics	
−2 Res Log Likelihood	21.3
AIC (smaller is better)	23.3
AICC (smaller is better)	25.3
BIC (smaller is better)	22.9

PARMS model likelihood ratio test		
DF	**Chi-square**	**Pr > ChiSq**
1	0.60	0.4402

Solution for fixed effects					
Effect	**Estimate**	**Standard error**	**DF**	**t value**	**Pr > |t|**
Intercept	2.0598	1.1360	2.27	1.81	0.1964

The intercept agrees with the random-effect estimate of the treatment effect produced by GenStat, but its SE is somewhat different. This discrepancy is associated with the specification of the Kenward–Roger method in the MODEL statement, which gives denominator d.f. $= 2.27$, whereas the method used by GenStat gave denominator d.f. $= 4$.

8.8　The multiple testing problem: Shrinkage of BLUPs as a defence against the Winner's Curse

It often happens in science that some striking phenomenon is reported – the effect of an experimental treatment or an association between two variables – but that when other researchers attempt to replicate the finding, the observed effect is much weaker or even entirely absent. This is partly due to the contribution of random variation to scientific results. One important function of meta-analysis is to reveal any such decline in the strength of the evidence, as illustrated in Figure 8.3. This problem is not too serious when each hypothesis tested by researchers has a strong basis in prior experience, but it became acute in contexts such as genetic and genomic research around the year 2000, when new high-throughput technologies first permitted testing the effect of hundreds of thousands of genetic variants on a phenotype, or the effect of some variable (experimental treatment, tissue type, etc.) on the expression of thousands of genes, with little or no distinctive expectation concerning individual variants or genes. In order to combat this embarrassing flow of false-positive results, very stringent significance criteria have been adopted in genetic association studies, while in gene expression studies there has been a strong emphasis on the control of the 'false discovery rate' (FDR).

The basic method for correcting the significance threshold to take account of multiple testing is that of Šidák (1967). Suppose that m mutually independent significance tests are performed, each with the significance threshold set at β, and that the null hypothesis is true in every case. In order to limit the probability that at least one test gives a significant result to α, we must specify

$$\beta = 1 - (1 - \alpha)^{\frac{1}{m}} \tag{8.12}$$

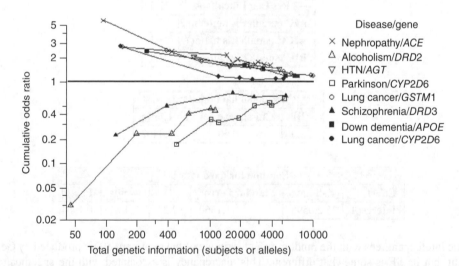

Figure 8.3　Evolution of the strength of an association as more information is accumulated.
Cumulative information from successive studies on eight topics in which the results of the first study or studies showed a significant effect ($p < 0.05$). An odds ratio of 1 indicates that there is no evidence of association between the gene and the disease. (Source: Reproduced from Ioannidis *et al.* (2001) with permission.)

which when β is small approximates to the better-known Bonferroni correction (Bonferroni, 1936),

$$\beta \approx \frac{\alpha}{m}. \tag{8.13}$$

There are several objections to this procedure for correction for multiple testing. Firstly, it is valid only if the significance tests are independent, whereas in practice there are usually correlations among them: for example, if a particular genetic variant is associated with a phenotype, neighbouring variants are more likely also to be associated with it. When such correlations are present, the effective number of tests is less than the nominal number and the Šidák or Bonferroni correction is conservative. Secondly, it is difficult to decide where to put the boundary of the 'family' of tests to be jointly considered. Even if one performs a correction so that one's probability of reporting a false-positive result in a particular research project is 0.05, other laboratories may be testing similar sets of hypotheses, in which case the probability of a false-positive result will accumulate. Thirdly, by framing the problem in terms of significance tests, it makes what may be a false distinction between 'true-' and 'false-positive' results. Finally, even on its own terms, it results in excessively stringent significance criteria. For example, it is common in human genetics to apply a criterion of genome-wide significance, arguing that a comprehensive set of tests of association with genetic variants distributed over the whole genome is equivalent to a million independent tests, and hence that a significance threshold of $0.05/1000000 = 5 \times 10^{-8}$ should be applied to each test. Only a small proportion of real genetic associations are likely to pass such a hurdle: the price of achieving a specified false-positive result is a high false-negative result, as many real associations go unreported. This dilemma is known as the *multiple testing problem* or the *multiplicity problem*.

An alternative approach to the identification of a subset of hypotheses that are worth investigating further is the specification of an FDR. The FDR can be limited to a specified value, q, by the 'step-up procedure' of Benjamini and Hochberg (1995), as follows. The values of p are first placed in ascending order so that

$$p_1 < p_2 < \ldots p_i < \ldots p_m$$

The largest value i for which

$$p_i \leq \frac{i}{m} \cdot q \tag{8.14}$$

is then identified. The null hypotheses corresponding to p_i and all smaller p-values are then rejected. The specification of a FDR is perhaps more reasonable than the Bonferroni or Šidák correction, because it is not based on the premise that the null hypothesis is universally true but accepts that true positive results exist, and tries to estimate what fraction they comprise of all positive results. However, it still tends to result in very conservative judgements when a low FDR is specified.

The investment of further research effort in published associations that turn out to be false positives is sometimes referred to as the *Winner's Curse*, a term originally applied to the tendency for the winner of an auction to pay somewhat more than the true worth of the item for sale (Thaler, 1988). This is most easily illustrated in the case of a 'blind' auction, in which each bidder submits a single bid in ignorance of the other bids. Assuming that the bids are randomly distributed around the true value of the item, the winning bid will tend to be higher than this value. This analogy suggests a remedy for the multiple testing problem, by viewing it not qualitatively, as the rejection of a null hypothesis that is true, but quantitatively, as the over-estimation of a model parameter. In this case, a shrinkage factor can be applied to the estimate.

In some contexts in which multiplicity is an issue, it is routinely dealt with in this way. For example, in crop breeding situations such as the field trial of barley breeding lines discussed in Chapter 5, the BLUE (Best Linear Unbiased Estimate) for each line is routinely shrunk to produce a BLUP (Best Linear Unbiased Predictor) as described in Section 5.2, effectively overcoming the Winner's Curse. It would be possible, for each breeding line, to test the null hypothesis that its yield does not differ from the mean yield of the population of lines and to compare the p-values obtained for the different lines, but plant breeders do not normally do this. Similarly, in a random-effects meta-analysis, estimates from a number of studies are shrunk towards the average value, as we have just seen (Section 8.4). However, in other contexts, including genetic and genomic studies, large numbers of p-values are routinely compared, and this practice should be questioned: shrunk estimates may provide a better basis for the comparison of the multiple entities (e.g. genetic variants or genes) under consideration.

The shrunk-estimate approach to the multiplicity problem can be illustrated using a data set that comprises the expression levels of a large number of genes in samples of bone marrow plasma cells from 559 patients diagnosed with multiple myeloma (NCBI Gene Expression Omnibus accession GSE24080). Most of the patients were enrolled on a treatment protocol named total therapy 2 (TT2), but some were on total therapy 3 (TT3). Details of these protocols are given by Pineda-Roman *et al.* (2008). The first and last few rows of a spreadsheet specifying the treatment received by each patient are presented in Table 8.3. The level of gene expression (i.e. the concentration of mRNA produced by the gene in question in the plasma cells, in arbitrary normalized units) is given for each of 1094 genes in each patient in another spreadsheet, the first and last few rows of which are presented in Table 8.4. (Data reproduced by kind permission of Leming Shi.) Each value of 'probeset' in this data set represents a particular gene. The genes included are a representative subset of those available, sufficient to illustrate the methods presented here. Further details of this data set are given on the web page http://www.ncbi.nlm.nih.gov/geo/query/acc.cgi?acc=GSE24080.

Table 8.3 Characteristics of cell samples used in a study of gene expression in bone marrow plasma in patients diagnosed with multiple myeloma.

Meanings of column labels: original order = arbitrary index value; PATID = patient ID; PROT = treatment protocol; PROT_coded = treatment protocol, coded as 0 or 1.

	A	B	C	D	E
1	Sample ID	original order	PATID	PROT	PROT_coded
2	GSM592391_P0002-01-FAKE03-U133Plus-2	1	8241	TT2	0
3	GSM592392_P0008-01-FAKE04-U133Plus-2	2	9843	TT2	0
4	GSM592393_P0009-01-FAKE05-U133Plus-2	3	9888	TT2	0
5	GSM592394_P0010-01-FAKE06-U133Plus-2	4	9835	TT2	0
6	GSM592395_P0015-01-A016-U133Plus-2	5	9997	TT2	0
	.				.
	.				.
	.				
553	GSM592947_P0995-02-D102-U133Plus-2	356	16626	TT3	1
554	GSM592948_P0996-04-D062-U133Plus-2	357	16563	TT3	1
555	GSM592949_P0997-02-D076-U133Plus-2	358	16593	TT3	1

Source: Data reproduced by kind permission of Leming Shi.

Table 8.4 Expression level of each of 1094 genes in each of the samples described in Table 8.3.

	A	B	C
1	probeset!	Sample	Expression
2	AFFX-BioB-5_at	GSM592391_P0002-01-FAKE03-U133Plus-2	5.05
3	AFFX-BioB-5_at	GSM592392_P0008-01-FAKE04-U133Plus-2	5.62
4	AFFX-BioB-5_at	GSM592393_P0009-01-FAKE05-U133Plus-2	4.06
5	AFFX-BioB-5_at	GSM592394_P0010-01-FAKE06-U133Plus-2	5.76
	.		.
	.		.
	.		.
606075	1570596_at	GSM592947_P0995-02-D102-U133Plus-2	2.31
606076	1570596_at	GSM592948_P0996-04-D062-U133Plus-2	2.31
606077	1570596_at	GSM592949_P0997-02-D076-U133Plus-2	2.31

Source: Data reproduced by kind permission of Leming Shi.

The relationship between the pattern of gene expression and the patients' disease outcome (duration of remission, duration of survival) in these data has been explored by Shaughnessy *et al.* (2007), and the use of this data set as a testing-ground for analysis techniques has been described by the MACQ Consortium (2010). Here, we will explore the direction, magnitude and significance of the difference in expression of each gene between patients receiving TT2 and those receiving TT3. In order to highlight the consequences of random variation in gene expression, we will confine the analysis to a representative subset of 110 samples.

The following GenStat statements read the data describing each sample (i.e. the plasma cells from a particular patient), and determine the number of samples:

```
IMPORT\
'IMM edn 2\\Ch 8\\MAQC2 Sample Info selected, training set.xlsx'; \
   SHEET = 'ClinInfo ordered'
CALCULATE nsample = NVALUES(SampleID)
```

The following GenStat statements read and organize the gene expression data:

```
IMPORT [FORDER = unsorted] \
   'IMM edn 2\\Ch 8\\gene expression subset, training set.xlsx'
CALCULATE nprobeset = NLEVELS(probeset)
PRINT nprobeset
VARIATE [NVALUES = nsample] exprn_set[1...nprobeset]
EQUATE OLDSTRUCTURES = expression; NEWSTRUCTURES = exprn_set
DELETE expression
```

The `IMPORT` statement reads the gene expression values into a single variate of length 110 samples × 1094 probe sets = 120340 values, which is then divided by the `VARIATE` and `EQUATE` statements into 110 variates of length 1094, each holding the values from an individual sample. The `DELETE` statement deletes the variate named 'expression' into which the data were initially read, to conserve storage space. This may be important when working with large data files.

The following statements produce a regression analysis of the effect of treatment protocol on the expression level of each probe set (gene). Because the treatment protocols are coded as 0 and 1, the regression coefficient is equal to the mean difference in expression level between the two protocols. Regression analysis is used here, in preference to more conventional methods for comparing means, in order to facilitate the extension of the method to the situation where the explanatory variable has more than two values on a numerical scale.

```
VARIATE [NVALUES = nprobeset] b, seb, df, ms
FOR i = 1¦nprobeset
    MODEL exprn_set[i]
    FIT [PRINT = *] PROT_coded
    RKEEP ESTIMATES = b_i; SE = seb_i; DF = df_i; MEANDEVIANCE = ms_i
    CALCULATE b$[i] = b_i$[2]
    CALCULATE seb$[i] = seb_i$[2]
    CALCULATE df$[i] = df_i
    CALCULATE ms$[i] = ms_i
ENDFOR
CALCULATE t = b/seb
```

The summary variates 'b', 'seb', 'df' and 'ms' are created to hold the results of the regression analysis on each probe set. The ESTIMATES parameter in the RKEEP statement creates a variate of length 2, b_i, in which the first value is the intercept and the second the slope (b, the regression coefficient). Similarly, the SE parameter creates a variate that holds the SEs of these estimates. The DF and MEANDEVIANCE parameters create scalar structures that hold, respectively, the residual degrees of freedom and the residual mean square from the regression analysis. The CALCULATE statements that follow copy the information retained from the regression analysis to the ith value of the appropriate summary variates: that is, the estimate of the slope for the ith probe set to the ith value of 'b' and so on.

There are seven probe sets for which $b = 0$ and $SE_b = 0$ as there is no variation in the expression value, and which are therefore excluded from further consideration, leaving 1087 probe sets. The distributions of b, SE_b and $t = b/SE_b$ are presented in Figure 8.4. Although there are a few extreme values of b, the distributions are sufficiently homogeneous for further analysis of the combined results to be reasonable.

If there are effects of treatment protocol on gene expression, it is likely that the expression level of some genes is higher under TT3 than under TT2, while that of others is lower, and it is therefore reasonable to conduct a two-tailed significance test in each regression analysis. The following statement obtains the appropriate p-values from the training sample set:

```
CALCULATE ptwotail = 2 * CUT(ABS(t); df)
```

The function ABS obtains the absolute value of its argument (i.e. it removes the minus sign from a negative value), and the function CUT() obtains the corresponding p-value as described in Section 4.4. On the null hypothesis that there is no effect of treatment protocol on the expression level of any gene, the p-values obtained are a random sample from a uniform distribution between 0 and 1. Departures from this hypothesis can therefore be detected by plotting the observed p-values, transformed to $-\log(p)$, against the corresponding quantiles

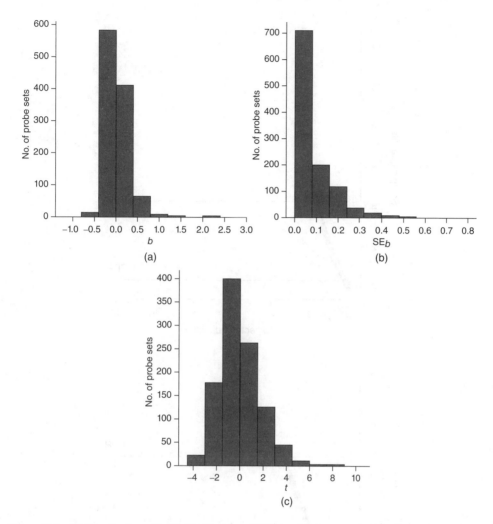

Figure 8.4 (a) Distribution of b, (b) distribution of SE_b, (c) distribution of t. Distributions, over probe sets, of summary statistics from the regressions of gene expression values on treatment protocol in data from bone marrow plasma cells of patients diagnosed with multiple myeloma.

from the uniform distribution (Figure 8.5). The points in this *quantile–quantile* (Q–Q) plot lie consistently above the line of unit slope passing through the origin, giving evidence that there is true association between treatment protocol and expression level for many genes. (For an introduction to Q–Q plots, see Wilk and Gnanadesikan (1968).) Setting $\alpha = 0.05$ and substituting n = number of probe sets under consideration = 1050 into Equation 8.12 gives a Šidák-corrected significance threshold of $\beta = 1 - (1 - 0.05)^{1/1050} = 0.0000488$, corresponding to $-\log(\beta) = 4.31$, and there are 24 probe sets that have lower p-values than this. However, it is likely that there are correlations in expression level among the genes, over the samples, even if the treatment protocol is held constant, in which case the effective number of independent significance tests is less than 1050 and the Šidák correction is conservative. Instead of using the Šidák correction, we may decide to limit the FDR to $q = 0.1$. The 127th-ranked

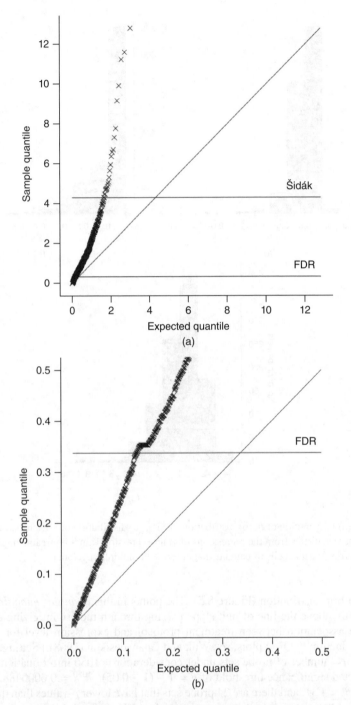

Figure 8.5 (a) Full range of quantiles and (b) enlargement of the region close to the origin. Q–Q plot of −log(p) for two-tailed tests of significance of association of gene expression values with treatment protocol.

Points above the horizontal lines represent associations that survive the Šidák correction for multiple testing, and associations that meet the FDR criterion, respectively.

p-value, 0.0120, corresponding to $-\log(p) = 1.92$ and associated with probe set 239838_at, satisfies Criterion 8.14 – that is,

$$0.0120 \leq \frac{127}{1050} \times 0.1$$

and it is the largest p-value that does so. Acceptance of the association of gene expression with this probe set as real, together with the 126 associations that give smaller p-values, is expected to achieve the specified FDR. But the Q–Q plot suggests that either of these criteria would eliminate many true associations from further consideration: the whole distribution of p-values is shifted towards smaller values than would be expected on the null hypothesis.

Turning attention from the p-values to the estimate of association obtained from each probe set, Figure 8.6a shows that regression coefficients that deviate strongly from the average, positively or negatively – that is, those that give large values of abs(b_i – mean(b)) – generally have larger SEs.[2] This indicates that extreme values of b_i are partly due to high levels of random variation for the probe set in question over the samples, and leads us to expect that such large values will not be consistently confirmed in future experiments. Therefore, instead of seeking to control the probability that a false association is selected or to control the FDR, we may shrink the estimates of association in order to avoid the over-optimistic expectations raised by the most extreme associations.

We will take an empirical-Bayesian approach to this shrinkage process, as outlined in Section 5.6, adapting Equation 5.14 (Section 5.3) to obtain a prior distribution for the measure of association:

$$B \sim N(\text{mean}(\beta), \text{var}(\beta)) \tag{8.15}$$

The term var(β) is the component of the variance among the regression coefficients that is due to real variation in the association with treatment protocol, excluding the component that is due to random variation among the samples, and our next task is to evaluate this term. The method used is a simplification of one presented by Smyth (2004), also in the context of microarray data. Smyth's approach is mathematically more rigorous than that presented here, and also differs in that

- it applies shrinkage to the t statistic for association rather than to the regression coefficient and

- it assumes that not all the genes studied, but only a specified proportion of them, are differentially expressed, so that for the remaining genes $\beta_i = 0$ and var(β) = 0.

The method is as follows. Random variation among the samples is a consequence of the random variation among observations within each sample, which will be designated σ_i^2, and an estimate of this, designated s_i^2, is given by MS_{Resid} for each sample. The value of σ_i^2 may itself vary from sample to sample, and a slope (b_i) associated with a large value of σ_i^2 should be more strongly shrunk than one associated with a small value. However, s_i^2 is itself an observation of a random variable, which should be shrunk towards its middle value before being used as the

---

[2] mean(b) here indicates the mean of the b_is. Analogous notation is used below for other functions and variables.

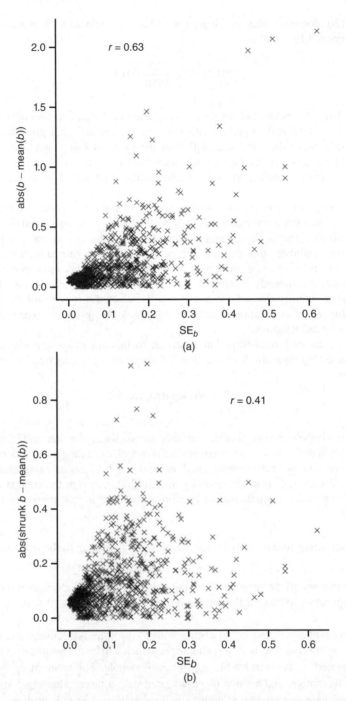

Figure 8.6   (a) Relationship of b and (b) relationship of shrunk b. The relationship of abs(b − mean(b)) and abs(shrunk b − mean(b)) with SE$_b$, over probe sets, for association of gene expression values with treatment protocol.[3]

---

[3]The correlation coefficient r is explained in a footnote in Section 7.5.

basis for shrinking b_i. In order to determine how much of the variation among the s_i is due to real variation among the σ_i^2, we make use of the fact that if

$$v = \text{the d.f. of } s_i^2,$$

then $(v/\sigma_i^2)s_i^2$ has a chi-square distribution, also with d.f. $= v$, and the variance of this distribution is $2v$. Hence if $\text{var}(\sigma^2) = 0$, that is, if

$$\sigma_1^2 = \ldots = \sigma_i^2 = \ldots = \sigma_m^2 = \sigma^2,$$

then

$$\text{var}(s^2) = \frac{\sigma^2}{v} \cdot \text{var}(\chi_v^2) = \frac{\sigma^2}{v} \cdot 2v = 2\sigma^2 \approx 2 \cdot \text{median}(s^2). \tag{8.16}$$

Median(s^2) is used in preference to mean(s^2) as an estimate of σ^2 in order to prevent a few probe sets with very large values of s_i^2 from dominating the calculation. Hence, approximately,

$$\text{var}(\sigma^2) = \max(\text{var}(s^2) - 2 \cdot \text{median}(s^2), 0) \tag{8.17}$$

Note that Equation 8.17 imposes the constraint $\text{var}(\sigma^2) \geq 0$, as it is assumed that the true value of this parameter cannot be negative. Hence, adapting Equation 5.14, a shrunk estimate of σ_i^2 is given by

$$\text{shrunk } s_i^2 = \text{median}(s^2) + (s_i^2 - \text{median}(s^2)) \cdot \frac{\text{var}(\sigma^2)}{\text{var}(s^2)}. \tag{8.18}$$

The idea that each b_i is an estimate of a true value β_i, which is imprecise due to the random variation among the samples, is expressed by the distribution

$$b_i \sim N(\beta_i, \sigma_{b_i}^2), \tag{8.19}$$

where

$$\sigma_{b_i}^2 = v_i \sigma_i^2. \tag{8.20}$$

The coefficient v_i is a function of the explanatory variables in the regression model. Depending on the regression model fitted, this function may be complicated, but in the present case it can be obtained by noting that σ_i^2 is estimated by s_i^2, and σ_{b_i} is estimated by the value of SE_{b_i} given in the output of the regression analysis. Hence v_i can be obtained by rearranging Equation 8.20 and substituting these estimates to give

$$v_i = \frac{\text{SE}_{b_i}^2}{s_i^2}. \tag{8.21}$$

We also require an estimate $\text{var}(\beta)$, and this is obtained from

$$\text{var}(b) \approx \text{var}(\beta) + \text{median}(\text{SE}_b^2) \tag{8.22}$$

and

$$\text{median}(\text{SE}_b^2) \approx \text{median}(v) \cdot \text{median}(s^2), \tag{8.23}$$

whence

$$\text{var}(\beta) \approx \text{var}(b) - \text{median}(v) \cdot \text{median}(s^2). \tag{8.24}$$

We are now in a position to substitute the following values into Equation 5.24 (Section 5.6):

$$\mu_\Delta = \text{mean}(\beta), \text{ adequately estimated by mean}(b)$$

$$\sigma_\Delta^2 = \text{var}(\beta)$$

$$d = b_i$$

$$\sigma_d^2 = \sigma_{b_i}^2 \approx v_i \cdot \text{shrunk } s_i^2$$

$$\text{posterior mean}(\Delta|\mu_\Delta, \sigma_\Delta^2, d, \sigma_d^2) =$$

$$\text{posterior mean}(\beta_i|\text{mean}(\beta), \text{var}(\beta), b_i, \sigma_d^2) = \text{shrunk } b_i.$$

Making these substitutions, we obtain

$$\text{shrunk } b_i = \text{mean}(b) + (b_i - \text{mean}(b)) \cdot \frac{\text{var}(\beta)}{\text{var}(\beta) + v_i \cdot \text{shrunk } s_i^2} \qquad (8.25)$$

Applying these formulae to the present results, we obtain

$$\text{median}(s^2) = 0.03696,$$

$$\text{var}(s^2) = 0.7514,$$

and substituting these values into Equation 8.17,

$$\text{var}(\sigma^2) = 0.7514 - 2 \times 0.03696 = 0.6775.$$

For the probe set with the smallest p-value, 212103_at,

$$s_i^2 = 0.623207$$

and substituting this value and those obtained above into Equation 8.18,

$$\text{shrunk } s_i^2 = 0.03696 + (0.623207 - 0.03696) \times \frac{0.6775}{0.7514} = 0.565538.$$

For this probe set,

$$SE_{b_i} = 0.155667,$$

and substituting this value and that of s_i^2 into Equation 8.21,

$$v_i = \frac{0.155667}{0.623207} = 0.038883.$$

In the present case, the value of the explanatory variable, and hence the value of v_i, is the same for every probe set, and

$$\text{median}(v) = 0.038883.$$

We obtain also

$$\text{var}(b) = 0.06146,$$

and substituting into Equation 8.24,

$$\text{var}(\beta) \approx 0.06290 - 0.038883 \times 0.03696 = 0.06146.$$

For the present results,

$$\text{mean}(b) = 0.05877$$

and for probe set 212103_at,

$$b_i = 1.31477.$$

Substituting into Equation 8.25,

$$\text{shrunk } b_i = 0.05877 + (1.31477 - 0.05877) \times \frac{0.06146}{0.06146 + 0.038883 \times 0.565538}.$$

$$= 0.983802$$

Figure 8.6b shows that the association of large SEs with large values of b is reduced, though not eliminated, by applying this shrinkage process to every b. Another way of assessing the effect of random variation on the distribution of the bs, and the effectiveness of shrinkage in compensating for it, is to compare the bs obtained from the 110 samples that we have studied so far, which we will call the training set, with those obtained by performing equivalent analyses on the remaining 444 samples, which we will call the validation set. (The terms 'training set' and 'validation set' are also used in the website describing the data and by the MACQ Consortium (2010) to distinguish the samples obtained from patients enrolled on TT2 and TT3, and that usage should not be confused with the one introduced here.) Large positive values of b in the training set are generally more positive than the corresponding values in the validation set, while large negative values of b are generally more negative: that is, there is a positive correlation between b_{training} and $(b_{\text{training}} - b_{\text{validation}})$ $(r = 0.491, \text{d.f.} = 1048, p < 0.001^4)$. By this measure, shrinking the bs has rather more than the desired effect: that is, there is a negative correlation between b_{training} and (shrunk $b_{\text{training}} - b_{\text{validation}})$ $(r = -0.359, \text{d.f.} = 1048, p < 0.001)$. This indicates that the shrunk b-values will give an over-pessimistic indication of the magnitudes of association with treatment that we can expect to observe for the same probe sets in a future study. However, they may serve as a useful corrective to the over-optimistic indication from the original b-values. Ideally, one would apply a shrinkage that would result in an indication neither optimistic nor pessimistic, but it is not obvious how a prior distribution for B that would reliably produce such an outcome could be specified.

A caveat must be mentioned concerning the use of shrunk estimates obtained in this way. In a conventional mixed-model analysis, in which all the BLUPs are based on a common estimate of within-group variance (e.g. the analysis of a plant breeding trial presented in Chapter 5 or the meta-analysis presented in Sections 8.1–8.7), it is expected that the shrunk estimates will not only avoid over-optimistic expectations concerning future observations, but will also be more closely correlated with the true values. Thus selection of entities (e.g. breeding lines) for further study on the basis of BLUPs will be more reliable than selection on the basis of BLUEs. In the present example, we cannot know the true values of associations between treatment protocol and gene expression for each probe set, but we should note that the original values of b from the training set are slightly more strongly correlated with the values

---

[4] The sample correlation coefficient r is an estimate of ρ, the true correlation coefficient in the population from which the sample is taken, and can be used as a test statistic for the null hypothesis $\rho = 0$. The relationship between r and p (obtainable from published tables or standard statistical software) depends on the degrees of freedom (d.f.), which in this simple case are given by $n - 2$, where $n =$ the number of pairs of observations contributing to r. The resulting p-value is two-sided if large values of r in either direction (positive or negative) are taken as evidence against the null hypothesis, and one-sided if values only in one direction are taken as such evidence. In the present case only values of $r > 0$ are taken as evidence against the null hypothesis.

of b from the validation set than are the shrunk values (original bs: $r = 0.856$, d.f. $= 1048$, $p < 0.001$; shrunk bs: $r = 0.848$, d.f. $= 1048$, $p < 0.001$). The values of SE_b are themselves estimates, subject to error, and the additional lack of precision that this introduces into the shrunk estimates has evidently offset the increase in correlation expected from the shrinkage, despite our attempt to allow for this lack of precision by also shrinking the values of s_i^2. A prudent course is therefore to use the original values of b to select probe sets for further study, but use the shrunk values as a guide to the magnitude of the values that can be expected in future work.

A related approach has been implemented for categorical data by Fram, Almenoff and DuMouchel (2003). In this approach, the relative risk (RR) for a particular association is shrunk towards the average RR for the class of associations under consideration using a criterion called the *empirical Bayes geometric mean*.

The multiple testing problem can arise either in the form of multiple responses to a single explanatory variable, as in the present case, or in the form of multiple explanatory variables for a single response, as in the study of the effect of genetic variants on a phenotype. In the latter situation either the individual effects of the explanatory variables on the response variable must be small, or the true explanatory variables must be few in number, and approaches to such data have therefore focused on *penalized regression* models which eliminate all but a few of the possible explanatory variables – in Bayesian terms, models in which the prior distribution for most effects has a large probability density at zero, so that the best estimate is zero unless the data provide strong evidence to the contrary. This is the basis of the Least Absolute Shrinkage and Selection Operator (LASSO) introduced by Tibshirani (1996): see also Leng, Lin and Wahba (2006), Malo, Libiger and Schork (2008) and Bühlmann and van de Geer (2011). This data structure also suggests that the effects of all the explanatory variables retained should be included in a single model, both to explain as fully as possible the variation in the response variable and to maximize the power to detect each effect by eliminating all the others from the residual variation. The models fitted by the LASSO method are of this kind.

8.9   Fitting of multiple models using R

The following commands read the sample data into R:

```
rm(list = ls())
sample_info <- read.table(
   "IMM edn 2\\Ch 8\\MAQC2 Sample Info selected, training set.txt",
   header=TRUE, sep = '\t')
names(sample_info)
attach(sample_info)
nsample <- length(PROT_coded)
```

The following commands read in the expression data and store them in a matrix, one column per probe set:

```
gene_expression <- read.table(
   "IMM edn 2\\Ch 8\\gene expression subset, training set.txt",
   header=TRUE, as.is = TRUE)
attach(gene_expression)
names(gene_expression)
```

```
print(gene_expression[1:20,])
fprobeset <- factor(probeset, levels = unique(probeset))
nprobeset <- nlevels(fprobeset)
exprn.mat <- matrix(data = expression, nrow = nsam-
ple, ncol = nprobeset)
expression <- NULL
```

The last command deletes the data from the table named 'expression' into which they were initially read, to conserve storage space. This may be important when working with large data files.

The following commands produce a regression analysis of the effect of treatment protocol on the expression level of each probe set and save the results:

```
b <- numeric()
seb <- numeric()
df <- numeric()
t <- numeric()
ptwotail <- numeric()
for (i in 1:nprobeset){
    exprn.model <- lm(exprn.mat[,i] ~ PROT_coded)
    b[i] <- summary(exprn.model)$coefficients[2,1]
    seb[i] <- summary(exprn.model)$coefficients[2,2]
    t[i] <- summary(exprn.model)$coefficients[2,3]
    ptwotail[i] <- summary(exprn.model)$coefficients[2,4]
    df <- summary(exprn.model)$df[2]
}
```

The function `numeric()` is used to specify a series of structures to hold the results. The `for` loop performs the required analyses for each probe set in turn. The function `lm()` is used to specify the regression models to be fitted. The function `summary()` is used to extract results from the regression analyses, and the component name 'coefficients' and the indices in square brackets that follow it indicate the results to be extracted. The index 'i' is used to ensure that the correct column of the matrices exprn.model.train and exprn.model.valid is used in the analyses for each probe set, and that the results are allocated to the correct position in the output structures.

There are several probe sets for which there is no variation in the expression values, and for which therefore $b_i = 0$ and $SE_{b_i} = 0$. GenStat automatically produces the missing t-values and p-values for these probe sets, but R calculates specious values from the infinitesimal values of b_i and SE_{b_i} that it has obtained. The following commands eliminate these values, identifying probe sets for which $abs(SE_{b_i}) < 10^{-10}$ and replacing all results for these probe sets with missing values:

```
seb[abs(seb) < 1.0e-10] <- NA
b[is.na(seb)] <- NA
t[is.na(seb)] <- NA
ptwotail[is.na(seb)] <- NA
```

The function `abs()` obtains the absolute value of its argument (i.e. it removes the minus sign from a negative number). The function `is.na()` identifies the positions in a structure that hold a missing value, represented by 'NA'. When the specious values have been removed, the values held in the structures produced by these statements are the same as those in the equivalently named structures produced in GenStat, and in the same order.

The execution of a `for` loop is relatively time-consuming in R, and the following statements produce the same results more efficiently, without explicit use of loops:

```
get.coefinfo <- function(x){
    exprn.model <- lm(x ~ PROT_coded)
    out1 <- summary(exprn.model)$coefficients[2,]
    out2 <- df <- summary(exprn.model)$df[2]
    out <- c(out1, out2)
}
coefinfo <- apply(exprn.mat, 2, get.coefinfo)
```

The syntax of these facilities is beyond the scope of this book, but is provided in the documentation of R (R Development Core Team, 2012, Section 3.2.2 and Chapter 4 for user-defined functions; http://stat.ethz.ch/R-manual/R-devel/library/base/html/apply.html for `apply()`). However, we can note that the keys to this *vectorized* specification are

- the capacity for user-defined functions, in this case `get.coefinfo`, to perform customized tasks and

- the function `apply()`, which invokes other functions repeatedly in a single step.

8.10   Fitting of multiple models using SAS

The following statements read the sample data into SAS:

```
PROC IMPORT OUT = sampleinfo DBMS = EXCELCS REPLACE
    DATAFILE =
    "&pathname.\IMM edn 2\Ch 8\MAQC2 Sample Info selected, training set.xlsx";
    SHEET = "ClinInfo ordered";
RUN;

DATA sampleinfo2; SET sampleinfo;
    index_var = _N_;
RUN;
```

The DATA step adds an index variable to the data set.

The following statements read the expression data:

```
PROC IMPORT OUT = geneexprn DBMS = EXCELCS REPLACE
    DATAFILE =
    "&pathname.\IMM edn 2\Ch 8\gene expression subset, training set.xlsx";
    SHEET = "for SAS";
RUN;
```

The following statements add the sample data to the file holding the expression data:

```
PROC SQL;
    CREATE TABLE geneexprn2 AS
    SELECT a.*, b.*
    FROM geneexprn AS a, sampleinfo2 AS b
    WHERE a.sample eq b.SampleID
    ORDER BY probeset, index_var;
QUIT;
```

The WHERE statement ensures that the data describing each sample are added to every line of the expression data relating to that sample. The ORDER statement ensures that the data are sorted by the probe set in ascending order, which is required for the subsequent analysis, and that within each probe set they are sorted by the index variable, that is, that the samples are kept in the order in which they are given in the original data set.

The following statements perform a regression analysis of the effect of treatment protocol on the expression level of each gene.

```
PROC GLM DATA = geneexprn2;
    BY probeset;
    MODEL expression = PROT_coded;
    ODS OUTPUT ParameterEstimates = esti;
RUN;

DATA esti2; SET esti;
    IF ABS(StdErr) < 10**(-10) THEN
        DO
            StdErr = . ;
            Estimate = .;
        END;
    IF Parameter = "PROT_coded" THEN OUTPUT;
RUN;
```

PROC GLM fits the required regression model for each probe set in turn, and writes the parameter estimates and associated information to the file 'esti'. The DATA step identifies the probe sets for which $SE_{b_i} = 0$ (as described in Section 8.8), and sets both b_i and SE_{b_i} to missing values for these probe sets. It then discards the information related to the intercept, retaining that related to the parameter of the explanatory variable PROT_coded in the model fitted (i.e. related to the slope), and writes the results retained to the file esti2. The first few rows of esti2 are as shown in Table 8.5. The values of b_i, SE_{b_i}, t and p are given in the columns headed

Table 8.5   First few rows of the data set 'esti2', produced by SAS, holding summary results of regression analysis on the samples described in Table 8.1.

probeset_	Dependent	Parameter	Estimate	StdErr	tValue	Probt
1552320_a_at	expression	PROT_coded	−0.001014493	0.001318548	−0.769401462	0.443334506
1552395_at	expression	PROT_coded	−0.005602686	0.007222448	−0.775732312	0.439602433
1552463_at	expression	PROT_coded	−0.002431955	0.001236143	−1.967373051	0.051704639
1552528_at	expression	PROT_coded	0.010957936	0.005275213	2.077250076	0.040150347

Estimate, StdErr, tValue and Probt, respectively. The values in these columns are the same as those in the corresponding structures produced by GenStat, though not in the same order.

8.11 Summary

Meta-analysis is the combined statistical analysis of the results of several studies, with the aim of obtaining clearer or firmer conclusions than can be obtained by considering them individually. When possible, meta-analysis should be performed on the basis of the individual observations used in each study.

In a fixed-effect meta-analysis, it is assumed that the true treatment effect is the same in every study. However, treatment effects may vary between studies, and it is then natural to regard the studies as a random sample, leading to a random-effect meta-analysis.

The main effect of treatment may be partitioned into a between-study and a within-study component. When the studies are randomized experiments (e.g. clinical trials), it is considered prudent to interpret only the within-study component. This is because the between-studies component may be biased by confounding with other variables, whereas the within-studies component is defended against such bias by the randomization process.

The within-studies estimate of the treatment effect may be obtained without explicit partitioning, by specifying study and treatment as fixed-effect terms, but the study × treatment interaction as a random-effect term. This is an exception to the rule that the interaction between two fixed-effect terms is also fixed (see Section 6.3).

When a random-effect meta-analysis is performed, shrunk estimates of the treatment effects in individual studies may be obtained.

Although meta-analysis based on individual observations is preferable, in most cases only summary data are available, for example, from the published results of each study. Statistical methods for fixed- and random-effect meta-analysis in this situation are presented. These methods can be applied directly to studies with a binary outcome (e.g. survived/died), whereas the methods presented for meta-analysis based on individual observations must be adapted (see Exercise 10.7).

When a strong effect or association is reported in scientific research, it often happens that in subsequent studies the effect is found to be weaker or disappears entirely. This is partly due to random variation, and is sometimes referred to as the *Winner's Curse*. It is a particular problem when large numbers of possible effects are considered, for example, in a study of genetic variation or gene expression.

To defend against such false-positive results, a stringent correction for multiple testing, the Šidák or Bonferroni correction, is often applied to the statistical significance criterion, or a threshold is set that ensures a low FDR. However, these procedures can result in a high false-negative result, as real associations go undetected.

Alternatively, shrunk estimates of the effects can be used to defend against unrealistic expectations for future results. This avoids the following problems with the Šidák or Bonferroni correction:

- The assumption that the tests are independent.

- The arbitrary boundary of the 'family' of tests to be jointly considered.

- The potentially false distinction between 'true-' and 'false-positive' results.

- Excessively stringent significance criteria.

The specification of an empirical-Bayesian prior probability distribution of effect sizes, in order to obtain such shrunk estimates, is explored.

8.12   Exercises

8.1   A multi-centre study to compare two anaesthetic agents (A and B) in patients undergoing short surgical procedures was reported by Whitehead (2002, Section 3.6, pp. 49–55). The response variable considered was the recovery time in minutes, transformed to logarithms. The results are presented in Table 8.6. (Data reproduced by kind permission of John Wiley & Sons, Ltd.) The individual-patient data from these studies are not available, but simulated data have been produced that give the values in Table 8.6, and the first and last few lines of this data set are presented in Table 8.7: the full data set is held in the file 'anaesthesia indiv patient.xlsx' on this book's website (see Preface).

(a)   Analyse these data, specifying both 'centre' and 'treatment' as fixed-effect terms, with no centre × treatment interaction term. State whether there is significant evidence of a difference between the effects of the two treatments, and if so, which one appears to be more effective in giving rapid recovery. State whether there is significant evidence of variation among the centres in the mean log(recovery time), averaged over the two treatments.

(b)   Extend your model so as to test whether there is evidence that the difference between the effects of the two treatments varies among centres.

(c)   Change your model so as to specify 'centre' as a random-effect term. Change other aspects of the model accordingly. Comment on the results from this model, and the ways in which they differ from those of the model fitted in Part (a).
   It is desired to partition the effect of treatment into between-centre and within-centre components.

(d)   Confirm that the variable 'b' holds the mean value of 'treatment' in each centre, when the treatments are coded as 0 and 1. Confirm that for each patient

$$w = \text{treatment} - b.$$

Table 8.6   Recovery time (minutes, log-transformed) after anaesthesia in a multi-centre study to compare two anaesthetics.

Centre	Treatment A			Treatment B		
	No. of patients	Mean	SD	No. of patients	Mean	SD
1	4	1.141	0.967	5	0.277	0.620
2	10	2.165	0.269	10	1.519	0.913
3	17	1.790	0.795	17	1.518	0.849
4	8	2.105	0.387	9	1.189	1.061
5	7	1.324	0.470	10	0.456	0.619
6	11	2.369	0.401	10	1.550	0.558
7	10	1.074	0.670	12	0.265	0.502
8	5	2.583	0.409	4	1.370	0.934
9	14	1.844	0.848	19	2.118	0.749

Source: Reproduced by permission of Wiley.

Table 8.7   Simulated individual-patient data from the clinical studies to compare two anaesthetics, summarized in Table 8.6.

	A	B	C	D	E	F
1	centre!	patient	treatment!	b	w	logrt
2	1	1	0	0.555556	−0.555556	0.4973
3	1	2	1	0.555556	0.444444	0.5399
4	1	3	1	0.555556	0.444444	−0.0413
5	1	4	1	0.555556	0.444444	0.5306
.						.
.						.
.						.
181	9	31	1	0.575758	0.424242	1.3515
182	9	32	0	0.575758	−0.575758	1.4634
183	9	33	1	0.575758	0.424242	1.6574

Treatment 0 = Treatment A; 1 = Treatment B.

'b' represents the between-study component of the treatment effect, and 'w' the within-centre component. Change the model fitted in Part c by removing the term 'treatment' and adding the terms 'b', 'w' and 'w.centre', allocating these terms to the fixed- and random-effect models as appropriate. Fit the resulting model, and comment on the results.

(e) Fit a model in which both 'centre' and 'treatment' are specified as fixed-effect terms but the centre × treatment interaction is specified as a random-effect term. Confirm that this gives equivalent results to the partitioning of the treatment effect in Part (d).

(f) Re-test the significance of the centre × treatment interaction term using the deviance accounted for by this term. Compare the result of this test with that performed in Part (b).

(g) Obtain the shrunk mean log(recovery time) for each treatment in each centre produced by the analysis performed in Part (e). Hence obtain a shrunk estimate of the difference between the effects of the two treatments in each centre. Make a graphical comparison between these shrunk differences and the differences obtained from the unshrunk means presented in Table 8.6.

(h) Perform a fixed-effect meta-analysis on the summary data presented in Table 8.6. Present the results of this analysis in a forest plot.

(i) Perform the corresponding random-effect meta-analysis on the summary data. Present the results of this analysis in a forest plot. Comment on the differences between the results of these two analyses and their relationship to the results of the corresponding analyses on the individual-patient data.

8.2 A meta-analysis of seven randomized clinical trials that studied the effect of aspirin when given to heart-attack patients was reported by Fleiss and Gross (1991). In each trial, the treatments were aspirin and placebo, and the outcome considered was death or survival: in each arm of each trial, the total number of patients and the number who died were reported. The spreadsheet holding the results is presented in Table 8.8. (Data reproduced by kind permission of Elsevier Ltd.)

Table 8.8 Results used in a meta-analysis to study the effect of aspirin when given to heart-attack patients.

	A	B	C	D	E	F	G	H	I	J	K	L	M	N
1	trial!	placebo_patients	placebo_deaths	aspirin_patients	aspirin_deaths	propn1	propn2	diff	se_diff	weight	e	v	logOR	SE(log(OR))
2	MRC-1	624	67	615	49	0.07967	0.10737	0.02770	0.01652	3665.4	57.5787	26.2835	−0.32639	0.19506
3	CDP	771	64	758	44	0.05805	0.08301	0.02496	0.01307	5852.7	53.5409	25.0911	−0.38025	0.19964
4	MRC-2	850	126	832	102	0.12260	0.14824	0.02564	0.01667	3599.3	112.7800	49.2678	−0.21880	0.14247
5	GASP	309	38	317	32	0.10095	0.12298	0.02203	0.02521	1574.0	35.4473	15.5406	−0.22182	0.25367
6	PARIS	406	52	810	85	0.10494	0.12808	0.02314	0.01977	2557.4	91.2582	27.0366	−0.23147	0.19232
7	AMIS	2257	219	2267	246	0.10851	0.09703	−0.01148	0.00903	12271.1	233.0139	104.3007	0.12451	0.09792
8	ISIS-2	8600	1720	8587	1570	0.18283	0.20000	0.01717	0.00600	27774.9	1643.7557	665.0536	−0.11090	0.03878

Source: Data reproduced by kind permission of Elsevier Ltd.

The variable 'propn1' records the proportion of placebo-treated patients who died in each study, calculated as r_1/n_1 where

$$r_1 = \text{number of placebo-treated patients who died and}$$

$$n_1 = \text{total number of placebo-treated patients.}$$

The variable 'propn2' records the proportion of aspirin-treated patients who died, similarly calculated. Then

$$\text{diff} = \text{propn2} - \text{propn1}$$

$$\text{se\_diff} = \sqrt{\frac{r_1(n_1 - r_1)}{n_1^3} + \frac{r_2(n_2 - r_2)}{n_2^3}} \tag{8.26}$$

and

$$\text{weight} = \frac{1}{\text{se\_diff}^2} \tag{8.27}$$

where

$$r_2 = \text{number of aspirin-treated patients who died and}$$

$$n_2 = \text{total number of aspirin-treated patients.}$$

On the null hypothesis that there is no difference between the effect of aspirin and placebo, the expected number of deaths of aspirin-treated patients is

$$e = \frac{(r_1 + r_2)n_1}{(n_1 + n_2)}. \tag{8.28}$$

Then the log odds ratio between the deaths of aspirin-treated and placebo-treated patients is

$$\log \text{OR} = \frac{(o - e)}{v} \tag{8.29}$$

where

$$o = \text{the observed number of deaths of aspirin-treated patients} = r_2$$

and

$$v = \frac{(r_1 + r_2)(n_1 + n_2 - (r_1 + r_2))n_1 n_2}{(n_1 + n_2)^3}. \tag{8.30}$$

Then

$$\text{SE(logOR)} = \sqrt{\frac{1}{v}}. \tag{8.31}$$

The variables 'diff' and 'logOR' provide estimates of the effect of aspirin on different scales. The SEs of these variables, se_diff and SE(logOR), respectively, are approximations based on the normal distribution.

(a) Perform fixed- and random-effect meta-analyses of the effect of aspirin as measured by 'diff'.

(b) Perform fixed- and random-effect meta-analyses of the effect of aspirin as measured by 'logOR'. Comment on the results and compare them with those obtained from the analysis of 'diff'.

(c) Comment on the relative merits of the difference between the proportions and the odds ratio as measures of the size of a treatment effect.

Alternative analyses that do not depend on the normal approximation are specified in Exercise 10.7.

8.3 Data have been generated that satisfy the equation

$$y_{ij} = \beta_{0i} + \beta_{1i}x_{ij} + \epsilon_{ij}. \tag{8.32}$$

The values in this equation are specified as follows:

- the β_{0i} are values of a random variable B_0 that has the distribution

$$B_0 \sim N(2.65, 0.28^2)$$

- the β_{1i} are values of a random variable B_1 that has the distribution

$$B_1 \sim N(0.044, 0.0072^2)$$

- the x_{ij} are observations of a random variable X that has the distribution

$$X \sim N(151, 41^2)$$

- the ϵ_{ij} are values of a random variable E that has the distribution

$$E \sim N(0, 0.89^2)$$

- the y_{ij} are observations of a variable Y.

All the random variables except Y are mutually independent. Values of B_0, B_1, X and Y for $i = 1 \ldots 1000$, $j = 1 \ldots 10$ are held in the file 'multiple y v x.xlsx' on this book's website (see Preface).

(a) By fitting a regression model to each sample (i.e. to the values of X and Y for a common value of i), obtain estimates of the β_{1i}, designated b_{1i} and their SEs, designated $SE_{b_{1i}}$. Obtain the degrees of freedom and the value of MS_{Resid} from each regression analysis.

(b) For each sample, obtain the statistic $= b_{1i}/SE_{b_{1i}}$. Plot histograms of b_1 (i.e. of all the b_{1i}), SE_{b_1} and t, and comment on the distributions of these statistics.

(c) Obtain the two-tailed p-value corresponding to each t statistic. Produce a Q–Q plot of the p-values, transformed to $-\log(p)$. Mark the Šidák-corrected significance threshold on this plot and determine the number of samples in which the association between X and Y survives this correction for multiplicity. Determine the largest value of p that is expected to give an FDR of 0.1 if accepted as evidence of a true association, and mark this threshold on the Q–Q plot. Determine how many associations meet this criterion. Comment on the appropriateness of the Šidák correction and the FDR criterion, in the light of your knowledge of the true distribution of B_1.

(d) For each sample, calculate abs($b_{1i} - $ mean(b_1)), where mean(b_1) indicates the mean of the b_is. Plot these values against $SE_{b_{1i}}$ and obtain the correlation coefficient (r) between these two variables. Obtain the one-sided p-value for the null hypothesis $\rho = 0$, with the alternative hypothesis $\rho > 0$, where ρ is the true correlation coefficient of which r is an estimate. Comment on these results.

(e) Use Equation 8.17 to obtain an approximate value of $\text{var}(\sigma^2)$. From the way in which the data were specified, what do you know of the true value of $\text{var}(\sigma^2)$? For each sample, use Equation 8.18 to obtain shrunk s_i^2, and Equation 8.21 to obtain v_i. Obtain median(v) and var(b), and hence from Equation 8.24 obtain $\text{var}(\beta)$. Hence from Equation 8.25 obtain shrunk b_i for each sample.

(f) For each sample, calculate abs(shrunk b_{1i} − mean(b_1)). Plot these values against $\text{SE}_{b_{1i}}$, and obtain the correlation coefficient between these two variables and the corresponding p-value. Compare these results with those obtained in Part (d).

(g) For all samples, plot b_{1i} against β_{1i}. Obtain the correlation coefficient (r) between these two variables and the p-value associated with r, specified in the same way as in Part (d).

(h) For all samples, plot shrunk b_{1i} against β_{1i}. Obtain the correlation coefficient between these two variables and the associated p-value, specified as in Part (d). Compare these results with those obtained in Part (g).

(i) Obtain var(b) and var(shrunk b) and compare them with $\text{var}(\beta)$. Comment on the relationships among these variances.

8.4 (a) The file 'multiple y v x, vary n.xlsx', available on this book's website (see Preface), contains data generated in the same way as those used in Exercise 8.3, except that the sample size varies: that is, the number of values of j varies between the values of i.

Repeat Exercise 8.3 using these data. N.B. The solution using GenStat or R requires programming techniques that go somewhat beyond those presented in this chapter, whereas that using SAS follows the pattern of Exercise 8.3 more closely.

References

Anzures-Cabrera, J. and Higgins, J.P.T. (2010) Graphical displays for meta-analysis: an overview with suggestions for practice. *Research Synthesis Methods*, **1**, 66–80.

Benjamini, Y. and Hochberg, Y. (1995) Controlling the false discovery rate: a practical and powerful approach to multiple testing. *Journal of the Royal Statistical Society*, **57**, 289–300.

Bonferroni, C.E. (1936) Teoria statistica delle classi e calcolo delle probabilità. *Pubblicazioni del R Istituto Superiore di Scienze Economiche e Commerciali di Firenze*, **8**, 3–62.

Bühlmann, P. and van de Geer, S. (2011) *Statistics for High-Dimensional Data: Methods, Theory and Applications*, Springer, Berlin, 556 pp.

DerSimonian, R. and Laird, N. (1986) Meta-analysis in clinical trials. *Controlled Clinical Trials*, **7**, 177–188.

Dwan, K., Altman, D.G., Arnaiz, J.A. *et al.* (2008) Systematic review of the empirical evidence of study publication bias and outcome reporting bias. *PLoS One*, **3**, e3081.

Fleiss, J.L. and Gross, A.J. (1991) Meta-analysis in epidemiology, with special reference to studies of the association between exposure to environmental tobacco smoke and lung cancer: a critique. *Journal of Clinical Epidemiology*, **44**, 127–139.

Fram, D.M., Almenoff, J.S. and DuMouchel, W. (2003) Empirical Bayesian data mining for discovering patterns in post-marketing drug safety, in *Proceedings of the Ninth ACM SIGKDD International Conference on Knowledge Discovery and Data Mining*, Association for Computing Machinery, New York, pp. 359–368.

Higgins, J.P. and Thompson, S.G. (2002) Quantifying heterogeneity in a meta-analysis. *Statistics in Medicine*, **21**, 1539–1558.

Ioannidis, J.P.A., Ntzani, E.E., Trikalinos, T.A. and Contopoulos-Ioannidis, D.G. (2001) Replication validity of genetic association studies. *Nature Genetics*, **29**, 306–309.

Kirsch, I., Deacon, B.J., Huedo-Medina, T.B. *et al.* (2008) Initial severity and antidepressant benefits: a meta-analysis of data submitted to the Food and Drug Administration. *PLoS Medicine*, **5**, e45. doi: 10.1371/journal. pmed.0050045

Leng, C., Lin, Y. and Wahba, G. (2006) A note on the lasso and related procedures in model selection. *Statistica Sinica*, **16**, 1273–1284.

Malo, N., Libiger, O. and Schork, N.J. (2008) Accommodating linkage disequilibrium in genetic-association analyses via ridge regression. *American Journal of Human Genetics*, **82**, 375–385.

MAQC Consortium (2010) The MicroArray Quality Control (MAQC)-II study of common practices for the development and validation of microarray-based predictive models. *Nature Biotechnology*, **28**, 827–841.

Pineda-Roman, M., Zangari, M., Haessler, J. *et al.* (2008) Sustained complete remissions in multiple myeloma linked to bortezomib in total therapy 3: comparison with total therapy 2. *British Journal of Haematology*, **140**, 625–634.

R Development Core Team (2012) R Language Definition, Version 2.15.0. Distributed with the R Software. ISBN: 3-900051-13-5.

Shaughnessy, J.D. Jr.,, Zhan, F., Burington, B.E. *et al.* (2007) A validated gene expression model of high-risk multiple myeloma is defined by deregulated expression of genes mapping to chromosome 1. *Blood*, **109**, 2276–2284.

Šidák, Z. (1967) Rectangular confidence regions for the means of multivariate normal distributions. *Journal of the American Statistical Association*, **62**, 626–633.

Simes, R.J. (1986) Publication bias: the case for an international registry of clinical trials. *Journal of Clinical Oncology*, **4**, 1529–1541.

Smyth, G.K. (2004) Linear models and empirical Bayes methods for assessing differential expression in microarray experiments. *Statistical Applications in Genetics and Molecular Biology*, **3** (1), 1–25, Article 3.

Spiegelhalter, D.J., Abrams, K.R. and Myles, J.P. (2004) *Bayesian Approaches to Clinical Trials and Health-Care Evaluation*, John Wiley & Sons, Ltd, Chichester, 391 pp.

Thaler, R.H. (1988) Anomalies: the Winner's Curse. *Journal of Economic Perspectives*, **2**, 191–202.

Tibshirani, R. (1996) Regression shrinkage and selection via the lasso. *Journal of the Royal Statistical Society, Series B*, **58**, 267–288.

Viechtbauer, W. (2010) Conducting meta-analyses in R with the metafor package. *Journal of Statistical Software*, **36** (3), 48 pp..

Whitehead, A. (2002) *Meta-Analysis of Controlled Clinical Trials*, John Wiley & Sons, Ltd, Chichester, 336 pp.

Wilk, M.B. and Gnanadesikan, R. (1968) Probability plotting methods for the analysis of data. *Biometrika*, **55**, 1–17.

9

The use of mixed models for the analysis of unbalanced experimental designs

9.1 A balanced incomplete block design

In Chapter 2, we saw that the analysis of the variance of a standard, balanced experimental design, the split plot design, could be viewed as the fitting of a mixed model, the treatment terms being the fixed-effect terms and the block terms being the random-effect terms. One of the major uses of mixed modelling is the analysis of experiments that cannot be tackled by the analysis of variance, because it has not been possible to achieve exact balance during the design phase.

A *balanced incomplete block design* presented by Cox (1958, Sections 11.1 and 11.2, pp. 219–230) provides the starting point for a simple illustration of this problem. The data are displayed in a spreadsheet in Table 9.1. (Data reproduced by kind permission of Wiley and Sons, Inc.) This experiment comprises five treatments, T1 to T5. It is expected that there will be random variation in the response from day to day, and 'day' is therefore to be included in the analysis as a block term. However, each treatment does not occur on every day: though there are five treatments, there are only three observations per day, so this is clearly not an ordinary randomized complete block design. Each block is *incomplete*. Such a compromise is often necessary: nature does not always provide groups of experimental units (areas of land, litters of animals) that are the right size to permit the application of every treatment to one unit within every group. Despite this, the design of the experiment is *balanced*. Each treatment occurs exactly six times, and any pair of treatments occurs on the same day exactly three times: for example, Treatments T1 and T2 occur together on Days 3, 6 and 10, whereas Treatments T3 and T4 occur together on Days 5, 7 and 8. Therefore every possible comparison between two treatments is made with the same precision. Consequently

Introduction to Mixed Modelling: Beyond Regression and Analysis of Variance, Second Edition. N. W. Galwey.
© 2014 John Wiley & Sons, Ltd. Published 2014 by John Wiley & Sons, Ltd.
Companion website: http://www.wiley.com/go/beyond_regression

Table 9.1   Data from an experiment with a balanced incomplete block design.

	A	B	C		A	B	C
1	day!	T!	response	17	6	T2	1.60
2	1	T4	4.43	18	6	T3	2.13
3	1	T5	3.16	19	6	T1	1.31
4	1	T1	1.40	20	7	T3	4.26
5	2	T4	5.09	21	7	T1	3.86
6	2	T2	1.81	22	7	T4	5.87
7	2	T5	4.54	23	8	T3	2.57
8	3	T2	3.91	24	8	T5	3.06
9	3	T4	6.02	25	8	T2	3.45
10	3	T1	3.32	26	9	T2	3.31
11	4	T5	4.66	27	9	T3	5.10
12	4	T3	3.09	28	9	T4	5.42
13	4	T1	3.56	29	10	T5	5.53
14	5	T3	3.66	30	10	T1	4.46
15	5	T4	2.81	31	10	T2	3.94
16	5	T5	4.66				

Source: Data reproduced by kind permission of Wiley and Sons, Inc.

this design can be analysed by the ordinary methods of analysis of variance, as is done by the following GenStat statements:

```
IMPORT 'IMM edn 2\\Ch 9\\incmplt block design.xlsx'; \
   SHEET = 'Sheet1'
BLOCKS day
TREATMENTS T
ANOVA [FPROB = yes] response
```

The output of the ANOVA statement is as follows:

Analysis of variance

Variate: response

Source of variation	d.f.	s.s.	m.s.	v.r.	F pr.
day stratum					
T	4	4.4503	1.1126	0.28	0.882
Residual	5	20.1348	4.0270	7.17	
day.*Units* stratum					
T	4	15.7533	3.9383	7.01	0.002
Residual	16	8.9915	0.5620		
Total	29	49.3298			

Information summary

Model term	e.f.	Non-orthogonal terms
day stratum		
T	0.167	
day.*Units* stratum		
T	0.833	day

Message: the following units have large residuals.

day 5 *units* 2	−1.42 approx. s.e. 0.55
day 8 *units* 3	1.18 approx. s.e. 0.55

Tables of means

Variate: response
Grand mean 3.73

T	T1	T2	T3	T4	T5
	2.88	2.90	3.60	4.82	4.46

Standard errors of differences of means

Table	T
rep.	6
d.f.	16
s.e.d.	0.474

As in the analysis of the split plot design (Section 2.2), the anova is divided into strata defined by the block term. However, whereas in the split plot design each treatment term was tested in only one stratum, in the present case the treatment term 'T' is tested both in the 'day' stratum and in the 'day.*Units*' stratum (the within-day stratum). This reflects the fact that each comparison between treatments is made partly among and partly within days. The value of $MS_{Residual}$ in the 'day' stratum is several times larger than that in the 'day.*Units*' stratum, confirming that the decision to specify days as blocks was justified. Consequently, the comparisons among treatments are made with considerably more precision in the 'day.*Units*' stratum, and it is only in this stratum that the effect of 'T' is significant according to the F test. The *efficiency factor* (e.f.) in the *information summary* shows how the information concerning the effects of 'T' is distributed between the strata. In the present simple case, the proportion of the information on 'T' given by comparison among observations within each day is given by

$$\frac{(\text{no. of treatments}) \times (\text{no. of units per block} - 1)}{(\text{no. of treatments} - 1) \times (\text{no. of units per block})} = \frac{5 \times (3 - 1)}{(5 - 1) \times 3} = 0.833. \quad (9.1)$$

The remaining proportion, 0.167, is given by comparison among the means for each day. The information summary also notes that 'T' is non-orthogonal to 'day' – that is, the effects of this term are estimated neither entirely within days, nor entirely among them.

The remainder of the output follows the pattern that we have seen in previous anovas. However, the methods for the calculation of the treatment means and of the $SE_{Difference}$ for comparisons between them are modified to take account of the incomplete block structure of the experiment. Each mean is not the simple mean of the observations for the treatment concerned, but is adjusted to allow for the effects of those blocks in which the treatment occurs, as follows (Cox, 1958). We first obtain the block totals, the sum of the totals for the blocks in which the treatment in question occurs. For example, for Treatment T1,

associated sum of blocks total = sum of totals for Blocks 1, 3, 4, 6, 7 and 10

$$= 8.99 + 13.25 + 11.31 + 5.04 + 13.99 + 13.93 = 66.51.$$

Then

adjusted mean

$= [(\text{no. of units per block}) \times (\text{treatment total}) - (\text{associated sum of block totals})]$

$$\times \frac{(\text{no. of treatments} - 1)}{(\text{total no. of units}) \times (\text{no. of units per block} - 1)} + (\text{overall mean of observations})$$

(9.2)

For example, for Treatment T1,

$$\text{adjusted mean} = [3 \times 17.91 - 66.51] \times \frac{(5 - 1)}{30 \times (3 - 1)} + 3.733 = 2.881.$$

The means obtained in this way are based only on the variation among observations within days, and the same is true of $SE_{\text{Difference}}$, which is given by

$$SE_{\text{Difference}} = \sqrt{\frac{2MS_{\text{Residual, day.*Units*stratum}}}{r \cdot (\text{efficiency factor, day.*Units*stratum})}} = \sqrt{\frac{2 \times 0.5620}{6 \times 0.833}} = 0.474 \quad (9.3)$$

where
$r = $ the number of replications of each treatment.

This approach to the treatment means of an incomplete block design was standard in the days when calculations were performed by hand, but by using statistical software it is straightforward to obtain alternative estimates of the means and $SE_{\text{Difference}}$, including the information about treatment effects that is contained in the among-days variation. These estimates with the *recovery of inter-block information* are produced by the statement

ANOVA [PRINT = cbmeans] response

The option setting 'PRINT = cbmeans' indicates that 'combined means', based on information from all strata in the anova, are to be presented (see Payne and Tobias, 1992). The output of this statement is as follows:

Tables of combined means

Variate: response

T	T1	T2	T3	T4	T5
	2.91	2.92	3.57	4.85	4.42

Standard errors of differences of combined means

Table	T
rep.	6
s.e.d.	0.463
effective d.f.	17.51

The combined means are slightly different from the adjusted means, though still not the same as the simple means. The value of $SE_{Difference}$ is reduced from 0.474 to 0.463, showing that a gain in precision has been obtained by the fuller use of the information available. Later (Section 9.2), we will see how the differences from the simple means are determined, and consider the assumptions that are made when combined means are used.

9.2   Imbalance due to a missing block: Mixed-model analysis of the incomplete block design

Now suppose that it were possible to conduct this experiment only on 9 days. The design would still be quite a good one: in the analysis without the recovery of inter-block information, only the comparisons between T1, T2 and T5 (the treatments applied on Day 10) would be made with slightly less precision. However, the design is no longer balanced, and no longer analysable by the standard analysis of variance techniques. When the ANOVA statement is applied to the data with Day 10 omitted, it produces the following output:

Fault 1, code AN 1, statement 1 on line 5

Command: ANOVA [FPROB = yes] response
Design unbalanced – cannot be analysed by ANOVA.
Model term T (non-orthogonal to term day) is unbalanced, in the day.*Units* stratum. Note, though, that the terms are nearly orthogonal (average e.f. = 1.0000). So it may be worth checking their factor values if you were expecting the design to be balanced.

This unbalanced design can, however, be analysed by mixed-modelling methods. This is done by the following statements:

```
VCOMPONENTS [FIXED = T] RANDOM = day
REML [PRINT = Wald, means] response
```

The REML statement produces the following output:

Tests for fixed effects

Sequentially adding terms to fixed model

Fixed term	Wald statistic	n.d.f.	F statistic	d.d.f.	F pr.
T	25.11	4	6.28	15.6	0.003

Dropping individual terms from full fixed model

Fixed term	Wald statistic	n.d.f.	F statistic	d.d.f.	F pr.
T	25.11	4	6.28	15.6	0.003

Message: denominator degrees of freedom for approximate F-tests are calculated using algebraic derivatives ignoring fixed/boundary/singular variance parameters.

Table of predicted means for Constant

3.593   Standard error   0.2926

Table of predicted means for T

T	T1	T2	T3	T4	T5
	2.670	2.814	3.465	4.759	4.259

Standard errors of differences

Average	0.5147
Maximum	0.5399
Minimum	0.4860

Average variance of differences: 0.2652

The F statistic indicates that the variation among the levels of 'T' is highly significant, as was indicated by the within-blocks F statistic in the anova of the complete experiment ($F_{4,16} = 7.01$, $p = 0.002$). The treatment means are similar, but not identical, to the combined means given by the anova on the complete experiment. Thus by using mixed modelling in place of analysis of variance, the requirement that the experiment should have exactly 10 blocks is overcome.

In order to compare the results of the analysis of variance and mixed-modelling approaches more closely, the complete (10-day) experiment can be analysed by mixed modelling. This is done with option settings in the REML statement that specify fuller output, as follows:

```
VCOMPONENTS [FIXED = T] RANDOM = day
REML [PRINT = model, components, Wald, means] response
```

The output of this REML statement is as follows:

REML variance components analysis

Response variate:	response
Fixed model:	Constant + T
Random model:	day
Number of units:	30

Residual term has been added to model
Sparse algorithm with AI (average information) optimization

Estimated variance components

Random term	Component	s.e.
Day	0.6889	0.4227

Residual variance model

Term	Model(order)	Parameter	Estimate	s.e.
Residual	Identity	Sigma2	0.558	0.1966

Tests for fixed effects

Sequentially adding terms to fixed model

Fixed term	Wald statistic	n.d.f.	F statistic	d.d.f.	F pr.
T	28.71	4	7.18	17.5	0.001

Dropping individual terms from full fixed model

Fixed term	Wald statistic	n.d.f.	F statistic	d.d.f.	F pr.
T	28.71	4	7.18	17.5	0.001

Message: denominator degrees of freedom for approximate F-tests are calculated using algebraic derivatives ignoring fixed/boundary/singular variance parameters.

Table of predicted means for Constant

3.733   Standard error   0.2958

Table of predicted means for T

T	T1	T2	T3	T4	T5
	2.906	2.925	3.570	4.849	4.415

Standard error of differences: 0.4627

The value of the F statistic in the mixed-model analysis ($F_{4,17.5} = 7.18$) is slightly larger, with larger denominator d.f. (degrees of freedom), than that for the effect of 'T' in the day.*Units* stratum of the anova ($F_{4,16} = 7.01$), indicating that there has been a slight gain in statistical power from the recovery of the inter-block information. This is to be expected, if the true effects of the treatments are the same in both strata of the experiment. Only if this is the case is it legitimate to combine within- and among-blocks information, and we should consider whether this assumption is reasonable. Many statisticians consider that it is, in the case of an incomplete block design, because the randomization process ensures that the treatment effects are unconfounded with any other source of variation among blocks, just as they are among observations within blocks. However, we have seen (Sections 8.3–8.4) that in the closely related situation of meta-analysis, where there is information on treatment effects both within and among studies, it is considered advisable to use only the within-study information, in order to avoid the danger of confounding. Note that when information is pooled over strata in this analysis, the denominator d.f. are not a whole number – a common phenomenon in mixed modelling, which we first encountered in Section 1.8.

Table 9.2 Comparison of types of means from an incomplete block design.

Simple mean = mean of observations; adjusted mean = mean based on within-block effects only; adjustment = adjusted mean − simple mean; combined mean = mean obtained by pooling within- and among-block information; and difference = combined mean − simple mean.

Treatment	Simple mean	Adjusted mean	Adjustment	Combined mean	Difference
T1	2.985	2.906	−0.079	2.906	−0.079
T2	3.003	2.925	−0.078	2.925	−0.078
T3	3.468	3.570	0.102	3.57	0.102
T4	4.940	4.849	−0.091	4.849	−0.091
T5	4.268	4.415	0.147	4.415	0.147

The treatment means obtained from mixed modelling are now identical to the 'combined means' obtained from analysis of variance (Section 9.1) and the SE (standard error) for differences between them is the same. They are compared with the simple treatment means, and the adjusted means based on within-block effects only, in Table 9.2.

The differences between the combined means and the simple means are based on the estimates of the day effects, which are obtained from the mixed-modelling analysis by the following statement:

```
REML [PRINT = effects; PTERMS = day; METHOD = Fisher] response
```

The output of this statement is as follows:

Table of effects for day

day	1	2	3	4	5
	−0.8348	−0.1965	0.6747	0.1097	−0.4473

day	6	7	8	9	10
	−1.1446	0.6995	−0.4803	0.6527	0.9668

Standard errors of differences

Average: 0.5591
Maximum: 0.5678
Minimum: 0.5547

Average variance of differences: 0.3126

The combined mean for each treatment is obtained from the formula

combined mean = simple mean − mean(effects of days on which the treatment occurs).

$$(9.4)$$

For example, Treatment T1 occurs on days 1, 3, 4, 6, 7 and 10. The mean of the effects of these days is

$$\frac{-0.8348 + 0.6747 + 0.1097 - 1.1446 + 0.6995 + 0.9668}{6} = 0.07855$$

and
$$\text{combined mean}_{T1} = 2.985 - 0.07855 = 2.906.$$

If 'day' is specified as a fixed-effect term – that is, if the VCOMPONENTS statement is changed from

```
VCOMPONENTS [FIXED = T] RANDOM = day
```

to

```
VCOMPONENTS [FIXED = T, day]
```

then the treatment means obtained are the adjusted means from the original anova, based only on the within-day variation. This is analogous to the specification of 'study' as a fixed-effect term in a meta-analysis, in order to obtain an estimate of the treatment effect based on within-study variation only (Section 8.4).

9.3 Use of R to analyse the incomplete block design

The following commands import the data into R, perform the appropriate analysis of variance and present the results:

```
rm(list = ls())
incomplete.blk <- read.table(
    "IMM edn 2\\Ch 9\\incmplt block design.txt",
    header=TRUE)
attach(incomplete.blk)
fday <- factor(day)
fT <- factor(T)
incomplt.blk.modelaov <-
    aov(response ~ fT + Error(fday))
summary(incomplt.blk.modelaov)
model.tables(incomplt.blk.modelaov, type = "means", se = TRUE)
```

The output of these commands is as follows:

```
Error: fday
          Df Sum Sq Mean Sq F value Pr(>F)
fT         4   4.45   1.113   0.276  0.882
Residuals  5  20.14   4.027

Error: Within
          Df Sum Sq Mean Sq F value  Pr(>F)
fT         4 15.753   3.938   7.008 0.00185 **
Residuals 16  8.992   0.562
```

```
---
Signif. codes:  0 '***' 0.001 '**' 0.01 '*' 0.05 '.' 0.1 ' ' 1

Error in FUN(X[[1 L]], ...) :
  eff.aovlist: non-orthogonal contrasts would give an
incorrect answer
```

The anova agrees with that produced by GenStat, but the function `model.tables()` is not able to produce the means from this incomplete block design.

When the data from Day 10 are omitted, the output produced by the function `summary()` is as follows:

```
Error: fday
          Df Sum Sq Mean Sq F value Pr(>F)
fT         4  10.35   2.588   0.902  0.538
Residuals  4  11.47   2.868

Error: Within
          Df Sum Sq Mean Sq F value Pr(>F)
fT         4 14.669   3.667    5.86 0.0055 **
Residuals 14  8.761   0.626
---
Signif. codes:  0 '***' 0.001 '**' 0.01 '*' 0.05 '.' 0.1 ' ' 1
```

R does not give a warning that the design is unbalanced.

The following commands produce and present a mixed-model analysis of the data with Day 10 omitted:

```
library(nlme)
incomplt.blk.modellme <- lme(response ~ fT,
   data = incomplete.blk, random = ~ 1|fday)
summary(incomplt.blk.modellme)
anova(incomplt.blk.modellme)
```

The output of these commands is as follows:

```
Linear mixed-effects model fit by REML
 Data: incomplete.blk
       AIC       BIC    logLik
  84.36937 92.00666 -35.18468

Random effects:
 Formula: ~1 | fday
         (Intercept)  Residual
StdDev:   0.7492358 0.7888804
```

```
Fixed effects: response ~ fT
                   Value Std.Error DF  t-value p-value
(Intercept) 2.6699719 0.4499793 14 5.933544  0.0000
fTT2        0.1441268 0.5399094 14 0.266946  0.7934
fTT3        0.7949600 0.5068698 14 1.568371  0.1391
fTT4        2.0895048 0.5068698 14 4.122370  0.0010
fTT5        1.5886669 0.5399094 14 2.942469  0.0107
 Correlation:
      (Intr) fTT2    fTT3    fTT4
fTT2 -0.600
fTT3 -0.622  0.533
fTT4 -0.622  0.533  0.540
fTT5 -0.600  0.500  0.533  0.533

Standardized Within-Group Residuals:
        Min            Q1          Med            Q3          Max
-2.05374371 -0.53906073  0.05276362  0.54670433  1.25581019

Number of Observations: 27
Number of Groups: 9

            numDF denDF   F-value p-value
(Intercept)     1    14 154.41387  <.0001
fT              4    14   6.27776  0.0041
```

The *F* value in the analysis of variance table agrees with that given by GenStat. In the estimates of the fixed effects, the mean for treatment T1 is used as the intercept, and its value, 2.66997, agrees with that given by GenStat. The effects of the other treatments are presented relative to this intercept, and the means are related to the effects by the formula

$$\text{mean}(\text{Treatment } i) = \text{intercept} + \text{effect}(\text{Treatment } i). \tag{9.5}$$

Thus
$$\text{mean}(\text{Treatment } 2) = 2.6699719 + 0.1441268 = 2.8140987.$$

These values agree with those given by GenStat. SEs of these means and the differences between them can be obtained by the methods described in Section 2.8: see also Exercise 9.3 at the end of this chapter.

9.4   Use of SAS to analyse the incomplete block design

No satisfactory way of using SAS's PROC ANOVA to analyse this incomplete block design is known to the present author. The following SAS statements import the data and perform a mixed-model analysis with Day 10 omitted:

```
PROC IMPORT OUT = incmplt DBMS = EXCELCS REPLACE
   DATAFILE = "&pathname.\IMM edn 2\Ch 9\incmplt block design.xlsx";
   SHEET = "for SAS, omit day 10";
RUN;
```

```
ODS RTF;
PROC MIXED ASYCOV NOBOUND;
   CLASS day T;
   MODEL response = T /DDFM = KR HTYPE = 1;
   RANDOM day;
   LSMEANS T;
RUN;
ODS RTF CLOSE;
```

Part of the output from PROC MIXED is as follows:

Convergence criteria met.

Covariance parameter estimates	
Cov Parm	**Estimate**
Day	0.5614
Residual	0.6223

Asymptotic covariance matrix of estimates			
Row	**Cov Parm**	**CovP1**	**CovP2**
1	day	0.1590	−0.01866
2	Residual	−0.01866	0.05471

Fit statistics	
−2 Res Log Likelihood	70.4
AIC (smaller is better)	74.4
AICC (smaller is better)	75.0
BIC (smaller is better)	74.8

Null model likelihood ratio test		
DF	**Chi-square**	**Pr > ChiSq**
1	4.42	0.0355

Type 1 tests of fixed effects				
Effect	**Num DF**	**Den DF**	**F value**	**Pr > F**
T	4	15.7	6.05	0.0038

Least squares means						
Effect	**T**	**Estimate**	**Standard error**	**DF**	***t* value**	**Pr > |*t*|**
T	T1	2.6700	0.4556	21.1	5.86	<0.0001
T	T2	2.8141	0.4556	21.1	6.18	<0.0001
T	T3	3.4649	0.4226	19.9	8.20	<0.0001
T	T4	4.7595	0.4226	19.9	11.26	<0.0001
T	T5	4.2586	0.4556	21.1	9.35	<0.0001

The F value and denominator d.f. are similar, but not identical to those given by GenStat, probably because the denominator d.f. are determined by the Kenward–Roger method in SAS and a somewhat different method in GenStat. The treatment means agree with those given by GenStat. The value of SE_{Mean} for Treatments T1, T2 and T3, which occurred in the omitted Block 10, is larger than that for the other treatments, as expected.

9.5   Relaxation of the requirement for balance: Alpha designs

In the early literature on experimental design, exemplified by Cochran and Cox (1957), there was a strong emphasis on balance, which was necessary in order to make the calculations required for statistical analysis feasible. However, the advent of electronic computers and the development of the mixed-modelling approach permitted the development of a range of designs in which the requirement for balance was relaxed, in order to allow the investigator to follow more closely the pattern of natural variation in his or her experimental material. A common problem, which occurs in many contexts, is that the number of experimental units that form a natural, homogeneous block is much smaller than the number of treatments to be compared. In this situation, a randomized complete block design will take account only of a small proportion of the natural variation, leaving an undesirably large residual variance. The merit of the balanced incomplete block design is that the number of experimental units per block is smaller than the number of treatments, which helps to overcome this problem. However, the possibility of finding an appropriate design of this type depends on the particular combination of

- the number of treatments,

- the number of replications and

- the number of units per block.

If the requirement for balance is relaxed, designs covering a much wider range of situations can be specified.

Hence the ability to apply mixed-model analyses to unbalanced designs is useful in many situations in which perfect balance is difficult to achieve. However, this does not exempt the investigator from trying to achieve balance. To illustrate this, consider two alternative designs for an experiment in which three treatments, A, B and C, are to be allocated to five blocks, each comprising three experimental units, as shown in Figure 9.1. Design 1 is a standard randomized block design, whereas in Design 2, the treatments have been randomly allocated to the units without regard to the block structure. If the block term is included in the model, Design 2 is unbalanced and cannot be analysed by the standard analysis of variance methods: however, it can still be analysed by mixed modelling. In this design, the treatment effects are estimated partly within, and partly among, blocks. For example, the difference between the means of Block 3 and Block 5 could be due either to natural variation among the blocks, or to a difference between the effects of Treatment B and Treatment C. If the variance component due to blocks is greater than zero – that is, if there is any real natural variation among the blocks – there will be a reduction in the precision with which the treatment effects are estimated, relative to that obtained from Design 1. No amount of sophistication in the analysis methods used can overcome this loss of efficiency, and the investigator should therefore seek an efficient experimental design, with balance or near-balance as one of the criteria, even though other designs are analysable.

Many extensions to the concept of the balanced incomplete block design have been devised to help the investigator in this task. One of these is *alpha designs*, which consist of incomplete blocks, each comprising only a small proportion of the treatments, but in which these blocks are grouped into complete replications. The treatments allocated to each block are chosen so that the design is nearly balanced. Alpha designs were originally devised for the analysis of plant breeding trials (Patterson and Williams, 1976), and a set of arrays for generating alpha designs with varying numbers of treatments, numbers of replications and block sizes was presented by Patterson, Williams and Hunter (1978). Computer software can be used to generate designs from these arrays. For example, the following GenStat statement produces a design with 23 treatments in three replicates, each replicate comprising five blocks:

```
AGALPHA [PRINT = design] LEVELS = 23; NREPLICATES = 3; \
    NBLOCKS = 5; SEED = 60594; \
    TREATMENTS = trtmnt; REPLICATES = rep; BLOCKS = blk; \
    UNITS = unt
```

The arbitrary value in the setting of the SEED parameter is used to initiate the randomization of the design. The output of this statement is as follows.

Treatment combinations on each unit of the design

rep	unt blk	1	2	3	4	5
1	1	16	15	22	–	14
	2	10	19	9	17	18
	3	7	23	6	3	1
	4	8	–	12	13	5
	5	11	4	20	21	2
2	1	14	7	2	13	9
	2	1	11	17	5	15
	3	16	6	20	–	10
	4	–	8	21	23	18
	5	3	19	4	22	12

3	1	16	17	3	21	13
	2	8	10	14	1	4
	3	11	6	9	–	12
	4	5	7	20	22	18
	5	–	2	23	19	15

Treatment factors are listed in the order: trtmnt.

Because the number of treatments is not an exact multiple of the number of blocks, some blocks contain five units and others four. The 'missing' unit in a four-unit block is indicated by a hyphen (-). The design is nearly balanced: for example, Treatments 15 and 16 occur in the same block in Replicate 1, but not in any other replicate, and the same is true of many other pairs of treatments. However, the balance is not perfect: some pairs of treatments, for example, Treatments 15 and 18, do not occur in the same block in any replicate. The number of blocks in which each pair-wise combination of treatments occurs is shown in the following matrix:

	1	2	3	4	5	6	7	8	9	10	11	12	13	14	15	16	17	18	19	20	21	22	23
1	0
2	0	0
3	1	0	0
4	1	1	1	0
5	1	0	0	0	0
6	1	0	1	0	0	0
7	1	1	1	0	1	1	0
8	1	0	0	1	1	0	0	0
9	0	1	0	0	0	1	1	0	0
10	1	0	0	1	0	1	0	1	1	0
11	1	1	0	1	1	1	0	0	1	0	0
12	0	0	1	1	1	1	0	1	1	0	1	0
13	0	1	1	0	1	0	1	1	1	0	0	1	0
14	1	1	0	1	0	0	1	1	1	1	0	0	1	0
15	1	1	0	0	1	0	0	0	0	0	1	0	0	1	0
16	0	0	1	0	0	1	0	0	0	1	0	0	1	1	1	0
17	1	0	1	0	1	0	0	0	1	1	1	0	1	0	1	1	0
18	0	0	0	0	1	0	1	1	1	1	0	0	0	0	0	0	1	0
19	0	1	1	1	0	0	0	0	1	1	0	1	0	0	1	0	1	1	0
20	0	1	0	1	1	1	1	0	0	1	1	0	0	0	0	1	0	1	0	0
21	0	1	1	1	0	0	0	1	0	0	1	0	1	0	0	1	1	1	0	1	0
22	0	0	1	1	1	0	1	0	0	0	0	1	0	1	1	1	0	1	1	1	0	0
23	1	1	1	0	0	1	1	1	0	0	0	0	0	0	1	0	0	1	1	0	1	0	0

Each combination occurs either in one block or in none: no combination occurs in more than one block. The number of treatments with which each treatment is combined is as shown in Table 9.3. The numbers are nearly equal – each treatment is combined with 10, 11 or 12 others.

	Design 1					Design 2			
	Block					Block			
	1	B	A	C		1	C	B	A
	2	C	B	A		2	B	C	A
	3	C	A	B		3	B	B	A
	4	B	C	A		4	B	A	C
	5	B	C	A		5	A	C	C

Figure 9.1   Alternative designs for an experiment with three treatments.

Table 9.3   Number of treatments with which each treatment is combined in the incomplete block design.

i = treatment number; N = number of treatments with which the ith treatment is combined.

i	N	i	N
1	12	13	11
2	11	14	11
3	12	15	10
4	12	16	10
5	11	17	12
6	10	18	11
7	12	19	11
8	10	20	11
9	11	21	11
10	11	22	11
11	11	23	10
12	10		

Thus the experiment is as balanced as is possible given the constraints of the design, or nearly so.

In addition to being presented in the GenStat output, the treatment, block and unit for each observation in the experiment are stored in the factors 'trtmnt', 'rep' and 'unt', respectively. These can be retrieved when the experiment is complete, to analyse the data.

There are some constraints on the numbers of treatments, replications and units per block that can be used when specifying an alpha design. If an invalid combination of values is specified, GenStat gives informative diagnostics.

In order to generate alpha designs in R, a package named 'agricolae' is required. This can be installed from the R website by the method described in Section 3.16. The following commands then generate an alpha design with the required specification:

```
rm(list = ls())
library(agricolae)
```

```
trtmnt.lab <- 1:25
alpha.design <- design.alpha(trt = trtmnt.lab,
    k = 5, r = 3, seed = 60594)
print(alpha.design$design)
```

The function `library()` loads the package 'agricolae'. A structure named 'trtmnt.lab' is then created to hold the treatment labels. R requires that the number of treatments be a multiple of the size of each block, so labels from 1 to 25 are generated: labels 24 and 25 can be discarded when the design is used. The function `design.alpha()` then generates the required design. The arguments `trt`, `k` and `r` specify a list of treatment labels, the size of each block and the number of replications, respectively. The argument `seed` provides an arbitrary starting value from which the randomization process can 'grow'. The results are stored in a structure named 'alpha.design'. The function `print()` produces the following output:

```
$rep1
      [,1] [,2] [,3] [,4] [,5]
[1,]  "1"  "13" "16" "25" "18"
[2,]  "8"  "20" "23" "19" "4"
[3,]  "2"  "15" "3"  "14" "11"
[4,]  "5"  "12" "17" "7"  "24"
[5,]  "22" "6"  "10" "9"  "21"

$rep2
      [,1] [,2] [,3] [,4] [,5]
[1,]  "2"  "16" "8"  "12" "6"
[2,]  "19" "11" "13" "10" "17"
[3,]  "24" "3"  "18" "21" "20"
[4,]  "15" "7"  "1"  "22" "4"
[5,]  "25" "5"  "14" "23" "9"

$rep3
      [,1] [,2] [,3] [,4] [,5]
[1,]  "15" "25" "17" "8"  "21"
[2,]  "20" "5"  "2"  "1"  "10"
[3,]  "4"  "6"  "24" "13" "14"
[4,]  "18" "12" "11" "23" "22"
[5,]  "19" "9"  "3"  "16" "7"
```

Within each replication, each row comprises a block. Note that although this alpha design meets the same criteria as that produced by GenStat, the allocation of treatments to experimental units is different: the same seed value does not produce the same randomization in different software systems. The structure 'alpha.design' also contains the treatment, block and unit for each observation in the experiment: this information is displayed by the following command:

```
print(alpha.design$book)
```

The first and last few lines of the output from this statement are as follows:

	plots	cols	block	trtmnt.lab	replication
1	1	1	1	1	1
2	2	2	1	13	1
3	3	3	1	16	1
4	4	4	1	25	1
5	5	5	1	18	1
6	6	1	2	8	1
7	7	2	2	20	1
.					.
.					.
.					.
73	73	3	15	3	3
74	74	4	15	16	3
75	75	5	15	7	3

SAS has no facilities for the convenient production of alpha designs, as far as is known to the present author. However, a design very similar to that specified above is produced by the following commands:

```
DATA trt_file;
    DO trtmnt = 1 TO 25;
        OUTPUT;
    END;
RUN;

ODS RTF;
PROC OPTEX DATA = trt_file SEED = 60594;
    CLASS trtmnt;
    MODEL trtmnt;
    BLOCKS STRUCTURE = (15)5;
    OUTPUT OUT = design;
RUN;
ODS RTF CLOSE;
```

The DATA step produces a file that contains a single variable, 'trtmnt', holding the values 1–25. PROC OPTEX is then used to specify the production of an optimized experimental design on the basis of this file. The option SEED provides an arbitrary starting value from which the randomization process can 'grow'. The CLASS statement indicates that 'trtmnt' is a categorical variable. The MODEL statement specifies the model formula for the experimental design to be generated – in this case, the single term 'trtmnt'. The BLOCKS statement specifies that the design is to comprise 15 blocks, each containing five experimental units. The OUTPUT statement indicates that the design generated is to be stored in the file named 'design'. Part of the output from these statements is as follows:

Design number	Treatment D-efficiency	Treatment A-efficiency	Block design D-efficiency
1	81.6497	80.0000	97.9796
2	81.6497	80.0000	97.9796
3	81.6497	80.0000	97.9796
4	81.6497	80.0000	97.9796
5	81.6497	80.0000	97.9796
6	81.6044	79.8581	97.9252
7	81.6044	79.8581	97.9252
8	81.6044	79.8581	97.9252
9	81.6044	79.8581	97.9252
10	81.5632	79.7290	97.8758

This table indicates the efficiency of the designs generated on different efficiency (or optimality) criteria, explained in the SAS/QC User's Guide (SAS Institute Inc., 2010, Chapter 10, pp. 889–890). D-efficiency is the criterion used by default for choosing designs, and design number 1 is the most efficient generated. The information in the data set 'design', which specifies this design, is presented in Table 9.4. It comprises 25 treatments, each replicated three times: experimental units containing Treatments 24 and 25 can be deleted to obtain the 23-treatment design required. As in the corresponding alpha design, the 75 units are arranged in 15 blocks of 5. However, this is not an alpha design as the blocks cannot necessarily be arranged to give three complete replications, each comprising five blocks.

9.6   Approximate balance in two directions: The alphalpha design

If each replication of an experiment is to be laid out in an array of rows and columns (as in the GenStat and R output above), an alpha design will take account of the natural variation among rows, but not of that among columns: this will contribute to residual variation and reduce the precision with which treatment effects are estimated. However, the concepts of alpha designs can be extended further, to give approximate balance in both dimensions. Results from an experiment with such a design (known as an *alphalpha design*) were presented by Mead (1997). The experiment was a field trial of 35 wheat genotypes, laid out in two replicates, the plots in each replicate being arranged in an array of five rows and seven columns. The grain yield from each plot was measured (units not given). The data are presented in the spreadsheet in Table 9.5. (Data reproduced by kind permission of Mike Talbot, Biomathematics and Statistics Scotland.)

In this experiment, if two treatments occur in the same row *or column* in Replicate 1, then in most cases they occur in different rows *and columns* in Replicate 2. However, the balance is not perfect and there are exceptions: for example, Genotypes 4 and 12 both occur in Row 1 in Replicate 1, and both in Row 2 in Replicate 2. The following statements import and analyse these data:

Table 9.4   Specification of a design with 25 treatments in 3 replications, in 15 blocks of 5 experimental units, generated by SAS.

	BLOCK	trtment		BLOCK	trtment
1	1	16	39	8	8
2	1	14	40	8	23
3	1	6	41	9	5
4	1	10	42	9	21
5	1	11	43	9	13
6	2	16	44	9	15
7	2	7	45	9	18
8	2	8	46	10	20
9	2	5	47	10	22
10	2	1	48	10	1
11	3	14	49	10	12
12	3	17	50	10	19
13	3	18	51	11	20
14	3	25	52	11	15
15	3	22	53	11	24
16	4	3	54	11	6
17	4	11	55	11	9
18	4	21	56	12	23
19	4	12	57	12	20
20	4	23	58	12	25
21	5	2	59	12	16
22	5	4	60	12	13
23	5	10	61	13	14
24	5	19	62	13	7
25	5	13	63	13	19
26	6	22	64	13	21
27	6	2	65	13	24
28	6	5	66	14	17
29	6	9	67	14	10
30	6	11	68	14	3
31	7	4	69	14	15
32	7	8	70	14	1
33	7	12	71	15	4
34	7	6	72	15	3
35	7	18	73	15	7
36	8	2	74	15	25
37	8	24	75	15	9
38	8	17			

Table 9.5 Yields of wheat genotypes, investigated in an alphalpha design.

	A	B	C	D	E
1	replicate!	row!	column!	genotype!	yield
2	1	1	1	20	3.77
3	1	1	2	4	3.21
4	1	1	3	33	4.55
5	1	1	4	28	4.09
6	1	1	5	7	5.05
7	1	1	6	12	4.19
8	1	1	7	30	3.27
9	1	2	1	10	3.44
10	1	2	2	14	4.30
11	1	2	3	16	*
12	1	2	4	21	3.86
13	1	2	5	31	3.26
14	1	2	6	6	4.30
15	1	2	7	18	3.72
16	1	3	1	22	3.49
17	1	3	2	11	4.20
18	1	3	3	19	4.77
19	1	3	4	26	2.56
20	1	3	5	29	2.87
21	1	3	6	15	1.93
22	1	3	7	23	2.26
23	1	4	1	24	3.62
24	1	4	2	25	4.52
25	1	4	3	5	4.23
26	1	4	4	32	3.76
27	1	4	5	2	3.61
28	1	4	6	27	3.62
29	1	4	7	8	4.01
30	1	5	1	17	3.81
31	1	5	2	9	3.75
32	1	5	3	3	4.81
33	1	5	4	34	3.69
34	1	5	5	13	4.61
35	1	5	6	35	2.68
36	1	5	7	1	4.15

	A	B	C	D	E
37	2	1	1	31	4.70
38	2	1	2	19	7.37
39	2	1	3	25	5.03
40	2	1	4	34	5.33
41	2	1	5	20	5.73
42	2	1	6	8	4.70
43	2	1	7	6	5.63
44	2	2	1	24	4.07
45	2	2	2	21	5.66
46	2	2	3	12	4.98
47	2	2	4	4	4.04
48	2	2	5	23	4.27
49	2	2	6	13	4.10
50	2	2	7	3	4.75
51	2	3	1	11	5.66
52	2	3	2	7	6.43
53	2	3	3	26	4.59
54	2	3	4	5	5.20
55	2	3	5	35	4.83
56	2	3	6	10	4.70
57	2	3	7	30	4.23
58	2	4	1	33	5.71
59	2	4	2	9	6.13
60	2	4	3	17	4.63
61	2	4	4	18	5.48
62	2	4	5	32	5.47
63	2	4	6	15	*
64	2	4	7	2	4.16
65	2	5	1	1	5.22
66	2	5	2	27	6.16
67	2	5	3	16	4.20
68	2	5	4	29	4.66
69	2	5	5	14	5.54
70	2	5	6	28	3.81
71	2	5	7	22	3.60

An asterisk (*) indicates a missing value.
Source: Data reproduced by kind permission of Mike Talbot, Biomathematics and Statistics Scotland.

```
IMPORT 'IMM edn 2\\Ch 9\\alphalpha design.xlsx'
VCOMPONENTS [FIXED = genotype] RANDOM = replicate/(row + column)
REML [PRINT = model, components, Wald, means] yield
```

The random-effects model,

$$\text{replicate}/(\text{row} + \text{column}) = \text{replicate} + \text{replicate.row} + \text{replicate.column},$$

indicates that both rows and columns are nested within replicates: that is,

- there is no main effect of 'row', and Row 1 in Replicate 1 is not the same row as Row 1 in Replicate 2;

- similarly, there is no main effect of 'column', and Column 1 in Replicate 1 is not the same column as Column 1 in Replicate 2.

The output of the REML statement is as follows:

REML variance components analysis

Response variate:	yield
Fixed model:	Constant + genotype
Random model:	replicate + replicate.row + replicate.column
Number of units:	68 (2 units excluded due to zero weights or missing values)

Residual term has been added to model
Sparse algorithm with AI optimization

Estimated variance components

Random term	component	s.e.
replicate	0.70356	1.05611
replicate.row	0.06387	0.04897
replicate.column	0.19265	0.09741

Residual variance model

Term	Model(order)	Parameter	Estimate	s.e.
Residual	Identity	Sigma2	0.0902	0.03685

Tests for fixed effects

Sequentially adding terms to fixed model

Fixed term	Wald statistic	n.d.f.	F statistic	d.d.f.	F pr.
genotype	166.52	34	5.02	12.1	0.002

Dropping individual terms from full fixed model

Fixed term	Wald statistic	n.d.f.	F statistic	d.d.f.	F pr.
genotype	166.52	34	5.02	12.1	0.002

Message: denominator degrees of freedom for approximate F-tests are calculated using algebraic derivatives ignoring fixed/boundary/singular variance parameters.

Table of predicted means for Constant

4.363   Standard error   0.6110

Table of predicted means for genotype

genotype	1	2	3	4	5	6	7	8
	4.814	3.915	5.098	3.521	4.395	5.409	5.085	4.603

genotype	9	10	11	12	13	14	15	16
	4.351	4.328	4.931	4.946	4.682	4.764	3.212	3.958

genotype	17	18	19	20	21	22	23	24
	4.154	4.565	5.669	4.320	4.593	4.011	3.423	3.888

genotype	25	26	27	28	29	30	31	32
	4.640	3.759	4.699	4.295	3.793	3.953	3.859	4.264

genotype	33	34	35
	4.914	4.298	3.602

Standard errors of differences

Average:	0.3854
Maximum:	0.5358
Minimum:	0.3511

Average variance of differences: 0.1498

The variance component estimate for the term 'replicate' is smaller than its SE, suggesting that the natural variation between the two replicates may be no greater than that among rows or columns within a replicate. However, the variance components for 'row' and 'column' are both larger than their respective SEs, suggesting that there is real natural variation among the rows and columns, and hence that the treatment effects are estimated with greater precision than they would have been if each replicate had been laid out in a randomized complete block. The F statistic indicates that there are highly significant differences among the genotype means. However, note that,

$$F \times DF_{\text{Numerator of } F} = 5.02 \times 34 = 170.68 \neq \text{Wald statistic,}$$

that is, this spatial analysis gives a slight exception to the relationship between the F and Wald statistics stated in Equation 1.5 (Section 1.8). As in the case of the balanced incomplete block design, the genotype means are slightly different from the corresponding simple means (e.g. the simple mean yield of Genotype 1 is $(4.15 + 5.22)/2 = 4.69$, whereas the value given by the mixed-model analysis is 4.814), and the adjustment is related to the effects of the rows and columns in which each treatment occurs.

Methods for producing two-dimensional incomplete block designs, and many other aspects of incomplete block designs, are discussed by John and Williams (1995). The software *CycDesigN* (distributed by VSN International: see website http://www.vsni.co.uk/software/cycdesign) can be used to generate such designs.

9.7   Use of R to analyse the alphalpha design

The following commands import the alphalpha design data into R, perform the appropriate mixed-model analysis and present the results:

```
rm(list = ls())
alphalpha <- read.table(
   "IMM edn 2\\Ch 9\\alphalpha design.txt",
   header=TRUE)attach(alphalpha)
freplicate <- factor(replicate)
frow <- factor(row)
fcolumn <- factor(column)
fgenotype <- factor(genotype)
library(lme4)
alphalpha.modellme <- lmer(yield ~ fgenotype +
   (1|freplicate) + (1|freplicate:frow) + (1|freplicate:fcolumn),
   data = alphalpha)
summary(alphalpha.modellme)
anova(alphalpha.modellme)
```

The output of these commands is as follows:

```
Linear mixed model fit by REML
Formula: yield ~ fgenotype + (1 | freplicate) +
(1 | freplicate:frow) +      (1 | freplicate:fcolumn)
   Data: alphalpha
   AIC    BIC  logLik deviance REMLdev
 151.3 237.8 -36.63    34.66    73.26
Random effects:
 Groups              Name        Variance Std.Dev.
 freplicate:fcolumn (Intercept) 0.192650 0.43892
 freplicate:frow    (Intercept) 0.063875 0.25273
 freplicate         (Intercept) 0.703555 0.83878
 Residual                       0.090172 0.30029
Number of obs: 68, groups: freplicate:fcolumn, 14;
freplicate:frow, 10; freplicate, 2
```

```
Fixed effects:
            Estimate Std. Error t value
(Intercept)   4.81429    0.66250   7.267
fgenotype2   -0.89883    0.37658  -2.387
fgenotype3    0.28332    0.36903   0.768
 .
 .
 .
fgenotype33   0.09998    0.36027   0.278
fgenotype34  -0.51583    0.36671  -1.407
fgenotype35  -1.21249    0.37184  -3.261

Correlation of Fixed Effects:
 .
 .
 .
<Here follows a large correlation matrix>
 .
 .
 .
Analysis of Variance Table
           Df Sum Sq Mean Sq F value
fgenotype 34 15.016 0.44164  4.8978
```

The estimates of the variance components agree with those given by GenStat, and the F value for 'fgenotype' agrees with the value of (Wald statistic)/DF$_{\text{Numerator of } F}$ given by GenStat, as expected from Equation 1.5 (Section 1.8), though not with the F value given by GenStat. As in the analysis of the balanced incomplete block design using R (Section 9.3), in the estimates of the fixed effects, the mean for the first level of the treatment factor is used as the intercept. Its value, 4.81429, agrees with that given by GenStat for the mean of Genotype 1. The effects of the other treatments are presented relative to this intercept, and the means are related to the effects by Equation 9.5. Thus

$$\text{mean(Genotype 2)} = 4.81429 - 0.89883 = 3.91546.$$

These values agree with those given by GenStat. SEs of these means and of the differences between them can be obtained by methods similar to those described in Section 2.8, though the details differ due to the use of the package 'lme4' and the function lmer() instead of 'nlme' and lme(). In brief, the fixed effects are extracted by the function coef(alphalpha. modellme)$'freplicate'[1,] and the covariance matrix is extracted by the function vcov(alphalpha.modellme). Full details are given in the solution to Exercise 9.4 at the end of this chapter.

9.8   Use of SAS to analyse the alphalpha design

The following statements import the alphalpha design data into SAS and perform the appropriate mixed-model analysis:

```
PROC IMPORT OUT = alphalpha DBMS = EXCELCS REPLACE
   DATAFILE = "&pathname.\IMM edn 2\Ch 9\alphalpha design.xlsx";
   SHEET = "for SAS";
RUN;

ODS RTF;
PROC MIXED ASYCOV NOBOUND;
   CLASS replicate row column genotype;
   MODEL yield = genotype /DDFM = KR HTYPE = 1;
   RANDOM replicate replicate*row replicate*column;
   LSMEANS genotype;
RUN;
ODS RTF CLOSE;
```

Part of the output from `PROC MIXED` is as follows:

Convergence criteria met.

Covariance parameter estimates	
Cov Parm	**Estimate**
replicate	0.7036
replicate*row	0.06387
replicate*column	0.1927
Residual	0.09017

Asymptotic covariance matrix of estimates					
Row	**Cov Parm**	**CovP1**	**CovP2**	**CovP3**	**CovP4**
1	replicate	1.1154	−0.00025	−0.00168	0.000110
2	replicate*row	−0.00025	0.002513	0.000166	−0.00052
3	replicate*column	−0.00168	0.000166	0.009516	−0.00061
4	Residual	0.000110	−0.00052	−0.00061	0.001384

Fit statistics	
−2 Res Log Likelihood	73.3
AIC (smaller is better)	81.3
AICC (smaller is better)	82.7
BIC (smaller is better)	76.0

Null model likelihood ratio test		
DF	**Chi-square**	**Pr > ChiSq**
3	45.01	<0.0001

Type 1 tests of fixed effects				
Effect	Num DF	Den DF	F value	Pr > F
genotype	34	11.8	4.84	0.0030

Least squares means						
Effect	genotype	Estimate	Standard error	DF	t value	Pr > |t|
genotype	1	4.8143	0.6664	1.38	7.22	0.0462
genotype	2	3.9155	0.6689	1.39	5.85	0.0602
genotype	3	5.0976	0.6698	1.4	7.61	0.0416
	⋮					
genotype	33	4.9143	0.6704	1.4	7.33	0.0436
genotype	34	4.2985	0.6660	1.38	6.45	0.0539
genotype	35	3.6018	0.6690	1.4	5.38	0.0671

The variance component estimates and the genotype means agree with those given by GenStat, and the F value for the effects of genotype is similar but not identical, probably due to the different methods for the determination of the denominator d.f. used by the two software systems (see Section 9.4).

9.9　Summary

A major use of mixed modelling is the analysis of experiments that cannot be tackled by the analysis of variance, because it has not been possible to achieve exact balance during the design phase.

A balanced incomplete block design provides the starting point for a simple illustration of this problem.

Information concerning the treatment effects is distributed over two strata of this design, the among-blocks and within-block strata. The efficiency factor indicates the proportion of the information in each stratum.

If one block is omitted from a balanced incomplete block design, it becomes unbalanced and cannot be analysed by the ordinary methods of analysis of variance. However, it can still be analysed by mixed modelling.

If a balanced incomplete block design is analysed by the methods of analysis of variance, the significance of the treatment term is tested both in the among-blocks stratum and the within-blocks stratum. If it is analysed by mixed modelling, with the blocks specified as a random-effects term, a single significance test, pooled over the two strata, is performed.

This pooling will usually give a gain in statistical power, if the true effects of treatments are the same in the two strata.

Two types of treatment means can be obtained for presentation with the anova of a balanced incomplete block design, namely

- adjusted means, obtained from the simple means by adding an adjustment so that they are based only on the within-block information;

- combined means, based on both within- and among-block information, obtained from the simple means by subtracting the mean of the block effects for the blocks in which the treatment occurs.

When an incomplete block design is analysed by mixed modelling, if the blocks are specified as a random-effect term, the combined treatment means are presented. If the blocks are specified as a fixed-effect term, the adjusted means based only on within-block information are presented.

The number of experimental units that form a natural, homogeneous block is often much smaller than the number of treatments to be compared. Balanced incomplete block designs can sometimes be used in such circumstances, but often no such design can be found that matches

- the number of treatments,

- the number of replications and

- the number of units per block.

If the requirement for balance is relaxed, designs covering a much wider range of situations can be specified.

However, imbalance always carries a penalty in the loss of efficiency, and the investigator should therefore still attempt to specify a design that is as nearly balanced as possible.

Many extensions of this kind to the concept of the balanced incomplete block design have been devised. One such extension is the alpha design, which consists of incomplete blocks, each comprising only a small proportion of the treatments, but in which these blocks are grouped into complete replications.

If each replication of an experiment is to be laid out in an array of rows and columns, an alpha design will take account of the natural variation among rows, but not of that among columns. A further extension is given by alphalpha designs, which give approximate balance over both rows and columns.

9.10   Exercises

9.1 Seven experimental treatments, A, B, C, D, E, F and G, are to be compared in a replicated experiment. However, the experimental units form natural groups of six, so one treatment must be omitted from each group.

(a) Devise a balanced incomplete block design within these constraints. How many replications are required to achieve balance?

(b) Invent values for the response variable from this experiment and perform the appropriate analysis of variance.

(c) Confirm that the efficiency factor in this analysis agrees with Equation 9.1 and that $SE_{Difference}$ for the treatment means agrees with Equation 9.3.

(d) Show that if one block is omitted from the experiment, it can no longer be simply analysed by analysis of variance, but that it can still be analysed by mixed modelling.

(e) Obtain the treatment means from the analysis of variance of the complete experiment:

(i) with recovery of inter-block information

(ii) without recovery of inter-block information.

(f) Obtain the same two sets of means from mixed-model analyses.

9.2 (a) Produce an alpha design with 37 treatments in four replicates, each replicate comprising seven blocks.

(b) Determine the number of blocks in which each pair-wise combination of treatments occurs. Confirm that each combination occurs either in one block or in none.

(c) Determine the number of treatments with which each treatment is combined. How closely do these values approach to the ideal outcome?

(d) Invent values for the response variable from this experiment and perform the appropriate mixed-model analysis.

9.3 If you are using the software R to solve these exercises, use the method given in Section 2.8 to obtain SEs for the following results from the incomplete block design with Day 10 omitted:

(a) the difference between the means for treatments 'TT3' and 'TT2'

(b) the difference between the means for treatments 'TT3' and 'TT1'.

9.4 If you are using the software R to solve these exercises, use the method given in Section 2.8, adapted as indicated in Section 9.7, to obtain SEs for the following results from the alphalpha design:

(a) the difference between the means for genotypes 3 and 2

(b) the difference between the means for genotypes 3 and 1.

References

Cochran, W.G. and Cox, G.M. (1957) *Experimental Designs*, 2nd edn, John Wiley & Sons, Inc., New York, 611 pp.

Cox, D.R. (1958) *Planning of Experiments*, John Wiley & Sons, Inc., New York, 308 pp.

John, J.A. and Williams, E.R. (1995) *Cyclic and Computer Generated Designs*, 2nd edn, Chapman & Hall, London, 255 pp.

Mead, R. (1997) Design of plant breeding trials, in *Statistical Methods for Plant Variety Evaluation* (eds R.A. Kempton and P.N. Fox), Chapman & Hall, pp. 40–67.

Patterson, H.D. and Williams, E.R. (1976) A new class of resolvable incomplete block designs. *Biometrika*, **63**, 83–92.

Patterson, H.D., Williams, E.R. and Hunter, E.A. (1978) Block designs for variety trials. *Journal of Agricultural Science, Cambridge*, **90**, 395–400.

Payne, R.W. and Tobias, R.D. (1992) General balance, combination of information and the analysis of covariance. *Scandinavian Journal of Statistics*, **19**, 3–23.

SAS Institute Inc. (2010) *SAS/QC 9.2: User's Guide*, 6th edn, SAS Institute Inc., Cary, NC, 2200 pp.

10

Beyond mixed modelling

10.1 Review of the uses of mixed models

Just as mixed modelling is an extension of the linear-modelling methods comprised in regression analysis and analysis of variance, mixed modelling itself can be further extended in several directions, to give even more versatile and realistic models. This chapter reviews the various contexts in which we have seen that mixed modelling is preferable to a simple regression or analysis of variance approach, then outlines the ways in which the concepts of mixed modelling can be developed further. Fuller accounts of such advanced uses of mixed modelling are given by Brown and Prescott (2006) and by Pinhero and Bates (2000). Brown and Prescott demonstrate the use of the statistical software SAS to fit the models, whereas Pinhero and Bates use the statistical computer language S (of which the software R is one implementation – see Section 1.11). Both books place much more emphasis on the underlying mathematical theory than is given here.

A mixed-model analysis provides a fuller interpretation of the data than a simple regression or analysis of variance approach, and permits wider inferences about the observations to be expected in future, in the following situations:

- When one or more of the factors in a regression model is a random-effect term, and should therefore contribute to the standard errors (SEs) of estimates of effects of other terms. Examples include

  - the variation in house prices among towns, which contributed to the SE of the estimated effect of latitude (Chapter 1);

  - the variation in bone mineral density among patients sampled at different hospitals, which contributed to the SEs of the estimated effects of gender, age, height and weight (Sections 7.2–7.4);

  - a meta-analysis of treatment effects in several studies, when study × treatment interaction effects are present (Sections 8.1–8.7).

Introduction to Mixed Modelling: Beyond Regression and Analysis of Variance, Second Edition. N. W. Galwey.
© 2014 John Wiley & Sons, Ltd. Published 2014 by John Wiley & Sons, Ltd.
Companion website: http://www.wiley.com/go/beyond_regression

- Where variance components are of intrinsic interest, for example, in the investigation of the sources of variation in the strength of a chemical paste (delivery, cask and sample – Sections 3.2–3.7). The relative magnitude of the different sources of variation can be estimated, with a view to their control by replication in subsequent investigation.

- When candidates from an exchangeable set of entities are to be identified, for example, in the identification of high-yielding breeding lines among the progeny of a cross between two barley varieties (Sections 3.8–3.17). The Best Linear Unbiased Predictor (BLUP), obtained from the mixed-model analysis, provides a more realistic – and more conserva-tive – prediction of the future performance of the selected candidates than is given by the simple mean performance of each candidate (or by the closely related Best Linear Unbiased Estimate (BLUE)) (Chapter 5; Sections 8.8–8.10).

- When it has not been possible to achieve exact balance in the design of an experiment, for example:

  – if one block has to be omitted from a balanced incomplete block design (Sections 9.1–9.4);

  – in an alpha or alphalpha design (Sections 9.5–9.8).

10.2 The generalized linear mixed model (GLMM): Fitting a logistic (sigmoidal) curve to proportions of observations

All the models that we have considered so far are *linear*: that is, they can be expressed in the form

$$y_k = \beta_0 + \beta_1 x_{1k} + \beta_2 x_{2k} + \cdots + \beta_p x_{pk} + v_1 z_{1k} + v_2 z_{2k} + \cdots + v_q z_{qk} + \varepsilon_k \qquad (10.1)$$

where
y_k = the kth observation of the response variable Y,
x_{ik} = the kth observation of the ith explanatory variable in the fixed-effect model, X_i,
p = the number of explanatory variables in the fixed-effect model,
z_{jk} = the kth observation of the jth explanatory variable in the random-effect model, Z_j,
q = the number of explanatory variables in the random-effect model,
ε_k = the kth value of the random variable E, which represents the residual variation in Y
     and $\beta_0 \ldots \beta_p$ and $v_1 \ldots v_q$ are parameters to be estimated.

When one or more of the explanatory variables are factors, some ingenuity is needed to express the model in this form. For example, the model used in Chapter 1 to relate house prices to latitude and town can be expressed in this form by setting
Y = log(house price)
X_1 = latitude
Z_1 = 1 for observations from Bradford, 0 otherwise
Z_2 = 1 for observations from Buxton, 0 otherwise
$\vdots$
Z_{11} = 1 for observations from Witney, 0 otherwise.

The variables Z_1 to Z_{11}, with their arbitrary values that indicate the category (the town) to which each observation belongs, are known as *dummy variables*. When the response variable and the explanatory variables are specified in this way, we find that

$$\beta_0 = \text{intercept}$$
$$\beta_1 = \text{effect of latitude}$$
$$v_1 = \text{effect of Bradford}$$
$$v_2 = \text{effect of Buxton}$$
$$\vdots$$
$$v_{11} = \text{effect of Witney.}$$

The estimates of $v_1 \ldots v_{11}$ are also the estimates of the parameters τ_1 to τ_{11}, the deviations of the town means from the regression line relating log(house price) to latitude, defined in Section 1.4. The decision to treat this model as a mixed model is equivalent to a decision to treat these parameters as values of a random variable, as described in Section 1.6.

In addition to being linear, all the models considered so far have had residuals that can reasonably be assumed to be normally distributed. There are many other regression models, relating a response variable to one or more explanatory variables, that do not have these properties. As an example of a situation in which neither a linear model nor a normal distribution of the residuals is adequate, we can consider the results of an experiment to determine the toxicity of ammonia to a species of beetle, *Tribolium confusum* (Finney, 1971, Section 9.1, p. 177). The experiment was performed in two batches, each comprising a series of samples. These batches will be represented by dummy variables, Z_1 and Z_2, and the explanatory variable X is $\log_{10}$(concentration of ammonia) applied to each sample. In any sample, the number of dead beetles, R, must be an integer between 0 and the number of beetles in the sample, N. The data are shown in the spreadsheet in Table 10.1. (Data reproduced by kind permission of Cambridge University Press.)

If the death of each individual is independent of that of every other individual, then the random variable R has a *binomial distribution*, the precise shape of which is determined by the value of N and by the probability that an individual beetle dies, designated by π. This statement can be written in symbolic shorthand as

$$R \sim \text{binomial}(N, \pi) \tag{10.2}$$

The value of π may depend on the value of X under consideration and on the batch, that is, π may be a function of X and 'batch'. This value can never be known, though it can be estimated from the data.

A brief digression on the binomial distribution is required here. This distribution is defined by the statement

$$P(R = r) = \frac{N!}{r!(N-r)!}\pi^r(1-\pi)^{(N-r)} \tag{10.3}$$

where

$$N! = N \times (N-1) \times (N-2) \ldots 3 \times 2 \times 1.$$

For example, if

$$\pi = 0.3,$$

then in a sample of 30 beetles, the probability that 8 die is given by

$$P(R = 8) = \frac{30!}{8! \times 22!} \times 0.3^8 \times 0.7^{22} = 0.1501.$$

Table 10.1   Mortality of the beetle *Tribolium confusum* at different concentrations of ammonia.

$X = \log_{10}$(concentration of ammonia), N=number of beetles in the sample, and R=number of dead beetles in the sample.

	A	B	C	D
		A	B	C	D
1	batch!	X	N	R
2	1	0.72	29	2
3	2	0.72	29	1
4	1	0.80	30	7
5	2	0.80	31	12
6	1	0.87	31	12
7	2	0.87	32	4
8	1	0.93	28	19
9	2	0.93	31	18
10	1	0.98	26	24
11	2	0.98	31	25
12	1	1.02	27	27
13	2	1.02	28	27
14	1	1.07	26	26
15	2	1.07	31	29
16	1	1.10	30	30
17	2	1.10	31	30

Source: Data reproduced by kind permission of Cambridge University Press.

Substituting each possible value of r from 0 to 30 into Equation 10.3, we obtain the distribution illustrated in Figure 10.1. For a fuller account of the binomial distribution, and why it occurs in such situations, see, for example, Snedecor and Cochran (1989, Sections 7.1–7.5, pp. 107–117) or Bulmer (1979, Chapter 6, pp. 81–90).

In a system of this kind, the relationship between the combined value of the explanatory variables ('batch' and X in this case) and π is often *sigmoidal* (S-shaped – see Figure 10.2). At one extreme of the range of the explanatory variables, the probability of the event under consideration (death in this case) is close to zero: at the other extreme it is close to one. There are two commonly used functions that specify a relationship of this form. One of these, the integral of the normal distribution, is the basis of a method called *probit analysis* (Finney, 1971). This method of fitting a sigmoid curve probably has the clearer conceptual basis: it is based on the assumption that an underlying variable, in this case the tolerance of the beetles to the toxin, is normally distributed, and that at any given dose, all individuals up to a certain level of tolerance are killed. The alternative is the logistic function, which we will use here as it has the advantage of being rather easier to express algebraically: in the present case, it is

$$p_{ij} = \frac{r_{ij}}{n_{ij}} = \frac{1}{1 + \exp\left[-\left(\beta_0 + \beta_1\left(x_j - \bar{x}\right) + v_1 z_{1i} + v_2 z_{2i} + v_3 z_{1i}\left(x_j - \bar{x}\right) + v_4 z_{2i}\left(x_j - \bar{x}\right)\right)\right]} + \varepsilon_{ij}$$

$$(10.4)$$

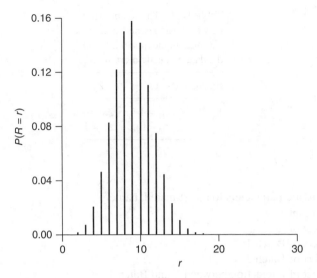

Figure 10.1 The binomial distribution with parameters $\pi = 0.3$ and $N = 30$.

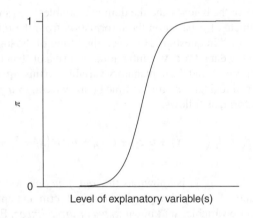

Figure 10.2 A sigmoidal relationship between an explanatory variable (or variables) and the probability (π) of a response.

where

r_{ij} = the value of R at the jth level of X in the ith batch, that is, the number of dead beetles in this sample (the ijth sample),

n_{ij} = the value of N, that is, the total number of beetles, in the ijth sample,

x_j = the jth level of X

z_{1i} = the value of the dummy variable Z_1 in the ith batch, indicating whether each observation was obtained from Batch 1

z_{2i} = the value of the dummy variable Z_2 in the ith batch, indicating whether each observation was obtained from Batch 2

ε_{ij} = the value of the random variable E in the ijth sample, which represents the residual effect on p_{ij},

$\bar{x} = \dfrac{\sum_{j=1}^{7} x_j}{7}$ = the mean value of X over all samples,

Table 10.2 The relationship between batches and dummy variables in the model fitted to the data on beetle mortality.

Batch	Z_1	Z_2
1	$z_{11} = 1$	$z_{21} = 0$
2	$z_{12} = 0$	$z_{22} = 1$

and the βs and υs are parameters to be estimated, namely

β_0 = constant
β_1 = effect of X
υ_1 = effect of Batch 1
υ_2 = effect of Batch 2
υ_3 = effect of interaction between X and Batch 1
υ_4 = effect of interaction between X and Batch 2.

The relationship between the batches and the dummy variables is as shown in Table 10.2. The value p_{ij} gives an estimate of π based on the information from the ijth sample, applicable to the batch and the level of X in question. However, the fitting of the logistic function will give estimates based on all the data, for any combination of levels of 'batch' and X.

The relationship between Y and the explanatory variables in this equation is not linear. However, if we ignore the residual term ε_{ij} for the time being, we can *transform* the equation to the familiar linear-model form, as follows:

$$\log_e \left(\frac{p_{ij}}{1 - p_{ij}} \right) = \beta_0 + \beta_1 \left(x_j - \bar{x} \right) + \upsilon_1 z_{1i} + \upsilon_2 z_{2i} + \upsilon_3 z_{1i} \left(x_j - \bar{x} \right) + \upsilon_4 z_{2i} \left(x_j - \bar{x} \right) \quad (10.5)$$

The function $\log_e \left(p_{ij} / \left(1 - p_{ij} \right) \right)$ is known as the *logit function*. Models of this type, which can be expressed in linear form by dropping the residual term and applying a suitable transformation to the response variable, are known as *generalized linear*. Every generalized linear model is characterized by a probability distribution and a *link function*: in the present case, the binomial distribution and the logit function. For each probability distribution, there is a particular link function known as the *canonical link* which has special mathematical properties, including the fact that when the residual term is the only random-effect term in the model, it always gives a unique set of parameter estimates, the *sufficient statistic* (McCullagh and Nelder, 1989, Sections 2.2.2–2.2.4, pp. 28–32). In the case of the binomial distribution, the logit function is the canonical link – another reason for preferring it to the *probit function*, the corresponding function used in probit analysis.

In the notation of Wilkinson and Rogers (1973; see Section 2.2), the model specified is

X*batch.

It is reasonable to specify 'X' as a fixed-effect term, and 'batch' as a random-effect term, from which it follows that 'X.batch' is also a random-effect term (see the guidelines in Section 6.3). We have then specified a *generalized linear mixed model* (GLMM).

This GLMM can be fitted to the data by the following GenStat statements:

```
IMPORT 'IMM edn 2\\Ch 10\\ammonia Tribolium.xlsx'
GLMM [PRINT = model, monitoring, components, vcovariance, effects; \
  DISTRIBUTION = binomial; LINK = logit; DISPERSION = *; \
  RANDOM = batch + X.batch; FIXED = X] Y = R; NBINOMIAL = N
VDISPLAY [PRINT = effects; PTERMS = batch/X; PSE = estimates]
```

In the GLMM statement, the DISTRIBUTION option specifies that the response variate 'R' follows the general form of the binomial distribution, though its variance may be greater or less than that of a binomial variable, as will be specified in a moment. The LINK option specifies that the logit function has been chosen as the link function. If the assumptions that each death is independent (conditional on π) and that R follows the binomial distribution in every respect were correct, it would follow that

$$\text{var}(R) = N\pi(1 - \pi) \tag{10.6}$$

and there would be no need to estimate the residual variance from the data. We could indicate that we were willing to make this assumption by setting the option 'DISPERSION = 1': instead, by setting this option to a missing value ('*') we indicate that the residual variance is to be estimated. The options RANDOM and FIXED specify the random-effect and fixed-effect model terms, respectively. It is possible that there is a correlation between the effects of 'batch' and those of 'X.batch', but with only two batches there is not enough information to estimate this and it cannot be included in the model. The parameter Y specifies the response variate, and the parameter NBINOMIAL specifies the variate that holds the number of observations in each sample, that is, the maximum value that each value of the response variate might take.

The output of the GLMM and VDISPLAY statements is as follows:

Generalized linear mixed-model analysis

Method:	c.f. Schall (1991) Biometrika
Response variate:	R
Binomial totals:	N
Distribution:	binomial
Link function:	logit
Random model:	batch + (batch.X)
Fixed model:	Constant + X

Dispersion parameter estimated

Monitoring information

Iteration	Gammas		Dispersion	Max change
1	0.02903	0.0002728	2.137	1.8314E+01
2	0.02653	0.000001091	2.238	1.0117E−01
3	0.02655	6.9016E−08	2.237	7.2607E−04
4	0.02657	4.3650E−09	2.237	4.1130E−05

Estimated variance components

Random term	Component	s.e.
batch	0.059	0.203
batch.X	0.000	bound

Residual variance model

Term	Model(order)	Parameter	Estimate	s.e.
Dispersn	Identity	Sigma2	2.237	0.877

Estimated variance matrix for variance components

batch	1	0.0410		
batch.X	2	0.0000	0.0000	
Dispersn	3	−0.0282	0.0000	0.7700
		1	2	3

Table of effects for constant

0.3314   Standard error   0.26538

Table of effects for X

17.83   Standard error   2.305

Table of effects for batch

batch	1	2
	0.1118	−0.1118

Standard error: 0.2167

Table of effects for batch.X

batch	1	2
	2.0310E−09	−2.0310E−09

Standard error: 0.00009882

The output first specifies the fitting method used and the model fitted. Next comes some monitoring information, indicating the estimated values of certain model parameters at successive iterations of the model-fitting process, leading to convergence, that is, successful fitting (see Section 10.5). Next come estimates of the variance components for the random-effect terms, with their SEs. The estimate for 'batch' is smaller than its SE, and that for 'batch.X' is zero, indicating that these terms could probably be dropped from the model. The estimate of the residual variance is presented in terms of the *dispersion*, which is given by the ratio

$$\frac{\text{observed residual variance}}{\text{residual variance expected if the response variate follows the distribution specified in the model}}$$

In the present case, the dispersion is substantially larger than 1, indicating that there is more residual variation from sample to sample than would be expected if the distribution were truly binomial. Next comes the matrix of covariances among these variance estimates. The values on the diagonal are simply the squares of the SEs above – the variances of the variance estimates. For example,

$$\text{var}(\hat{\sigma}^2_{\text{batch}}) = 0.203^2 \approx 0.0410.$$

The off-diagonal values indicate the extent to which the estimate of one variance component is associated with that of another.

Next come the estimated effects of each model term. These can be substituted into the original model (Equation 10.4), together with the value

$$\bar{x} = 0.9363,$$

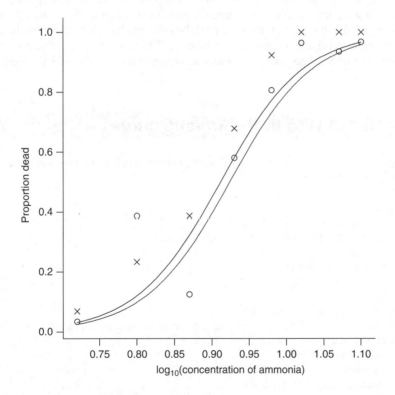

Figure 10.3 Fitted curves from a logistic model relating the proportion of *T. confusum* individuals killed to the concentration of ammonia applied.
——fitted, Batch 1; ------- fitted, Batch 2; × observed, Batch 1; ○ observed, Batch 2.

to provide an estimate of π, the true probability that an insect dies, for any combination of batch and X, thus

$$\hat{\pi}_{ij} = \frac{1}{1 + \exp\left[-\left(\begin{array}{l}0.3314 + 17.83 \times (x_j - 0.9363) + 0.1118 \times z_{1i} - 0.1118 \times z_{2i} + \\ 2.0310 \times 10^{-9} \times z_{1i}(x_j - 0.9363) - 2.0310 \times 10^{-9} \times z_{2i}(x_j - 0.9363)\end{array}\right)\right]}$$

This function is displayed over the range of the data, together with the observed values, in Figure 10.3.

Overall, the curves give a reasonable fit to the data, permitting a realistic estimate of the proportion of insects that would be killed by a particular concentration of ammonia. The effect of batch corresponds to the horizontal displacement between the two curves. This is small relative to the scatter of the data, confirming that the batch has little if any effect. The 'X.batch' interaction effect corresponds to the difference in average slope between the two curves, which is much too slight to be detected by eye. The corresponding variance component estimate is zero, and it might be preferable to drop these terms from the model, but they are included in the substitution into Equation 10.4 in order to show how the equation is constructed. The fact that the dispersion is larger than 1 indicates that the scatter of the observations about the curves is wider than would be expected if the deaths of individuals within each sample were independent: that is, there is evidence of some *heterogeneity* in the conditions of each sample, even after allowing for the effects of ammonia and batch. Because the 'X.batch' interaction effect is negligible, the decision to use a mixed model has had little effect on the precision of the parameter estimates in this case. However, if this effect were substantial, it would be important to take it into account in order to obtain realistic values for the SEs of $\hat{\beta}_0$, $\hat{\beta}_1$ and $\hat{\beta}_2$.

10.3   Use of R to fit the logistic curve

The following commands import the data into R, convert batch to a factor and fit the model in Equation 10.4 to the data:

```
rm(list = ls())
ammonia.tribolium <- read.table(
  "IMM edn 2\\Ch 10\\ammonia tribolium.txt",
  header=TRUE)
attach(ammonia.tribolium)
fbatch <- factor(batch)
library(lme4)
responses <- cbind(R, N - R)
ammonia.tribolium.glmer <- glmer(responses ~ X + (1|fbatch)+
  (X|fbatch), family = binomial(link = "logit"),
  data = ammonia.tribolium)
summary(ammonia.tribolium.glmer)
coef(ammonia.tribolium.glmer)
```

The function cbind() combines the number of individuals that respond (R) in each sample and the number that do not respond ($N - R$) into a single structure, 'responses'. The model specified in the glmer() function indicates that

- 'responses' holds the information on the responses;

- 'X' is a fixed-effect term;

- the main effect of batch is a random-effect term, indicated by (1|batch);

- the variation in the response to 'X' among batches is a random-effect term, indicated by (X|batch).

The family option in this function indicates the distribution of the response variable and the link function. The output of the function summary() is as follows:

```
Generalized linear mixed model fit by the Laplace approximation
Formula: responses ~ X + (1 | fbatch) + (X | fbatch)
 Data: ammonia.tribolium
 AIC BIC logLik deviance
 47.04 51.67 -17.52 35.04
Random effects:
 Groups Name Variance Std.Dev. Corr
 fbatch (Intercept) 0.000000 0.00000
 X 0.053433 0.23116 NaN
Number of obs: 16, groups: fbatch, 2

Fixed effects:
             Estimate Std. Error z value Pr(>|z|)
(Intercept)   -15.919     1.410   -11.29   <2e-16 ***
X              17.951     1.559    11.51   <2e-16 ***
---
Signif. codes:  0 '***' 0.001 '**' 0.01 '*' 0.05 '.' 0.1 ' ' 1

Correlation of Fixed Effects:
   (Intr)
X -0.990

$fbatch
  (Intercept) X
1 -15.91874 18.11988
2 -15.91874 17.77798
```

The estimates of effects are generally different from those obtained from GenStat, because the GenStat fits the model centred on the mean value of X, whereas R fits it without centring. One consequence of this is that it is now the main effect of batch that has a variance component estimate of zero, while the X.batch interaction has a positive variance component estimate. The function summary() reports the intercept and the coefficient of X for the two levels of fbatch directly, rather than as departures from the fixed-effect estimates as specified in Equation 10.4. Adapting this equation accordingly, and substituting the coefficient estimates, the probability that an insect dies is estimated by

$$\hat{\pi}_{ij} = \frac{1}{1 + \exp\left[-\left(-15.91874 \times z_{1i} - 15.91874 \times z_{2i} + 18.11988 \times z_{1i}x_j + 17.77798 \times z_{2i}x_j\right)\right]}$$

The curves produced by this function do not agree closely with those produced by GenStat, but agree closely with those produced by SAS, and fit the data fairly well. No straightforward method of obtaining an estimate of the dispersion parameter from R is known to the author.

10.4 Use of SAS to fit the logistic curve

The following commands import the data into SAS and fit the model in Equation 10.4 to the data:

```
PROC IMPORT OUT = ammonia DBMS = EXCELCS REPLACE
    DATAFILE = "&pathname.\IMM edn 2\Ch 10\ammonia Tribolium.xlsx";
  SHEET = "for SAS";
RUN;


ODS RTF;
PROC GLIMMIX DATA = ammonia METHOD = RSPL ASYCOV;
    CLASS batch;
    MODEL R/N = X/
        DISTRIBUTION = BINOMIAL LINK = LOGIT SOLUTION;
    RANDOM batch batch * X/ SOLUTION;
    RANDOM _RESIDUAL_;
RUN;
ODS RTF CLOSE;
```

The logistic model is fitted by the procedure GLIMMIX. The option NOBOUND of this procedure is not set, so variance components estimates are constrained to be positive: without this constraint, the model-fitting process fails to converge. The MODEL statement indicates that the response variable is the number of individuals responding, R, out of a total N, and gives the fixed-effect model. In this statement, the DISTRIBUTION option specifies the distribution of the response variable, the LINK option specifies the link function and the SOLUTION option specifies that the regression coefficients for the fixed-effect terms are to be displayed in the output. The first RANDOM statement indicates the terms in the random-effects model. The second RANDOM statement indicates that the dispersion parameter, related to the residual variance, is to be estimated, not fixed at 1.

Part of the output from PROC GLIMMIX is as follows:

Covariance para meter estimates		
Cov Parm	Estimate	Standard error
batch	0	
X*batch	0.09689	0.2797
Residual (VC)	2.2192	0.8704

Asymptotic covariance matrix of covariance parameter estimates			
Cov Parm	**CovP1**	**CovP2**	**CovP3**
batch			
X*batch		0.07824	−0.03379
Residual (VC)		−0.03379	0.7576

Solutions for fixed effects					
Effect	**Estimate**	**Standard error**	**DF**	**_t_ value**	**Pr > |_t_|**
Intercept	−15.8489	2.0920	1	−7.58	0.0835
X	17.8700	2.3105	1	7.73	0.0819

Type III tests of fixed effects				
Effect	**Num DF**	**Den DF**	**_F_ value**	**Pr > _F_**
X	1	1	59.82	0.0819

Solution for random effects						
Effect	**batch**	**Estimate**	**Std Err Pred**	**DF**	**_t_ value**	**Pr > |_t_|**
batch	1	0				
batch	2	0				
X*batch	1	0.1548	0.2700	12	0.57	0.5771
X*batch	2	−0.1548	0.2700	12	−0.57	0.5771

The estimates of effects are generally different from those obtained from GenStat, because the GenStat fits the model centred on the mean value of X, whereas SAS fits it without centring. One consequence of this is that it is now the main effect of batch that has a variance component estimate of zero, while the X.batch interaction has a positive variance component estimate. PROC GLIMMIX reports the intercept and the coefficient of X for the two levels of fbatch directly, rather than as departures from the fixed-effect estimates as specified in Equation 10.4. Adapting Equation 10.4 to take account of the absence of centring, and substituting the coefficient estimates into the model, the probability that an insect dies is estimated by

$$\hat{\pi}_{ij} = \frac{1}{1 + \exp\left[- \left(-15.8489 + 17.8700x_j + 0z_{1i} + 0z_{2i} + 0.1548z_{1i}x_j - 0.1548z_{2i}x_j\right)\right]}.$$

The curves produced by this function do not agree closely with those produced by GenStat, but agree closely with those produced by R and fit the data fairly well.

10.5   Fitting a GLMM to a contingency table: Trouble-shooting when the mixed modelling process fails

There are several other types of data that cannot be realistically represented by an ordinary linear model with normally distributed residual variation, but which do fulfil the criteria for fitting a GLMM, namely

- they can be represented by a model that can be converted to the general-linear form by omitting the residual term and applying an appropriate transformation;

- an appropriate probability distribution can be specified for the response variable.

Hence the use of GLMMs permits a wide extension to the range of situations in which the concepts of mixed modelling can be applied. Another important case in which generalized linear models can be used is the analysis of *contingency tables*. These are data sets in which events of a particular type are counted and are classified by factors that indicate the combination of circumstances, or contingency, in which each event occurred. This type of data set is illustrated by an example concerning the frequency of damage caused by waves to the forward sections of cargo-carrying ships (McCullagh and Nelder, 1989, Section 6.3.2, pp. 204–208). Each occurrence of damage is classified by the type of ship to which it occurred (A to E), the year of construction of the ship and its period of operation. For each category defined by these three factors – that is, each contingency – the number of incidents of damage observed was recorded, along with the number of months of service over which observations were available. The first and last few rows of the data are shown in the spreadsheet in Table 10.3. The null hypothesis to be tested (H_0) is that none of the factors influenced the frequency of incidents, and in this case the number of incidents in each category is expected to be proportional to the number of months of service. (Data reproduced by kind permission of Chapman and Hall.)

We can represent the number of incidents of damage to ships of the ith type, constructed during the jth range of years, during the kth period of operation (the ijkth category) by the symbol r_{ijk}. If each damage incident is independent of all the others, then it can be shown that r_{ijk} is an observation of a random variable R_{ijk} which has a *Poisson distribution*, the mean of which is given by $N\pi_{ijk}$ where

N = the expected total number of incidents of damage given the total number of months of
     observation and their distribution over the categories

and

π_{ijk} = the true probability that an individual damage incident falls in the ijkth category.

This statement can be written in symbolic shorthand as

$$R_{ijk} \sim \text{Poisson}\left(N\pi_{ijk}\right) \tag{10.7}$$

Again a brief digression is required, this time on the Poisson distribution. This is defined by the statement that if

$$R \sim \text{Poisson}(\mu), \tag{10.8}$$

Table 10.3 Number of incidents of damage to ships, classified by ship type, year of construction and period of operation.

	A	B	C	D	E
	type!	constrctn_y!	operatn_period!	service_months	damage_incidents
1					
2	A	1960–64	1960–74	127	0
3	A	1960–64	1975–79	63	0
4	A	1965–69	1960–74	1095	3
5	A	1965–69	1975–79	1095	4
6	A	1970–74	1960–74	1512	6
7	A	1970–74	1975–79	3353	18
8	A	1975–79	1960–74		
9	A	1975–79	1975–79	2244	11
10	B	1960–64	1960–74	44882	39
11	B	1960–64	1975–79	17176	29
.					.
.					.
.					.
34	E	1960–64	1960–74	45	0
35	E	1960–64	1975–79	0	0
36	E	1965–69	1960–74	789	7
37	E	1965–69	1975–79	437	7
38	E	1970–74	1960–74	1157	5
39	E	1970–74	1975–79	2161	12
40	E	1975–79	1960–74		
41	E	1975–79	1975–79	542	1

Source: Data reproduced by kind permission of Chapman and Hall.

where
μ = the mean number of incidents in an observation period,

then

$$P(R = r) = \frac{\mu^r}{r!}e^{-\mu} \tag{10.9}$$

For example, if the expected number of incidents of damage in a particular category (the mean number over an infinite hypothetical population of data sets similar to the present data set) is 8, then the observed number of incidents will be distributed as shown in Figure 10.4. For a fuller account of the Poisson distribution, and why it occurs in this context, see, for example, Snedecor and Cochran (1989, Section 7.14, pp. 130–133) or Bulmer (1979, pp. 90–97).

In the present case, if H_0 is true, then

$$\pi_{ijk} = a_{ijk} \tag{10.10}$$

where
a_{ijk} = the proportion of months of service that fall in the ijkth category.

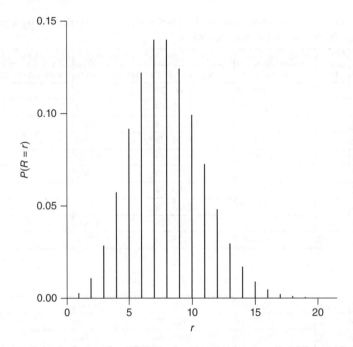

Figure 10.4 The Poisson distribution with parameter (mean value) $\mu = 8$.

In this case, the expected number of incidents in the ijkth category is Na_{ijk}, and

$$r_{ijk} = Na_{ijk} + \varepsilon_{ijk} \tag{10.11}$$

where

ε_{ijk} = the value of the random variable E in the ijkth category, which represents the residual effect on r_{ijk}.

We can add terms to this model to represent the possibility that the factors type of ship, year of construction and period of operation and their interactions influence the probability of a damage incident, as follows:

$$r_{ijk} = Na_{ijk}q_{i..}q_{.j.}q_{..k}q_{ij.}q_{i.k}q_{.jk}q_{ijk} + \varepsilon_{ijk} \tag{10.12}$$

where

$q_{i..}$ = the main effect of the ith type of ship,
$q_{.j.}$ = the main effect of the jth range of years of construction,
$q_{..k}$ = the main effect of the kth period of operation,
$q_{ij.}$ = the effect of interaction between the ith type and the jth range of years, and so on.

N and the qs are parameters to be estimated from the data. This model, like that in Equation 10.4, can be transformed to the linear-model form by ignoring the residual term ε_{ijk} and applying a suitable transformation, in this case the logarithmic transformation, namely

$$\log\left(r_{ijk}\right) = \log\left(N\right) + \log\left(a_{ijk}\right) + \log\left(q_{i..}\right) + \log\left(q_{.j.}\right) + \log\left(q_{..k}\right) + \log\left(q_{ij.}\right)$$
$$+ \log\left(q_{i.k}\right) + \log\left(q_{.jk}\right) + \log\left(q_{ijk}\right) \tag{10.13}$$

However, the term $\log(a_{ijk})$ does not have to be estimated: it is a separate variable supplied with the data. It is equivalent to a term $\beta \log(a_{ijk})$ when the value of the parameter β is fixed at 1. Such a term is called an *offset*. In the notation of Wilkinson and Rogers (1973), the model specified here (excluding the offset term) is

$$\text{type*constrctn\_y*operatn\_period.}$$

The following statements import the data and fit this model, specifying all terms as fixed-effect terms for the time being:

```
IMPORT 'IMM edn 2\\Ch 10\\ship damage.xlsx'
CALCULATE logservice = LOG(service_months/SUM(service_months))
MODEL [DISTRIBUTION = poisson; LINK = log; DISPERSION = 1; \
 OFFSET = logservice] damage_incidents
TERMS type*constrctn_y*operatn_period
FIT [PRINT = *] type
ADD [PRINT = *] constrctn_y
ADD [PRINT = *] operatn_period
ADD [PRINT = *] type.constrctn_y
ADD [PRINT = *] type.operatn_period
ADD [PRINT = model, accumulated; FPROBABILITY = yes]\
 constrctn_y.operatn_period
```

The CALCULATE statement obtains the natural logarithm of the proportion of months of service in each category, transforming this variable to the scale on which it will be required in the model. A message in the output (not shown here) warns that in the case of Unit 34 (Row 35 of the spreadsheet) an attempt has been made to obtain the logarithm of zero, and that the result is a missing value. Consequently this unit is omitted from the analysis, which is appropriate, as it represents a category of ship that spent no time at sea, and therefore was not exposed to the risk of damage. In the MODEL statement, the DISTRIBUTION option specifies that the response variate ('damage_incidents') follows the general form of the Poisson distribution, though its variance may be greater or less than that of a Poisson variable, unless we constrain it using the DISPERSION option (see below). The LINK function specifies the function required to transform the model to the linear form, just as the same option did in the GLMM statement in the previous example (Section 10.2). If the assumption that each damage incident is independent and that R follows the Poisson distribution in every respect is correct, then it follows that

$$\text{var}(R) = \mu \qquad (10.14)$$

and there is no need to estimate the residual variance from the data. (It is peculiarity of the Poisson distribution that its variance is equal to its mean.) The option setting 'DISPERSION = 1' indicates that we are willing to make this assumption. The offset term is specified by the OFFSET option.

In order to obtain an analysis in which the deviance accounted for by each model term is shown separately, it is necessary to specify each of the terms in a separate statement. The FIT statement specifies the model on H_0, comprising only the constant term $\log(N)$ and the offset term $\log(a_{ijk})$. The option setting 'PRINT = *' indicates that no output is to be produced from this initial model. The main effect and interaction terms are added by a succession of ADD statements, again with no printing, until the final ADD statement specifies that the complete model and the *accumulated analysis of deviance* are to be printed. The *analysis of deviance* is a method closely related to analysis of variance: in the special case where the residual variation is normally distributed (i.e. in all the models considered prior to the present chapter), the two are equivalent. The term *accumulated* indicates that the deviance accounted for by adding each term to the model successively is to be presented (cf. Section 1.4, where an anova constructed on the same basis is presented). The FPROBABILITY option specifies that the analysis of deviance table is to include a *p*-value for the significance of each term in the model. Note that it is not necessary to fit the three-way interaction term 'type.constrctn_y.operatn_period' explicitly, as there is only one observation for each combination of these factors, and this term is therefore the residual term.

The output of the final ADD statement is as follows:

Message: term constrctn_y.operatn_period cannot be fully included in the model because 1 parameter is aliased with terms already in the model.
(constrctn_y 1975–79 .operatn_period 1975–79) = (constrctn_y 1975–79)

Regression analysis

Response variate: damage_incidents
   Distribution: Poisson
   Link function: Log
   Offset variate: logservice
   Fitted terms: Constant + type + constrctn_y + operatn_period + type.constrctn_y + type.operatn_period + constrctn_y.operatn_period

Accumulated analysis of deviance

Change	d.f.	deviance	mean deviance	deviance ratio	approx chi pr
+type	4	55.4391	13.8598	13.86	<0.001
+constrctn_y	3	41.5341	13.8447	13.84	<0.001
+operatn_period	1	10.6601	10.6601	10.66	0.001
+type.constrctn_y	12	24.1079	2.0090	2.01	0.020
+type.operatn_period	4	6.0661	1.5165	1.52	0.194
+constrctn_y.operatn_period	2	1.6643	0.8322	0.83	0.435
Residual	7	6.8568	0.9795		
Total	33	146.3283	4.4342		

Message: ratios are based on dispersion parameter with value 1.

A message first warns that the term 'constrctn_y.operatn_period' cannot be fully included in the model. This is because of the presence of missing values, as a result of which not all combinations of levels of the three factors are represented in the data. The model fitted and the fitting method used are then specified. The accumulated analysis of deviance then shows how the variation among the values of 'damage_incidents' (after adjusting for 'logservice') is distributed among the terms in the model. Roughly speaking, the deviance per degree of freedom – the mean deviance – gives a measure of the amount of variation accounted for by each term, when the terms are added successively to the model. Thus the mean deviance corresponds to the mean square in an analysis of variance. It is divided by the dispersion parameter to give the deviance ratio: since the dispersion parameter has been fixed at 1 (i.e. 'damage_incidents' has been assumed to follow a Poisson distribution in every respect, including the value of the variance), the two are identical. If the dispersion parameter had not been specified, it would be estimated by the residual mean deviance: hence the deviance ratio is equivalent to the variance ratio (the F statistic) in an analysis of variance. If the assumption that 'damage_incidents' follows a Poisson distribution is correct, and if H_0 is true, then the deviance for each term is distributed approximately as χ^2 with the degrees of freedom (d.f.) indicated. Thus each deviance provides a significance test for the term in question, and the column headed 'approx chi pr' gives the corresponding p-value. These p-values indicate that the main effects of the three factors are highly significant, and that the 'type.constructn_y' interaction is also significant, but that the other interactions are not.

It is of interest to obtain estimates of the mean frequency of damage to ships of each type, but it is not possible to do so from the model fitted above, because some pair-wise combinations of factor levels are not represented in the data. However, we can omit the non-significant two-way interaction terms from the model and obtain estimates based on the simpler model comprising only the main effects of the three factors and the type.constructn_y interaction. This is done by the following statements:

```
FIT [PRINT = *] type * constrctn_y + operatn_period
PREDICT [PRINT = description, prediction, se, sed] type
```

The output of the PREDICT statement is as follows:

Predictions from regression model

These predictions are estimated mean values, formed on the scale of the response variable, adjusted with respect to some factors as specified below.

The predictions have been formed only for those combinations of factor levels for which means can be estimated without involving aliased parameters.

The predictions are based on the mean of the offset variate: logservice −4.956

The predictions have been standardized by averaging over the levels of some factors:

Factor	Weighting policy	Status of weights
operatn_period	Marginal weights	Constant over levels of other factors
constrctn_y	Marginal weights	Constant over levels of other factors

The standard errors are appropriate for the interpretation of the predictions as summaries of the data rather than as forecasts of new observations.

Response variate: damage_incidents

type	Prediction	s.e.
A	3.415	0.5738
B	2.463	0.1741
C	2.098	0.7250
D	2.902	0.7430
E	6.131	1.2224

Standard errors of differences of predictions

type A	1	*			
type B	2	0.600	*		
type C	3	0.924	0.746	*	
type D	4	0.936	0.763	1.037	*
type E	5	1.353	1.235	1.422	1.434	*
		1	2	3	4	5

Message: s.e's, variances and lsd's are approximate, since the model is not linear.
Message: s.e's are based on dispersion parameter with value 1.

The note that the estimated means are formed on the scale of the response variable indicates that they must be back-transformed, using the inverse of the link function, in order to obtain them on the original scale, namely the number of incidents of damage. For example, the mean value for ships of Type A corresponds to

$$e^{3.415} = 30.42 \text{ incidents.}$$

The accompanying SE indicates that the true value is likely to lie in the range from

$$e^{(3.415-0.5738)} = 17.14$$

to

$$e^{(3.415+0.5738)} = 53.99 \text{ incidents.}$$

Another note states that these estimates are based on a value of -4.956 for the offset variate 'logservice'. This means that they are based on the value

$$\log_e A = -4.956, \tag{10.15}$$

where
$A = $ the proportion of months of service that is assumed to fall in the category under consideration.

Although A varies from category to category in the data, it is held constant for the purpose of prediction, in order to provide a valid basis for comparison between risks in the different categories. From this decision, it follows that

$$\begin{array}{l} \textit{number} \text{ of months of} \\ \text{service assumed to fall} \\ \text{in the category under} \\ \text{consideration for the} \\ \text{purpose of prediction} \end{array} = TA \tag{10.16}$$

where
$T = $ total number of months of service.

Totalling the values of 'service_months' in the data, we obtain

$$T = 163574, \tag{10.17}$$

rearranging Equation 10.15 we obtain

$$A = e^{-4.956}, \tag{10.18}$$

and substituting from Equations 10.17 and 10.18 into Equation 10.16, we find that the predicted numbers of incidents are based on

$$163574e^{-4.956} = 1151.7 \text{ months of service.}$$

Another note indicates that the calculation of this mean has required averaging over the levels of the other factors, and that in this process *marginal weights* have been applied: that is, operation periods and construction years that are more heavily represented in the data contribute more heavily to the mean. However, it is noted that the status of weights is constant over levels of other factors: this means that a *combination of* 'operatn_period' and 'constrctn_y' that is more heavily represented in the data than the marginal weights would lead one to predict does *not* contribute more heavily to the mean. Another note states that the SEs are not appropriate for the forecasting of new observations: that is, they indicate the precision *of the means themselves*, not the amount of variation among individual observations. As noted earlier (Section 4.6), SEs of differences between means are usually of more interest than SEs of the means themselves, and these are provided, for all pair-wise comparisons, on the transformed scale.

If 'constrctn_y' and 'operatn_period' are representative of a broader population of construction and operation dates, it may be reasonable to specify these factors as random-effect terms. However, it may be reasonable to retain 'type' as a fixed-effect term representing particular methods of construction that are of individual interest, and that may be amenable to choice or control: ship builders can decide what type of hull to construct, and insurers can decide what premiums to offer on each type of ship. According to the guidelines given in Section 6.3, it follows that the remaining terms in the model should then be specified as random-effect terms. The following statements specify the fitting of this model:

```
GLMM [DISTRIBUTION = poisson; LINK = log; DISPERSION = 1; \
  OFFSET = logservice; \
  RANDOM = constrctn_y * operatn_period + type.constrctn_y + \
  type.operatn_period; FIXED = type] Y = damage_incidents
```

The structure of this statement corresponds to that of the GLMM statement in Section 10.2: the only new feature is the OFFSET option, which has the same meaning as that in the MODEL statement earlier in this section. However, the output from this statement begins with the following diagnostic messages:

Warning 4, code VC 31, statement 163 in procedure GLMM

Command: REML [PRINT = *; WORKSPACE = WORKSPACE;
MVINCLUDE = #MVINCLUDE] TRANS
Unsuccessful update for variance parameters - try smaller step.

Warning 5, code VC 17, statement 163 in procedure GLMM

Command: REML [PRINT = *; WORKSPACE = WORKSPACE;
MVINCLUDE = #MVINCLUDE] TRANS
The iterative process has not converged.

It is therefore the safest to ignore the rest of the output. Whereas the methods of ordinary regression analysis and analysis of variance can be applied without arithmetic problems to almost any data set, the same is not true of mixed modelling, and certainly not of GLMM. We will now examine the kinds of problems that can be encountered, in order to interpret these warning messages.

In ordinary regression analysis and analysis of variance, analytical formulae are applied which give, in a single step, the best estimates, according to a certain criterion, of the parameters of the model being fitted. The criterion used to identify the 'best' parameter estimates is that they should be the values that maximize the probability of the observed data, and hence minimize the value of the deviance (a concept introduced in Section 3.12). These are said to be the parameter estimates with the highest *likelihood*: as noted in Section 3.12,

$$\text{deviance} = -2\log_e(\text{likelihood}). \tag{10.19}$$

When all the random-effect terms are assumed to be normally distributed (as is the case in all the examples considered in earlier chapters), these estimates are also those that minimize the estimate of the residual variance. However, in mixed modelling, the maximum-likelihood estimates cannot generally be obtained in a single step: a *search* must be made for them. In broad terms, the search strategy is to start from an arbitrary set of parameter values, then determine a set of changes that can be made to these that will increase the value of the likelihood. This process is repeated, the change made getting smaller at each iteration, until no change that produces a further increase can be found. At this point, the model-fitting process is said to have *converged*. The process it analogous to trying to climb a mountain in fog, following the rule 'always walk uphill'. The process will lead to the summit (analogous to the maximum-likelihood estimates), provided that

- one starts on the flank of the mountain, not in some other part of the landscape,

- one takes steps that are not too large and

- the mountain has one, and only one, peak.

If these criteria are not met, one may walk uphill indefinitely (failure of convergence), or the process may lead to a small peak on the flank of the likelihood 'mountain', not its true summit (convergence to a local maximum). The latter problem may be recognized by unrealistic parameter estimates and a poor fit to the data. Sometimes these problems can be overcome by a careful choice of initial parameter values for the fitting process, giving the process a better chance of 'climbing the right peak'.

In the present case, the model-fitting process has encountered severe difficulties. Generally, such problems occur because the model being fitted does not represent the pattern of variation in the data well: a review of the model may suggest terms that can be dropped, or others that should be added. In the output above, the message that negative variance components are present suggests that the difficulties may have occurred because some of the model terms have little or no effect: if such terms are dropped, model-fitting may be more successful. Inspection of the results produced when all terms were specified as fixed-effect terms indicates that the non-significant terms 'type.operatn_period' and 'constrctn_y.operatn_period' are candidates for omission. Fitting of the resulting model is specified by the following statement:

```
GLMM [DISTRIBUTION = poisson; LINK = log; DISPERSION = 1; \
 OFFSET = logservice; \
 RANDOM = constrctn_y + oper-
atn_period + type.constrctn_y; FIXED = type; \
 PRINT = model, monitoring, components, vcovariance, means, \
 backmeans, waldtests] Y = damage_incidents
```

The output produced by this statement is as follows:

Generalized linear mixed-model analysis

Method:	c.f. Schall (1991) Biometrika
Response variate:	damage_incidents
Distribution:	poisson
Distribution:	binomial
Link function:	logarithm
Offset:	logservice
Random model:	constrctn_y + operatn_period + (constrctn_y.type)
Fixed model:	Constant + type

Dispersion parameter fixed at value 1.000

Warning 17, code UF 2, statement 236 in procedure GLMM

Missing values generated in weights/working variate.

Monitoring information

Iteration	Gammas			Dispersion	Max change
1	0.06317	0.07271	0.1455	1.000	8.6857E−01

Warning 18, code UF 2, statement 236 in procedure GLMM

Missing values generated in weights/working variate.

| 2 | 0.07559 | 0.07117 | 0.1366 | 1.000 | 1.2420E−02 |

Warning 19, code UF 2, statement 236 in procedure GLMM

Missing values generated in weights/working variate.

⋮

\<similar messages from subsequent iterations\>

⋮

| 6 | 0.07661 | 0.07151 | 0.1438 | 1.000 | 1.0626E−04 |

Warning 23, code UF 2, statement 236 in procedure GLMM

Missing values generated in weights/working variate.

| 7 | 0.07661 | 0.07151 | 0.1439 | 1.000 | 7.1114E−05 |

Estimated variance components

Random term	Component	s.e.
constrctn_y	0.077	0.129
operatn_period	0.072	0.111
constrctn_y.type	0.144	0.130

Residual variance model

Term	Model(order)	Parameter	Estimate	s.e.
Dispersn	Identity	Sigma2	1.000	fixed

Estimated variance matrix for variance components

constrctn_y	1	0.016543			
operatn_period	2	−0.000168	0.012291		
constrctn_y.type	3	−0.004545	−0.000003	0.016841	
Dispersn	4	0.000000	0.000000	0.000000	0.000000
		1	2	3	4

Tests for fixed effects

Sequentially adding terms to fixed model

Fixed term	Wald statistic	n.d.f.	F statistic	d.d.f.	F pr.
type	9.75	4	2.42	8.0	0.133

Dropping individual terms from full fixed model

Fixed term	Wald statistic	n.d.f.	F statistic	d.d.f.	F pr.
type	9.75	4	2.42	8.0	0.133

Tables of means with standard errors

calculated at offset zero

Table of predicted means for type

type	A	B	C	D	E
	6.349	5.763	5.610	6.131	6.753

Standard errors of differences

Average:	0.4279
Maximum:	0.5056
Minimum:	0.3450

Average variance of differences:	0.1853

Back-transformed Means (on the original scale)

calculated at offset zero

type
A	571.8
B	318.3
C	273.0
D	460.0
E	856.3

This simpler model has been successfully fitted. The output follows the same general form as that from the GLMM fitted in Section 10.2. The statement of the fitting method used and the model fitted is followed by a warning about the missing values in the data. Next comes the monitoring information on the fitting process, which shows that convergence has been achieved: this is indicated by the values of the *gammas*, statistics closely related to the variance components for the three random-effect terms (see Section 10.10), which hardly change between the sixth and seventh iteration. Next come the estimates of the variance components. Although the main effects of 'constrctn_y' and 'operatn_period' were significant in the fixed-effects-only analysis, the variance component estimates for these terms are smaller than their respective SEs. It should be remembered that in the fixed-effects-only analysis, each term was tested against the residual deviance: the significance of these terms may have been due to other variance components that contribute to their deviance. Several variance components can contribute to the deviance for a particular model term, just as several can contribute to an expected mean square in an anova (see Section 3.3). Next comes the variance–covariance matrix among these variance component estimates: the interpretation of such a matrix is explained in Section 10.2. Next come the tests of the significance of the fixed-effect term 'type'. The F statistic shows that this falls short of significance. Nevertheless, the estimated means for each level of type is then presented, on the transformed scale, that is, the scale after the link function has been applied to the model: in this case, the logarithmic scale. These means are somewhat different from those obtained from the fixed-effects model: consistently larger, but less variable. However, the ranking of the five types of ship is the same, Type C giving the fewest incidents of damage and Type E the most. The SEs of differences between these means are consistently smaller than those from the fixed-effects model. Overall, the variation among the means is somewhat less, relative to the SEs of differences, than when the fixed-effects model is used. This is perhaps to be expected, as the mixed model recognizes 'constrctn_y.type' as a random-effect term, which contributes to the SEs of differences between levels of 'type'. Finally, the back-transformed means on the original scale (count of incidents) are presented. These are obtained from the means on the transformed scale using the inverse of the link function, namely the exponential function: for example,

$$\text{back-transformed mean(Level A)} = e^{6.349} = 571.8.$$

10.6   The hierarchical generalized linear model (HGLM)

In the GLMMs fitted above (Sections 10.2 and 10.5), it is assumed that although the residual variation may not be normally distributed, the other random-effect terms do follow the normal distribution. However, this is not always the most natural assumption. For example, it has been suggested (Lee and Nelder, 2001) that when a Poisson distribution and a logarithmic link function are specified for the residual variation (as in the model fitted in Section 10.5), a more appropriate assumption for the other random-effect terms might be a gamma distribution and a logarithmic link function. Such models, in which a probability distribution and a link function can be specified for each random-effect term, have been called *hierarchical generalized linear models (HGLMs)*. However, this name is slightly misleading, as the random-effect terms do not have to be nested to form a hierarchy: they may be crossed, as in the model in Section 10.5. An alternative name, taking account of this possibility, would be stratified generalized linear models.

A system to fit HGLMs has been developed (Lee and Nelder, 1996, 2001), and is available in GenStat. The following statements use this system to fit the same GLMM as was fitted in Section 10.5, with no change to the distributions or link functions:

```
CALCULATE logservice1 = LOG(service_months)
SUBSET [logservice1 .NE. !s(*)] \
 type,constrctn_y,operatn_period,damage_incidents,logservice1
HGFIXEDMODEL [DISTRIBUTION = poisson; LINK = logarithm; \
 OFFSET = logservice1] type
HGRANDOMMODEL [DISTRIBUTION = normal; LINK = identity] \
 constrctn_y + operatn_period + type.constrctn_y
HGANALYSE [LMETHOD = eql] damage_incidents
HGPLOT METHOD = histogram, fittedvalues, normal, halfnormal
HGPLOT [RANDOMTERM = type.constrctn_y] \
 METHOD = histogram, fittedvalues, normal, halfnormal
```

Note that for presentation to the HGLM fitting system, the offset variable must be specified as 'LOG(service_months)' not 'LOG(service_months/SUM(service_months))'. Note also that HGLM fitting system is not able to cope with the missing values of the offset variate, so the SUBSET statement is used to exclude these from consideration, together with the corresponding values of the other variates and factors used. The HGFIXEDMODEL and HGRANDOMMODEL statements specify the fixed-effect and random-effect models, respectively. The HGFIXEDMODEL statement also specifies the distribution and link function for the residual term, and the offset variable, as was done in the equivalent GLMM statement (Section 10.5). The HGRANDOMMODEL statement further indicates that the other random-effect terms are to have a normal distribution and the *identity* link function, that is, no transformation, as was assumed implicitly in the GLMM statement. The HGANALYSE statement indicates that the response variable is 'damage_incidents', and that the model is to be fitted using a criterion known as *extended quasi-likelihood*, rather than the default criterion of exact likelihood. This specification produces output (not shown) that is numerically equivalent to that from the GLMM statement with regard to the fixed-effect estimates. The two HGPLOT statements produce diagnostic plots of the random effects, for the residual term and the term 'type.constrctn_y', respectively (Figure 10.5). The other two random-effect terms do not have enough levels to justify the production of diagnostic plots.

The distributions used in the model fitted here are not both normal, and the reference to normal plots and normal quantiles in the labelling of these diagnostic plots is therefore imprecise. However, the interpretation of the plots is the same as in models that assume a normal distribution (Section 1.10). In the case of the residual term, the histogram is reasonably bell-shaped

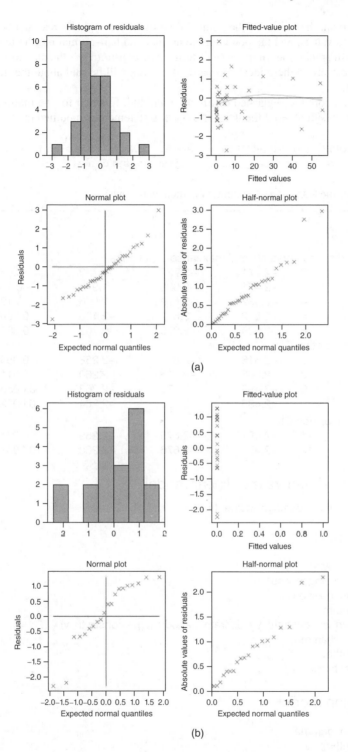

Figure 10.5   Diagnostic plots of the distribution of random effects from the mixed model relating damage to ships to their type, year of construction and period of operation, fitted with the normal distribution and the identity link function. (a) Effects in the residual term and (b) effects in the term 'type.constrctn_y'.

and symmetrical, the scatter of the points in the fitted-value plot is reasonably even over the range of fitted values, and the points in the normal and half-normal plot lie reasonably close to a straight diagonal line. In the case of term 'type.constrctn_y', the diagnostic plots do not conform so closely to these ideals: in particular, more will be said about the fitted-value plot in a moment.

In order to adopt the suggested distribution and link function for the random-effect terms other than the residual term, the HGRANDOMMODEL statement is modified to

```
HGRANDOMMODEL [DISTRIBUTION = gamma; LINK = logarithm] \
  constrctn_y + operatn_period + type.constrctn_y
```

The output of the HGANALYSE statement is then as follows:

Monitoring

Cycle no., disp. components and max. absolute change

2	−2.271	−2.727	−1.729	0.5663
3	−2.471	−2.688	−1.973	0.2440
4	−2.532	−2.679	−2.096	0.1228
5	−2.542	−2.677	−2.163	0.06693
6	−2.535	−2.676	−2.202	0.03880

Aitken extrapolation OK

7	−2.516	−2.675	−2.258	0.05584
8	−2.505	−2.675	−2.260	0.01117
9	−2.499	−2.675	−2.263	0.005888
10	−2.496	−2.676	−2.264	0.003227

Aitken extrapolation OK

11	−2.491	−2.676	−2.269	0.004577
12	−2.491	−2.676	−2.268	0.0003058

Hierarchical generalized linear model

Response variate: damage_incidents

Mean model

Fixed terms: type
Offset variate: logservice1
Distribution: poisson
Link: logarithm
Random terms: constrctn_y + operatn_period + (type.constrctn_y)
Distribution: gamma
Link: logarithm
Dispersion: fixed

Dispersion model

Distribution: gamma
Link: logarithm

Estimates from the mean model

	estimate	s.e.	$t(29)$
Constant	−5.624	0.3423	−16.427
type B	−0.579	0.3082	−1.878
type C	−0.729	0.4297	−1.697
type D	−0.150	0.3937	−0.381
type E	0.439	0.3699	1.188

Estimates from the dispersion model

Estimates of parameters

Parameter	estimate	s.e.	$t(*)$	Antilog of estimate
lambda constrctn_y	−2.49	1.05	−2.38	0.08283
lambda operatn_period	−2.68	1.46	−1.83	0.06885
lambda constrctn_y.type	−2.268	0.632	−3.59	0.1035

Likelihood statistics

$-2 \times h(y\|v)$	124.561
$-2 \times h$	119.851
$-2 \times P_v(h)$	153.904

$-2 \times P_{\beta,v}(h)$	156.249
$-2 \times EQD(y\|v)$	124.879
$-2 \times EQD$	119.747
$-2 \times P_v(EQD)$	153.800
$-2 \times P_{\beta,v}(EQD)$	156.145

Fixed parameters in mean model	5
Random parameters in mean model	26
Fixed dispersion parameters	3
Random dispersion parameters	0

Scaled deviances

Random term	deviance	d.f.
units	26.69	21.66
constrctn_y	1.82	1.77
operatn_period	0.93	0.91
constrctn_y.type	5.01	4.66
Total	34.46	29.00

The monitoring information at the beginning of the output shows the value of the dispersion parameter λ_i for each random-effect term (see below) at each stage of the model-fitting process, and the largest absolute change in this parameter value at each step. The largest absolute

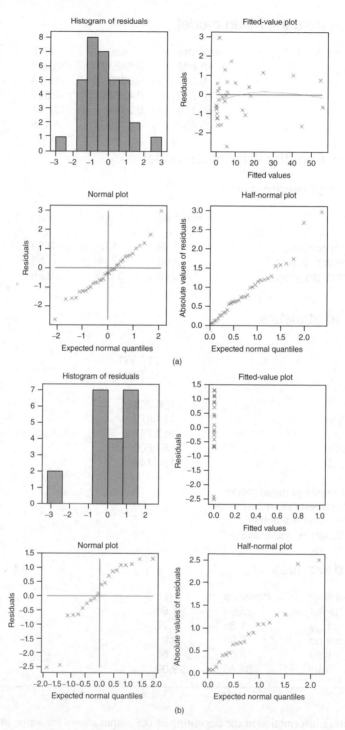

Figure 10.6 Diagnostic plots of the distribution of random effects from the mixed model relating damage to ships to their type, year of construction and period of operation, fitted with the gamma distribution and the logarithmic link function. (a) Effects in the residual term and (b) effects in the term 'type.constrctn_y'.

change between the final two cycles is very small, confirming that the model-fitting process has converged. The message 'Aitken extrapolation OK' indicates points at which the use of this method to accelerate convergence is acceptable. The model fitted is then specified. Next come the estimates of the constant and of the effects of 'type'. These are calculated using 'A' as the reference level of 'type', so the corresponding means are given by

$$\text{Mean}_{\text{Type A}} = \text{constant},$$

and

$$\text{Mean}_{\text{Type } i} = \text{constant} + \text{Mean}_{\text{Type } i}$$

for the other types. The values obtained are presented in Table 10.4. They differ from those given by the GLMM by approximately $\log_e(T) = \log_e(163574) = 12.005$, because the offset of the HGLM was specified as $\log_e(\text{service months})$ whereas that of the GLMM was specified as $\log_e(\text{service months/sum(service months)})$. Hence they are based on $A = e^{-4.956} = 0.0070410$ months of service.

The parameter estimates from the dispersion model, $\hat{\lambda}_1$, $\hat{\lambda}_2$ and $\hat{\lambda}_3$, indicate the deviance due to the random-effect terms 'constrctn_y', 'operatn_period' and 'constrctn_y.type', respectively. These estimates are given relative to the dispersion: that is,

$$\hat{\lambda}_i = \frac{\text{deviance(term } i)}{\text{dispersion}}. \tag{10.20}$$

In the present case, we have specified that

$$\text{dispersion} = 1, \tag{10.21}$$

so rearranging Equation 10.20 and substituting the numerical values from the output and from Equation 10.21, we obtain, for example,

$$\text{deviance(constrctn\_y)} = -2.49 \times 1 = -2.49.$$

The deviances for the random-effect terms can be interpreted and compared in roughly the same way as estimates of variance components. For example, each is compared in the output with its own SE, to obtain a t statistic that gives a tentative indication of whether any variation is accounted for by the term (cf. Section 3.11, where the analogous statistic based on estimates of variance components is referred to as a z statistic). In the present case, all three dispersion parameter estimates are negative, although GLMM gave small positive values for the corresponding variance component estimates. Thus according to the HGLM models, no variation is accounted for by these terms, and the estimates of all random effects, except the residual effects, are shrunk to zero. This is why the fitted values in the fitted-value plots for 'type.constrctn_y' are all zero (Figures 10.5b and 10.6b).

Table 10.4   Predicted number of damage incidents suffered by each type of ship, per 0.0070410 months of service, transformed to natural logarithms, obtained from an HGLM.

		Type		
A	B	C	D	E
−5.624	−6.203	−6.353	−5.774	−5.185

The diagnostic plots obtained from the residual term in this model (Figure 10.6a) is very little changed from that obtained previously (Figure 10.5a) but that for term 'type.constrctn_y' (Figure 10.6b) is considerably *less* satisfactory than when the normal distribution and the identity link function were used (Figure 10.5b). Thus, in this particular case, the use of the HGLM system has not improved the outcome of the modelling process. This may be because the parameters of a non-normal distribution are more difficult to fit, even when such a distribution is more appropriate on theoretical grounds (D. Hedderly, personal communication).

10.7   Use of R to fit a GLMM and a HGLM to a contingency table

The following commands import the ship-damage data into R, convert 'type', 'constrctn_y' and 'operatn_period' to factors and calculate the offset variable 'logservice':

```
rm(list = ls())
ship.damage <- read.table(
  "IMM edn 2\\Ch 10\\ship damage.txt",
  header=TRUE)
attach(ship.damage)
ftype <- factor(type)
fconstrctn_y <- factor(constrctn_y)
foperatn_period <- factor(operatn_period)
logservice <- log(service_months/sum(service_months, na.rm = TRUE))
logservice[logservice == -Inf] <- NA
```

Note that in categories where there are no months of service, and hence 'logservice' is minus infinity, this is replaced by a missing value.

The following commands fit to these data the model in which all terms are specified as fixed-effect terms:

```
ship.damage.glm.1 <-
  glm(damage_incidents ~ ftype * fconstrctn_y * foperatn_period,
  family = poisson(link = "log"),
  offset = logservice, data = ship.damage)
anova(ship.damage.glm.1)
```

The output from the function anova() is as follows:

	Df	Deviance	Resid. Df	Resid. Dev
NULL			33	146.328
ftype	4	55.439	29	90.889
fconstrctn_y	3	41.534	26	49.355
foperatn_period	1	10.660	25	38.695
ftype:fconstrctn_y	12	24.108	13	14.587
ftype:foperatn_period	4	6.066	9	8.521
fconstrctn_y:foperatn_period	2	1.664	7	6.856
ftype:fconstrctn_y:foperatn_period	7	6.856	0	0.000

The deviances and their d.f. agree with those obtained from GenStat.

The following commands fit the model from which the non-significant two-way interaction terms have been omitted:

```
ship.damage.glm.2 <-
  glm(damage_incidents ~ ftype * fconstrctn_y + foperatn_period,
  family = poisson(link = "log"),
  offset = logservice, data = ship.damage)
anova(ship.damage.glm.2)
```

Again, the deviances and their d.f. agree with those obtained from GenStat. However, the author has not been able to obtain meaningful predictions of the ship type means, or the differences between them, from this model.

The combination of the random-effect model 'constrctn_y*operatn_period + type.constrctn _y + type.operatn_period' with the fixed-effect model 'type', which produced non-convergence in GenStat, is successfully fitted by R, but for purposes of comparison, the model subsequently fitted in GenStat will also be fitted here. This is done by the following commands:

```
library(lme4)
ship.damage.glmer.2 <- glmer(damage_incidents ~ ftype +
  (1|fconstrctn_y) + (1|foperatn_period) +
  (1|ftype:fconstrctn_y),
  family = poisson(link = "log"), offset = logservice)
summary(ship.damage.glmer.2)
```

The function glmer() fits a generalized linear mixed model, the response variable, fixed-effect model and random-effect model being specified as described in Section 10.3. The family option indicates the distribution of the response variable and the link function, and the offset option indicates the offset variable.

Part of the output from the function summary() is as follows:

```
Generalized linear mixed model fit by the Laplace approximation
Formula: damage_incidents ~ ftype +         (1 | fconstrctn_y) +
    (1 | foperatn_period) + (1 | ftype:fconstrctn_y)
   Data; ship.damage
   AIC BIC logLik deviance
  71.02 83.23 -27.51 55.02
Random effects:
 Groups               Name         Variance Std.Dev.
 ftype:fconstrctn_y (Intercept)  0.030047 0.17334
 fconstrctn_y        (Intercept)  0.087829 0.29636
 foperatn_period     (Intercept)  0.045956 0.21437
Number of obs: 34, groups: ftype:fconstrctn_y, 20;
fconstrctn_y, 4; foperatn_period, 2
```

```
Fixed effects:
              Estimate Std. Error z value Pr(>|z|)
(Intercept)     6.3177      0.2889  21.865   <2e-16 ***
ftypeB         -0.5723      0.2272  -2.519   0.0118 *
ftypeC         -0.7197      0.3650  -1.972   0.0487 *
ftypeD         -0.1161      0.3270  -0.355   0.7226
ftypeE          0.3736      0.2853   1.310   0.1903
---
Signif. codes: 0 '***' 0.001 '**' 0.01 '*' 0.05 '.' 0.1 ' ' 1
```

The variance component estimates do not agree with those obtained from GenStat: the reason for this discrepancy is not known to the author. The estimates of the effect of the type are calculated relative to the estimated mean for Type A, which is the intercept. Thus, for example, the estimated mean for Type B on the transformed (logarithmic) scale is

$$\text{Intercept} + \text{effect of Type B} = 6.3177 - 0.5723 = 5.7454.$$

These estimates do not agree closely with those obtained from GenStat, but they rank the ship types in the same order.

At the time of writing, no straightforward method for specifying in R the HGLM fitted in GenStat, with several random-effect terms, is known to the author. A simpler HGLM with a single random-effect term is therefore fitted, using the package 'hglm'. This package can be installed from the R website by the method described in Section 3.16. It is not able to deal with an offset variate containing missing values, and a new data frame (see Section 1.11 for an explanation of this concept) in which the corresponding observations are excluded from all variables is therefore created by the following command:

```
ship.damage.short <-
  data.frame(damage_incidents, ftype, fconstrctn_y,
  foperatn_period, logservice)[is.na(logservice) == FALSE,]
```

The following commands then fit a HGLM that estimates the effects of 'ftype' and 'fconstrctn_y', but does not take account of 'foperatn_period':

```
library(hglm)
ship.damage.hglm.1 <- hglm(fixed = damage_incidents ~ ftype,
  random = ~ 1|fconstrctn_y,
  family = poisson(link = "log"), rand.family = gaussian
      (link = "identity"),
  offset = logservice, data = ship.damage.short)
summary(ship.damage.hglm.1)
```

In the function hglm(), the fixed and random options specify the fixed-effect and random-effect models, respectively. The family option specifies the distribution and the link function in the residual stratum, and the rand.family option specifies the distribution and link function for the term in the random-effect model – in the present case, the Gaussian (normal) distribution and the identity link function. The data option indicates that the analysis is to be performed using the data frame from which observations with missing values have been excluded.

The output from the `summary()` function is as follows:

```
Call:
hglm.formula(family = poisson(link = "log"), rand.family = gaus-
sian(link = "identity"),
  fixed = damage_incidents ~ ftype, random = ~1 | fconstrctn_y,
  data = ship.damage.short, offset = logservice)


- - - - - - - - - -
MEAN MODEL
- - - - - - - - - -


Summary of the fixed-effects estimates:

            Estimate Std. Error t-value Pr(>|t|)
(Intercept)  6.37823    0.29498  21.623   <2e-16 ***
ftypeB      -0.60312    0.24254  -2.487   0.0196 *
ftypeC      -0.70288    0.45216  -1.555   0.1322
ftypeD      -0.07265    0.39888  -0.182   0.8569
ftypeE       0.30599    0.32403   0.944   0.3537
- - -
Signif. codes: 0 '***' 0.001 '**' 0.01 '*' 0.05 '.' 0.1 ' ' 1
Note: P-values are based on 26 degrees of freedom

Summary of the random effects estimates:

                      Estimate Std. Error
fconstrctn_y1960-64   -0.5130     0.2384
fconstrctn_y1965-69    0.1367     0.2276
fconstrctn_y1970-74    0.3125     0.2309
fconstrctn_y1975-79    0.0639     0.2598
- - - - - - - - - - - - - - - -
DISPERSION MODEL
- - - - - - - - - - - - - - - -


WARNING: h-likelihood estimates through EQL can be biased.

Dispersion parameter for the mean model:
[1] 1.893956

Model estimates for the dispersion term:

Link = log

Effects:
  Estimate Std. Error
    0.6387     0.2748

Dispersion = 1 is used in Gamma model on deviances to
calculate the standard error(s).
```

```
Dispersion parameter for the random effects:
[1] 0.1525

Dispersion model for the random effects:

Link = log

Effects:
fconstrctn_y1975-79
  Estimate Std. Error
    -1.8806       0.8916

Dispersion = 1 is used in Gamma model on deviances to
calculate the standard error(s).

EQL estimation converged in 3 iterations.
```

The output first reproduces the HGLM model specification. It then shows the effect estimates for the fixed-effect model 'type', calculated relative to Type A as before. The ship type means obtained from these effect estimates agree fairly closely with those given by GenStat. BLUPs are then presented for the levels of the term in the random-effect model, 'fconstrctn_y'. A fuller account of the function hglm() is given by Rönnegård, Shen and Alam (2010), and a somewhat broader range of HGLMs can be fitted using the newer function hglm(2) (L. Rönnegård, personal communication).

Diagnostic plots of residuals from this model are produced by the following commands:

```
windows()
par(mfrow=c(2,2))
plot(ship.damage.hglm.1)
```

Part of the output from the plot() function is shown in Figure 10.7. The plot of studentized residuals against fitted values (Plot a) shows that larger fitted values are associated with more variable residuals, indicating that the model fitted should ideally be refined to take account of this heterogeneity of variance. This pattern is confirmed by the corresponding plot of the absolute values of the residuals (Plot b), which shows a clear positive trend. The Q–Q plot (Plot c) shows somewhat more residuals with very large positive or negative values than expected. However, the histogram of residuals (Plot d) is as close to a normal distribution as can reasonably be expected from this fairly small sample. Further diagnostic plots are produced, but they are less informative than those shown here. In particular, the plots that related to the random effects of 'fconstrctn_y' give very little information, being based on only four values.

The HGLM can be modified to specify a gamma distribution for the effects of 'fconstrctn_y' and the logarithmic link function for this stratum by changing the setting of the rand.family option to 'rand.family = Gamma(link = "log")'. This causes some change to the estimates of the effects of 'type', but has very little effect on the diagnostic plots.

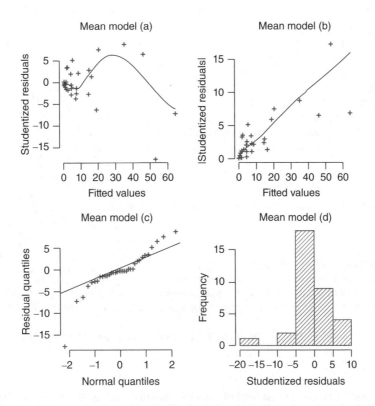

Figure 10.7 (a–d) Diagnostic plots of the distribution of random effects from the mixed model relating damage to ships to their type and year of construction, fitted with the normal distribution and the identity link function.

10.8 Use of SAS to fit a GLMM to a contingency table

The following SAS statements import the ship-damage data, calculate the offset variable 'logservice' and fit the model specifying all terms as fixed-effect terms:

```
PROC IMPORT OUT = ship DBMS = EXCELCS REPLACE
   DATAFILE = "&pathname.\IMM edn 2\Ch 10\ship damage.xlsx";
   SHEET = "for SAS";
RUN;

PROC UNIVARIATE DATA = ship NOPRINT;
   VAR service_months;
   OUTPUT OUT = ship_summ SUM = serv_sum;
RUN;

DATA ship_summ_2; SET ship_summ;
   dummy = 1;
RUN;
```

```
DATA ship_2; SET ship;
    dummy = 1;
RUN;

PROC SQL;
    CREATE TABLE ship_3 AS
    SELECT a.*, b.*
    FROM ship_2 AS a, ship_summ_2 AS b
    WHERE a.dummy EQ b.dummy;
QUIT;

DATA ship_4; SET ship_3;
    logservice = log(service_months/serv_sum);
RUN;

ODS RTF;
PROC GENMOD DATA = ship_4;
    CLASS type constrctn_y operatn_period;
    MODEL damage_incidents = type constrctn_y operatn_period
        type * constrctn_y type * operatn_period constrctn_y * oper-
atn_period /
        DIST = POISSON LINK = LOG OFFSET = logservice TYPE1;
RUN;
ODS RTF CLOSE;
```

PROC UNIVARIATE obtains the total of 'service_months', and the DATA steps and use of PROC SQL that follow place this total in an additional column in the data set and calculate 'logservice'. PROC GENMOD fits the model. In the MODEL statement, the DIST, LINK and OFFSET options specify the distribution of the response variable, the link function and the offset variable, respectively, and the TYPE1 option specifies Type I hypothesis tests, which are those performed in an accumulated analysis of deviance. Part of the output from PROC GENMOD is as follows:

LR statistics for Type 1 analysis				
Source	Deviance	DF	Chi-square	Pr > ChiSq
Intercept	146.3283			
type	90.8893	4	55.44	<0.0001
constrctn_y	49.3552	3	41.53	<0.0001
operatn_period	38.6951	1	10.66	0.0011
type*constrctn_y	14.5869	12	24.11	0.0197
type*operatn_period	8.5208	4	6.07	0.1943
constrctn*operatn_pe	6.8565	2	1.66	0.4351

The chi-square statistics and their d.f. agree with the deviances in the accumulated analysis of deviance produced by GenStat.

The following statements fit the GLMM with 'type' as the fixed-effect model and all other terms in the random-effect model, with the omission of the terms 'type.operatn_period' and 'constrctn_y.operatn_period':

```
ODS RTF;
PROC GLIMMIX DATA = ship_4 METHOD = RSPL ASYCOV NOBOUND;
   CLASS type constrctn_y operatn_period;
   MODEL damage_incidents = type /
      DISTRIBUTION = POISSON LINK = LOG OFFSET = logservice;
   RANDOM constrctn_y operatn_period constrctn_y * type;
   LSMEANS type;
RUN;
ODS RTF CLOSE;
```

In PROC GLIMMIX, the option setting 'METHOD = RSPL' indicates that the default RSPL method of estimation will be used. The MODEL statement specifies the response variable and the fixed-effect term, and the RANDOM statement specifies the random-effect terms. The LSMEANS statement indicates that the mean for each level of 'type' is to be estimated. Part of the output from PROC GLIMMIX is as follows:

Covariance parameter estimates		
Cov Parm	Estimate	Standard error
constrctn_y	0.07661	0.1329
operatn_period	0.07152	0.1109
type*constrctn_y	0.1440	0.1372

Asymptotic covariance matrix of covariance parameter estimates			
Cov Parm	CovP1	CovP2	CovP3
constrctn_y	0.01767	−0.00035	−0.00454
operatn_period	−0.00035	0.01230	−0.00002
type*constrctn_y	−0.00454	−0.00002	0.01883

Type III tests of fixed effects				
Effect	Num DF	Den DF	F value	Pr > F
type	4	12	2.43	0.1043

Type least squares means					
Type	Estimate	Standard error	DF	t value	Pr > |t|
A	6.3487	0.3639	12	17.45	<0.0001
B	5.7631	0.3114	12	18.51	<0.0001
C	5.6096	0.4411	12	12.72	<0.0001
D	6.1312	0.4175	12	14.68	<0.0001
E	6.7527	0.3853	12	17.53	<0.0001

The covariance parameter estimates agree with the estimated variance components produced by GenStat, and the covariances between the estimates are similar though not identical. The F test for 'type' agrees fairly closely with that given by GenStat, but the denominator d.f. is different, being obtained by a somewhat different method, and the p-value is therefore also different, though still non-significant. The estimated means for types of ship (on the logarithmic scale) agree with those given by GenStat.

At the time of writing, no straightforward method of fitting an HGLM in SAS is known to the author.

10.9 The role of the covariance matrix in the specification of a mixed model

In all the mixed models we have examined so far, each random-effect term consists of a set of factor levels. Any two observations either come from the same level of the factor, in which case they have the same random effect for the term in question, or from different levels, in which case their random effects are independent. In the residual stratum, each observation comprises a separate level, and its random effect is independent of those of all other observations. However, the relationship between the random effects in each term need not be so simple. In order to explore more elaborate relationships among the random effects, we need first to establish a notation in which to display them. Such a notation is provided by the *covariance matrix*, which is specified as follows.

Consider a simple data set comprising four groups of three observations of a response variable (Table 10.5). The natural model to fit to these data is

$$y_{ij} = \mu + \gamma_i + \varepsilon_{ij} \tag{10.22}$$

Table 10.5   A simple data set comprising a grouping factor and a response variable.

	A	B
1	group!	y
2	1	1.97
3	1	1.01
4	1	4.53
5	2	0.76
6	2	1.60
7	2	3.52
8	3	4.37
9	3	4.05
10	3	5.17
11	4	6.94
12	4	7.50
13	4	5.62

Table 10.6 Anova of a simple data set comprising a grouping factor and a response variable.

Source of variation	DF	MS	F	p
Group	3	13.875	8.44	0.007
Residual	8	1.644		
Total	11			

where
y_{ij} = the jth observation of the response variable Y in the ith group (the ijth observation),
μ = the overall mean,
γ_i = the effect of the ith group,
ε_{ij} = the residual effect on the ijth observation of Y.

We will make the usual assumption that the ε_{ij} are values of a random, normally distributed variable E, that is,

$$E \sim N(0, \sigma_E^2). \tag{10.23}$$

The anova corresponding to this model is shown in Table 10.6, and the estimate of σ_E^2 is given by $\mathrm{MS_{Residual}}$,

$$\hat{\sigma}_E^2 = 1.644.$$

In order to express the idea that the 12 values ε_{ij}, $i = 1 \ldots 4$, $j = 1 \ldots 3$, are independent, we need to think of each of them as a value of a different variable, E_{ij}. Similarly, each value of the response variable, y_{ij}, is thought of as an observation of a different variable, Y_{ij}. The relationships among the 12 variables E_{ij} can then be expressed in a covariance matrix, namely:

i'			1	1	1	2	2	2	3	3	3	4	4	4
i	j	j'	1	2	3	1	2	3	1	2	3	1	2	3
1	1		σ_E^2	0	0	0	0	0	0	0	0	0	0	0
1	2		0	σ_E^2	0	0	0	0	0	0	0	0	0	0
1	3		0	0	σ_E^2	0	0	0	0	0	0	0	0	0
2	1		0	0	0	σ_E^2	0	0	0	0	0	0	0	0
2	2		0	0	0	0	σ_E^2	0	0	0	0	0	0	0
2	3		0	0	0	0	0	σ_E^2	0	0	0	0	0	0
3	1		0	0	0	0	0	0	σ_E^2	0	0	0	0	0
3	2		0	0	0	0	0	0	0	σ_E^2	0	0	0	0
3	3		0	0	0	0	0	0	0	0	σ_E^2	0	0	0
4	1		0	0	0	0	0	0	0	0	0	σ_E^2	0	0
4	2		0	0	0	0	0	0	0	0	0	0	σ_E^2	0
4	3		0	0	0	0	0	0	0	0	0	0	0	σ_E^2

where
i = the group to which the first value in a particular comparison compared belongs,
j = the position within group i of the first value,
i' = the group to which the second value in the comparison belongs and
j' = the position within group i' of the second value.

The value of cov($E_{ij},E_{i'j'}$), the covariance between E_{ij} and $E_{i'j'}$, is given in the cell in the ijth row and the $i'j'$th column of this matrix. Since E_{ij} is the only random term that contributes to Y_{ij},

$$\text{cov}(Y_{ij}, Y_{i'j'}) = \text{cov}(E_{ij}, E_{i'j'}) \tag{10.24}$$

The variance of each of the variables E_{ij} – its covariance with itself – is given along the *leading diagonal* of the matrix (the 12 cells from top left to bottom right). All other covariances are zero, reflecting the assumption that all the variables are mutually independent. Because

$$\text{cov}(E_{ij}, E_{i'j'}) = \text{cov}(E_{i'j'}, E_{ij}) \tag{10.25}$$

by definition, the matrix is symmetrical about the leading diagonal: hence to improve clarity, the top right-hand half can be omitted.

Now suppose that we decide to treat the groups as a random-effect term. We then make the assumption that the γ_i, $i = 1 \dots 4$, are values of a variable Γ, and that

$$\Gamma \sim N(0, \sigma_\Gamma^2). \tag{10.26}$$

The anova can now be interpreted as shown in Table 10.7, and the estimate of σ_Γ^2 is

$$\hat{\sigma}_\Gamma^2 = \frac{\text{MS}_{\text{Group}} - \text{MS}_{\text{Residual}}}{3} = \frac{13.875 - 1.644}{3} = 4.077.$$

In order to express the idea that the four values γ_i, $i = 1 \dots 4$, are independent, we think of each of them as a value of a different variable, Γ_i. The variable representing the random part of Y_{ij} is now no longer E_{ij} but $\Gamma_i + E_{ij}$, and the covariances among these 12 variables are as follows:

i j	i' $\;$ 1 $\;$ j' $\;$ 1	1 2	1 3	2 1	2 2	2 3	3 1	3 2	3 3	4 1	4 2	4 3
1 1	$\sigma_\Gamma^2 + \sigma_E^2$											
1 2	σ_Γ^2	$\sigma_\Gamma^2 + \sigma_E^2$										
1 3	σ_Γ^2	σ_Γ^2	$\sigma_\Gamma^2 + \sigma_E^2$									
2 1	0	0	0	$\sigma_\Gamma^2 + \sigma_E^2$								
2 2	0	0	0	σ_Γ^2	$\sigma_\Gamma^2 + \sigma_E^2$							
2 3	0	0	0	σ_Γ^2	σ_Γ^2	$\sigma_\Gamma^2 + \sigma_E^2$						
3 1	0	0	0	0	0	0	$\sigma_\Gamma^2 + \sigma_E^2$					
3 2	0	0	0	0	0	0	σ_Γ^2	$\sigma_\Gamma^2 + \sigma_E^2$				
3 3	0	0	0	0	0	0	σ_Γ^2	σ_Γ^2	$\sigma_\Gamma^2 + \sigma_E^2$			
4 1	0	0	0	0	0	0	0	0	0	$\sigma_\Gamma^2 + \sigma_E^2$		
4 2	0	0	0	0	0	0	0	0	0	σ_Γ^2	$\sigma_\Gamma^2 + \sigma_E^2$	
4 3	0	0	0	0	0	0	0	0	0	σ_Γ^2	σ_Γ^2	$\sigma_\Gamma^2 + \sigma_E^2$

Table 10.7 Expected mean squares in the anova of a simple data set comprising a grouping factor and a response variable.

Source of variation	DF	Expected MS
Group	3	$3\sigma_\Gamma^2 + \sigma_E^2$
Residual	8	σ_E^2
Total	11	

The variance of each observation of the response variable is now $\sigma_\Gamma^2 + \sigma_E^2$. The covariance between observations from different groups – for example, between Y_{12} and Y_{23} – is zero, as before. However, the covariance between observations from the same group – for example, between Y_{12} and Y_{13} – is now σ_Γ^2, because they share the same group effect (Γ_1) though they have different individual-observation effects (E_{12} and E_{13}). This covariance matrix can be expressed in the notation of *matrix algebra* as

$$
\begin{pmatrix}
1 & & & & & & & & & & & \\
1 & 1 & & & & & & & & & & \\
1 & 1 & 1 & & & & & & & & & \\
0 & 0 & 0 & 1 & & & & & & & & \\
0 & 0 & 0 & 1 & 1 & & & & & & & \\
0 & 0 & 0 & 1 & 1 & 1 & & & & & & \\
0 & 0 & 0 & 0 & 0 & 0 & 1 & & & & & \\
0 & 0 & 0 & 0 & 0 & 0 & 1 & 1 & & & & \\
0 & 0 & 0 & 0 & 0 & 0 & 1 & 1 & 1 & & & \\
0 & 0 & 0 & 0 & 0 & 0 & 0 & 0 & 0 & 1 & & \\
0 & 0 & 0 & 0 & 0 & 0 & 0 & 0 & 0 & 1 & 1 & \\
0 & 0 & 0 & 0 & 0 & 0 & 0 & 0 & 0 & 1 & 1 & 1
\end{pmatrix}
\sigma_\Gamma^2 +
\begin{pmatrix}
1 & & & & & & & & & & & \\
0 & 1 & & & & & & & & & & \\
0 & 0 & 1 & & & & & & & & & \\
0 & 0 & 0 & 1 & & & & & & & & \\
0 & 0 & 0 & 0 & 1 & & & & & & & \\
0 & 0 & 0 & 0 & 0 & 1 & & & & & & \\
0 & 0 & 0 & 0 & 0 & 0 & 1 & & & & & \\
0 & 0 & 0 & 0 & 0 & 0 & 0 & 1 & & & & \\
0 & 0 & 0 & 0 & 0 & 0 & 0 & 0 & 1 & & & \\
0 & 0 & 0 & 0 & 0 & 0 & 0 & 0 & 0 & 1 & & \\
0 & 0 & 0 & 0 & 0 & 0 & 0 & 0 & 0 & 0 & 1 & \\
0 & 0 & 0 & 0 & 0 & 0 & 0 & 0 & 0 & 0 & 0 & 1
\end{pmatrix}
\sigma_E^2.
$$

In the first term in this expression, all rows and all columns representing observations in the same group are identical. (To see this, it may be helpful to fill in the top right-hand half of the matrix.) Hence the information in this term can be expressed more concisely in the form

$$
\text{covariance matrix} =
\begin{pmatrix}
1 & & & \\
0 & 1 & & \\
0 & 0 & 1 & \\
0 & 0 & 0 & 1
\end{pmatrix}
\sigma_\Gamma^2.
\tag{10.27}
$$

10.10   A more general pattern in the covariance matrix: Analysis of pedigrees and genetic data

In all the mixed models we have considered so far, the covariance matrix for each term is of the form shown in Equation 10.27 – an *identity matrix*, having 1s along the leading diagonal and 0s elsewhere, multiplied by the variance component for the term in question. However, the covariance matrix need not be so simple: we have already seen an instance of non-zero covariance between two random-effect terms, namely the slopes and intercepts in the random-coefficient model (Sections 7.5–7.7; see also Exercise 10.3 at the end of this chapter). For an example of non-zero covariances within a single term, consider the pedigree in Figure 10.8, which represents three generations of animals. Earlier generations are represented in higher rows of the pedigree: animals represented in the bottom row are the most recent generation, those in the row above are their parents and those in the top row their grandparents.

Suppose that a continuous variable Y (e.g. weight) is measured on each animal in this pedigree, and that this variable comprises a genetic effect Γ and an environmental effect E, so that

$$
Y = \mu + \Gamma + E
\tag{10.28}
$$

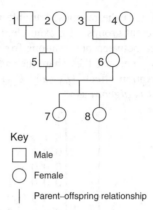

Key

☐ Male

◯ Female

| Parent–offspring relationship

Figure 10.8   A pedigree representing three generations of animals.

where
μ = the mean value of Y.

Then in the population from which this pedigree is randomly sampled, the variance of Y among unrelated animals (such as the grandparents 1, 2, 3 and 4) has two components, namely σ_Γ^2, the genetic component and σ_E^2, the environmental component, the total variance of Y being $\sigma_\Gamma^2 + \sigma_E^2$. The values of Γ in any two unrelated animals are independent, as are the values of E, so the covariance of Y between such animals is zero. However, Animal 7 shares half her genes with her sister, Animal 8, so the values of Γ in these two animals are not independent: the covariance of Y between them is $\frac{1}{2}\sigma_\Gamma^2$. Conversely, the variance among siblings is less than that among unrelated animals: it is $\frac{1}{2}\sigma_\Gamma^2 + \sigma_E^2$. Animal 7 also shares a quarter of her genes with her paternal grandmother, Animal 2, so that the covariance of Y between these animals is $\frac{1}{4}\sigma_\Gamma^2$. Similar arguments can be applied to every pair of animals, giving the following covariance matrix:

$$
\begin{array}{c|cccccccc}
i\ i' & 1 & 2 & 3 & 4 & 5 & 6 & 7 & 8 \\
\hline
1 & \sigma_\Gamma^2 + \sigma_E^2 \\
2 & 0 & \sigma_\Gamma^2 + \sigma_E^2 \\
3 & 0 & 0 & \sigma_\Gamma^2 + \sigma_E^2 \\
4 & 0 & 0 & 0 & \sigma_\Gamma^2 + \sigma_E^2 \\
5 & \frac{1}{2}\sigma_\Gamma^2 & \frac{1}{2}\sigma_\Gamma^2 & 0 & 0 & \sigma_\Gamma^2 + \sigma_E^2 \\
6 & 0 & 0 & \frac{1}{2}\sigma_\Gamma^2 & \frac{1}{2}\sigma_\Gamma^2 & 0 & \sigma_\Gamma^2 + \sigma_E^2 \\
7 & \frac{1}{4}\sigma_\Gamma^2 & \frac{1}{4}\sigma_\Gamma^2 & \frac{1}{4}\sigma_\Gamma^2 & \frac{1}{4}\sigma_\Gamma^2 & \frac{1}{2}\sigma_\Gamma^2 & \frac{1}{2}\sigma_\Gamma^2 & \sigma_\Gamma^2 + \sigma_E^2 \\
8 & \frac{1}{4}\sigma_\Gamma^2 & \frac{1}{4}\sigma_\Gamma^2 & \frac{1}{4}\sigma_\Gamma^2 & \frac{1}{4}\sigma_\Gamma^2 & \frac{1}{2}\sigma_\Gamma^2 & \frac{1}{2}\sigma_\Gamma^2 & \frac{1}{2}\sigma_\Gamma^2 & \sigma_\Gamma^2 + \sigma_E^2
\end{array}
$$

$$
= \begin{pmatrix}
1 \\
0 & 1 \\
0 & 0 & 1 \\
0 & 0 & 0 & 1 \\
\frac{1}{2} & \frac{1}{2} & 0 & 0 & 1 \\
0 & 0 & \frac{1}{2} & \frac{1}{2} & 0 & 1 \\
\frac{1}{4} & \frac{1}{4} & \frac{1}{4} & \frac{1}{4} & \frac{1}{2} & \frac{1}{2} & 1 \\
\frac{1}{4} & \frac{1}{4} & \frac{1}{4} & \frac{1}{4} & \frac{1}{2} & \frac{1}{2} & \frac{1}{2} & 1
\end{pmatrix} \sigma_\Gamma^2 + \begin{pmatrix}
1 \\
0 & 1 \\
0 & 0 & 1 \\
0 & 0 & 0 & 1 \\
0 & 0 & 0 & 0 & 1 \\
0 & 0 & 0 & 0 & 0 & 1 \\
0 & 0 & 0 & 0 & 0 & 0 & 1 \\
0 & 0 & 0 & 0 & 0 & 0 & 0 & 1
\end{pmatrix} \sigma_E^2
$$

where
i = the first animal in a particular pair-wise comparison and
i' = the second animal in the comparison.

The matrix $\begin{pmatrix} 1 & & & & & & & \\ 0 & 1 & & & & & & \\ 0 & 0 & 1 & & & & & \\ 0 & 0 & 0 & 1 & & & & \\ \frac{1}{2} & \frac{1}{2} & 0 & 0 & 1 & & & \\ 0 & 0 & \frac{1}{2} & \frac{1}{2} & 0 & 1 & & \\ \frac{1}{4} & \frac{1}{4} & \frac{1}{4} & \frac{1}{4} & \frac{1}{2} & \frac{1}{2} & 1 & \\ \frac{1}{4} & \frac{1}{4} & \frac{1}{4} & \frac{1}{4} & \frac{1}{2} & \frac{1}{2} & \frac{1}{2} & 1 \end{pmatrix}$ is called the *relationship matrix* among the animals in the

pedigree, and is said to hold the *variance structure* among them. The covariance matrix can be applied to the observations of Y, designated as y_i, $i = 1 \ldots 8$, using the methods of mixed modelling, to obtain estimates of the variance components, $\hat{\sigma}_\Gamma^2$ and $\hat{\sigma}_E^2$. From these, an estimate of the heritability of Y,

$$h^2 = \frac{\hat{\sigma}_\Gamma^2}{\hat{\sigma}_\Gamma^2 + \hat{\sigma}_E^2}, \tag{10.29}$$

(see also Equation 3.33) can be obtained. For a fuller account of genetic covariance between relatives and heritability, see Falconer and Mackay (1996, Chapters 9 and 10, pp. 145–183).

This method of analysis can be applied to a set of simulated data described by Goldin *et al.* (1997). (Data reproduced by kind permission of John Blangero, Tom Dyer, Jean MacCluer and Marcy Speer. Data presented at the 10th Genetic Analysis Workshop, supported by grant R01 GM31575 from the National Institute of General Medical Sciences.) The first and last few lines of the spreadsheet holding the pedigree data and the phenotypic data (i.e. the observations on the individuals in the pedigrees) are shown in the spreadsheet in Table 10.8: the full data set is held in the file 'GAW10 ped phen.xlsx' on this book's website (see Preface). The first three columns specify the relationships among the individuals, assumed here to belong to a species of domesticated animal. Thus in Row 19 we are told that the father and mother of Individual 1378 are Individuals 922 and 924, respectively. The information on these two parents is held in Rows 5 and 7, where it is confirmed that 922 is a male (sex = 1) and 924 is a female (sex = 2). Q4 is a quantitative trait, the inheritance of which is to be studied.

These data are imported and analysed by the following statements:

```
IMPORT 'IMM edn 2\\Ch 10\\GAW10 ped phen.xlsx'
VPEDIGREE INDIVIDUALS = ID; MALE = father; FEMALE = mother; \
    INVERSE = ainv
VCOMPONENTS [FIXED = sex * age] RANDOM = ID
VSTRUCTURE [TERM = ID] MODEL = fixed; INVERSE = ainv
REML [PRINT = model, monitor, component, Wald, effects] Q4
```

In the VPEDIGREE statement, the parameter INDIVIDUALS specifies the factor that identifies each individual to be considered, and the parameters MALE and FEMALE specify factors that identify the parents of each individual. From this information is produced not the relationship matrix described above but its *inverse* – a matrix derived from it contains the same information. (The details of the relationship between a matrix and its inverse need not concern us here.) The parameter INVERSE indicates the data structure in which this inverse matrix is to be stored, namely 'ainv'. The VCOMPONENTS statement specifies the terms in the fixed-effect and random-effect parts of the model in the usual way. In the present case, sex, age and their interaction are to be fitted as fixed-effect terms. This indicates that sex and age may have effects on the phenotypic trait studied, which should be taken into account when assessing the degree of similarity between relatives. The 'sex.age' interaction term indicates that the effect of age may differ between the sexes. The only random term is the individual observations. However, this term does not merely comprise a set of independent residual effects: it is structured by the pedigree relationships among the individuals, and this is indicated in the

Table 10.8   Pedigree data indicating the relationships among a set of individuals, with observations of a quantitative phenotypic trait.

The numbering of the individuals has been changed from that in the original version of this data set, in order to meet GenStat's requirement that every individual, including those who appear under 'father!' or 'mother!' but not under 'ID!', should have a unique number.

	A	B	C	D	E	F
	ID!	father!	mother!	sex!	age	Q4
1						
2	919	1	460	1	56	
3	920	2	461	2	59	
4	921	3	462	2	80	
5	922	4	463	1	80	
6	923	5	464	1	67	
7	924	6	465	2	77	
8	925	7	466	2	61	
9	926	8	467	2	63	
10	927	9	468	1	64	
11	928	10	469	2	37	13.1086
12	929	11	470	1	40	10.1638
13	930	12	471	1	40	10.3412
14	931	13	472	1	49	10.9942
15	932	14	473	1	64	10.685
16	933	15	474	1	65	11.4764
17	934	16	475	2	66	11.3533
18	935	17	476	1	67	10.1688
19	1378	922	924	2	80	
20	1379	922	924	1	80	
21	1380	922	924	1	80	
22	1381	927	926	2	63	
23	1382	927	926	1	72	
24	1596	919	1378	2	64	11.6905
25	1597	923	1381	1	58	10.5241
26	1598	923	1381	1	53	
27	1599	923	1381	2	64	11.3787

(continued overleaf)

Table 10.8 (*continued*)

	A	B	C	D	E	F
1484	2035	1588	1585	1	49	11.6033
1485	2329	2024	1590	2	37	12.1533
1486	2330	2024	1590	2	47	10.9051
1487	2331	2024	1590	2	40	9.9029
1488	2332	2024	1590	1	42	12.1348
1489	2333	2024	1590	2	46	10.8755
1490	2334	2024	1590	2	45	12.3281
1491	2335	2025	1373	2	34	10.8796
1492	2409	1365	2331	1	19	10.5293
1493	2410	1370	2334	2	24	10.7028
1494	2411	1370	2334	1	16	11.9798
1495	2412	1374	2329	1	16	11.0855
1496	2413	2035	2333	2	26	9.3804
1497	2414	2035	2333	2	24	11.3734
1498	2415	2035	2333	1	22	9.6838

Source: Data reproduced by kind permission of John Blangero, Tom Dyer, Jean MacCluer and Marcy Speer. Data presented at the 10th Genetic Analysis Workshop, supported by grant R01 GM31575 from the National Institute of General Medical Sciences.

VSTRUCTURE statement that follows. The option setting 'TERM = ID' indicates the term in the random-effect model for which a variance structure is to be specified. In the present case, there is only one term, ID, but if the random-effect model contained several terms, a different variance structure (or none) might be specified for each of them. The parameter setting 'MODEL = fixed' indicates that the variance structure is fully specified prior to the model-fitting process – in this case, by the VPEDIGREE statement. In the next example (Section 10.11), we shall see a situation in which some aspects of the variance structure are estimated during the model-fitting process. The parameter setting 'INVERSE = ainv' identifies the data structure that holds the information defining the variance structure. Finally, the REML statement specifies that the response variable to which the model is to be fitted is Q4. The PRINT option of this statement specifies that the output should comprise a statement of the model fitted, information monitoring the fitting process and estimates of the variance components.

The output of the REML statement is as follows:

REML variance components analysis

Response variate: Q4
Fixed model: Constant + sex + age + sex.age
Random model: ID
Number of units: 1000 (497 units excluded due to zero weights or missing values)

Residual term has been added to model

Sparse algorithm with AI (average information) optimization

All covariates centred

Covariance structures defined for random model

Covariance structures defined within terms:

Term	Factor	Model	Order	No. rows
ID	ID	Fixed matrix ainv (inverse)	1	2415

Convergence monitoring

```
Cycle    Deviance    Current variance parameters: gammas, sigma2, others
  0      877.367     0.468573    1.00000
  1      877.202     0.464133    1.02253
  2      876.839     0.451823    1.08776
  3      876.592     0.437181    1.17101
  4      876.538     0.426116    1.23839
  5      876.538     0.426581    1.23548
  6      876.538     0.426546    1.23569
  7      876.538     0.426549    1.23568
```

Estimated parameters for covariance models

Random term(s)	Factor	Model(order)	Parameter	Estimate	s.e.
ID	ID	Fixed matrix	Scalar	1.236	0.298

Note: the covariance matrix for each term is calculated as G or R where var(y) = Sigma2(ZGZ′ + R), that is, relative to the residual variance, Sigma2.

Residual variance model

Term	Model(order)	Parameter	Estimate	s.e.
Residual	Identity	Sigma2	0.427	0.0513

Tests for fixed effects

Sequentially adding terms to fixed model

Fixed term	Wald statistic	n.d.f.	F statistic	d.d.f.	F pr.
sex	0.71	1	0.71	969.7	0.400
age	6.09	1	6.09	801.8	0.014
sex.age	1.76	1	1.76	957.1	0.186

Dropping individual terms from full fixed model

Fixed term	Wald statistic	n.d.f.	F statistic	d.d.f.	F pr.
sex.age	1.76	1	1.76	957.1	0.186

Message: denominator degrees of freedom for approximate F tests are calculated using numerical derivatives ignoring fixed/boundary/singular variance parameters.

Table of effects for Constant

11.44 Standard error: 0.056

Table of effects for sex

sex	1	2
	0.00000	−0.05684

Standard error of differences: 0.05725

Table of effects for age

0.006357	Standard error	0.0023813

Table of effects for sex.age

sex	1	2
	0.000000	−0.004416

Standard error of differences: 0.003333

First comes the usual information about the model fitted – the response variable, the fixed-effect and random-effect models and so on – with some additional information about the variance structure specified (here referred to as the *covariance structure*). It is noted that the inverse of the relationship matrix, 'ainv', has 2415 rows, corresponding to the 2415 individuals (including parents) represented in the data set. Next comes the information on the convergence of the fitting process, which shows that after four cycles of parameter estimation there is little or no change in the deviance from the fitted model. Convergence has been achieved, and the process terminates successfully after seven cycles. Next come the estimates of the variance components. The first of these is the variance accounted for by the covariance structure specified for factor ID – that is, the genetic component that follows the pattern of relationships among the individuals, σ_Γ^2. This is followed by the residual variance component that affects each individual independently, σ_E^2. The estimate of the residual variance component is expressed straightforwardly as

$$\hat{\sigma}_E^2 = 0.427.$$

However, it is noted that the genetic component is calculated *relative to* the residual variance – that is, if we define

$$\gamma_\Gamma = \frac{\sigma_\Gamma^2}{\sigma_E^2} \tag{10.30}$$

then the estimate of γ_Γ is

$$\hat{\gamma}_\Gamma = 1.236,$$

and

$$\hat{\sigma}_\Gamma^2 = \hat{\gamma}_\Gamma \hat{\sigma}_E^2 = 1.236 \times 0.427 = 0.5278. \tag{10.31}$$

(Note that the ratio $\gamma_\Gamma = \sigma_\Gamma^2 / \sigma_E^2$ is not the same thing as the variable Γ introduced in Equations 10.22 and 10.28: the letter 'gamma' is used for both purposes.) Substituting the expression for $\hat{\sigma}_\Gamma^2$ in Equation 10.31 into Equation 10.29,

$$h^2 = \frac{\hat{\gamma}_\Gamma \hat{\sigma}_E^2}{\hat{\gamma}_\Gamma \hat{\sigma}_E^2 + \hat{\sigma}_E^2} = \frac{\hat{\gamma}_\Gamma}{\hat{\gamma}_\Gamma + 1}, \tag{10.32}$$

and in the present case,

$$h^2 = \frac{1.236}{1.236 + 1} = 0.553.$$

This tells us that the trait Q4 is heritable, but also considerably influenced by non-heritable sources of variation.

This estimate of heritability may be an overestimate, as it depends upon the assumption that all similarity between relatives is due to shared genes, not shared environment. This assumption may be reasonable in the case of domestic animals in an experimental or commercial environment. (It is never likely to be so in the case of human beings.) More elaborate methods of estimation can be used to attempt to correct for the effect of shared environment. Knowledge of the heritability of traits provides the basis for selective breeding of domestic animals: if Q4 were a trait of commercial importance in an animal species, it would be amenable to improvement by selection, though not easily so.

The F tests show that trait Q4 is significantly influenced by the animal's age, though neither by its sex nor by a 'sex.age' interaction effect. The coefficient for the effect of age, 0.006357, shows that the value of this trait tends to increase with age.

In order to choose the animals from which to breed in a genetic improvement programme, it is desirable to estimate the *breeding value* of each animal, the amount by which the mean value of its progeny will differ from the population mean. The simplest estimate of an animal's breeding value is its phenotypic value: the observed value of the trait in question. However, a better estimate is provided by the estimated *genetic effect* on the trait for each animal, the estimated value of the variable Γ. In the present case, this is obtained by the following statement:

```
REML [PTERMS = ID; PRINT = effects] Q4
```

The output of this statement is as follows:

Warning 30, code VC 67, statement 1 on line 8

Command: REML [PTERMS = ID; PRINT = effects] Q4
SEDs not available for all terms when sparse algorithm used.

Table of effects for ID

ID	1	2	3	4	5	6
	0.0508	0.0016	−0.1929	−0.0702	0.0065	−0.0702

ID	7	8	9	10	11	12
	−0.1748	−0.0842	−0.0842	0.6077	−0.4457	−0.3764

.
.
.

<similar output for other individuals>

.
.
.

ID	2407	2408	2409	2410	2411	2412
	0.3380	0.1222	−0.6363	−0.2330	0.2872	0.0768

ID	2413	2414	2415
	−1.0287	−0.2661	−0.8934

Standard errors of differences are not available.

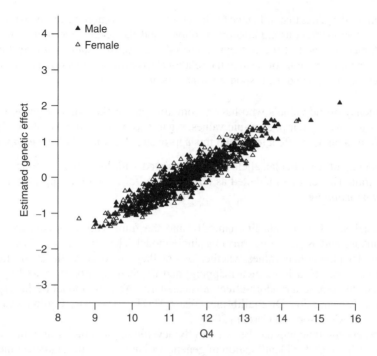

Figure 10.9   Relationship between phenotypic values and estimated genetic effects for trait Q4, obtained from pedigree data.

In the present case, the estimated genetic effect is an improvement on the phenotypic value as a predictor not only because it is adjusted for the sex and age of the animal (the fixed-effect terms in the mixed model), but also because it incorporates information from the individual's relatives. If it seems counter-intuitive that these can provide more reliable information concerning an animal's breeding value than observation of the animal's own phenotype, consider the extreme case of the selection of dairy bulls. Each of these has the potential to produce many daughters, so the choice of the best animal is crucial, yet no observation of milk yield can be made on any of the bulls available. They are compared entirely on the basis of the yield of their female relatives. The relationship between the phenotypic values and the genetic effects in the present data set (for those individuals for which both values are available) is shown in Figure 10.9. A breeder seeking to increase the value of trait Q4 would choose many of the same animals for mating on either basis, but there would be some discrepancies, and hence some increase in the progress made if selection were based on the estimated genetic effect.

Because the genetic effect of each animal is a random effect – an estimated value of the random variable Γ – it is a BLUP or shrunk estimate (see Chapter 5). This partly accounts for the narrow range of values of the estimated genetic effect (max − min = 2.11 − (−1.39) = 3.50) relative to the range of Q4 (max − min = 15.56 − 8.57 = 6.99). (The adjustment for age and sex also narrows the range of values of the estimated genetic effect.) This application of BLUPs is also considered by Robinson (1991, Section 6.2).

If information from genetic markers is available for the individuals in a pedigree, or for a sample from a population of organisms, the techniques of mixed modelling can be further applied to the genetic mapping of phenotypic traits. A genetic map estimates the positions on the chromosomes (the linear structures in each cell of an organism that carry its hereditary information) of the genes that influence the trait under consideration. Such a map is constructed by using genetic markers as explanatory variables and the phenotypic trait as the response variable. A clear account of the concepts involved is given by Sham (1998, Section 5.3, pp. 197–219). The information concerning the relationship between a marker at a particular position (known as a *locus*) and the response variable is of two types:

- The tendency of individuals who have a common variant (known as a *common allele*) at the locus to have similar phenotypic values, regardless of the particular allele. This can be modelled as a random-effect term, and this approach is known as *linkage mapping*.

- The tendency of a particular allele to be associated with high or low values of the phenotypic trait. This can be modelled as a fixed-effect term, and this approach is known as *association mapping*.

Thus the analysis of such data fits naturally into the framework of mixed models, which combine linkage and association terms in a single model. The tendency of related individuals to have similar phenotypic values whether or not they share an allele at the locus under consideration is an obstacle to such mapping, and precision can be increased by including the pedigree structure as a random-effect nuisance term. An early model of this type was the Quantitative Transmission Disequilibrium Test (QTDT) (Abecasis, Cardon and Cookson, 2000; Abecasis, Cookson and Cardon, 2000).

However, genetic mapping has been a rapidly developing field of research in recent years, both in the technology for identification of genetic variants and in the statistical methods and software for the analysis of the data, and an important refinement of this approach is now possible. A *genome-wide scan* can provide information on the genetic variants present in each individual at a large sample of loci fairly evenly distributed over the whole genome (the organism's whole complement of chromosomes), and this information can be used not only to seek a relationship between each locus and the phenotype, but also to estimate the degree of relatedness between every pair of individuals studied. For the latter purpose, it has the following advantages over pedigree records:

- It can be applied not only to recorded pedigrees but also to population-based samples, in which it provides information both on low levels of relatedness among large groups (known as *population structure*) and higher levels of relatedness among smaller groups (known as *cryptic kinship*) (Astle and Balding, 2009).

- It estimates the actual relatedness of the two individuals, whereas the pedigree record gives only the average for the relationship in question (full siblings, first cousins, etc.).

- It is not vulnerable to the errors that commonly occur in pedigree records, especially records of paternity.

Estimates of heritability based on genome-wide scans are vulnerable to bias due to uneven *linkage disequilibrium* – that is, variation in the extent to which variants at neighbouring loci are correlated – but methods have been developed that largely overcome this bias (Yang *et al.*, 2010; Speed *et al.*, 2012). Recently developed software systems for the use of the mixed model in heritability analysis include GCTA, 'a tool for Genome-wide Complex Trait Analysis', available from website http://www.complextraitgenomics.com/software/gcta/, and LDAK, from http://dougspeed.com/ldak/. Software systems for genetic mapping with mixed models include FaST-LMMs, 'Factored Spectrally Transformed Linear Mixed Models', from http://fastlmm.codeplex.com/.

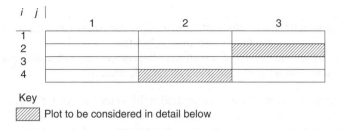

Key

▨ Plot to be considered in detail below

Figure 10.10 Layout of a small field experiment.

At the time of writing, no straightforward method of using a covariance matrix obtained from a pedigree to fit a mixed model in R or SAS is known to the author.

10.11 Estimation of parameters in the covariance matrix: Analysis of temporal and spatial variation

There are many other contexts in which it is possible, and valuable, to specify the structure of a covariance matrix. We will next consider the situation in which observations are obtained from a regular array of experimental units. This may be a one-dimensional array, such as a series of observations at regular time intervals, or a two-dimensional array, such as the spatial arrangement of plots in a field experiment on a crop species. In either case, a structured covariance matrix can be used to model the natural variation over the array, in order to improve the precision with which treatment effects are estimated. Consider a very small field experiment, comprising a 4×3 array of plots (Figure 10.10).

Provided that the experimental treatments are randomly allocated to the plots, the contribution of natural variation among plots to the response variable Y can be treated as a set of independent values of a random variable, E, which is usually assumed to be normally distributed: that is,

$$E \sim N(0, \sigma_E^2) \tag{10.33}$$

and the covariance matrix among the 12 values of E, ε_{ij}, $i = 1 \ldots 4, j = 1 \ldots 3$, is assumed to be

| i | j | i' | 1 | 1 | 1 | 2 | 2 | 2 | 3 | 3 | 3 | 4 | 4 | 4 |
		j'	1	2	3	1	2	3	1	2	3	1	2	3
1	1		σ_E^2											
1	2		0	σ_E^2										
1	3		0	0	σ_E^2									
2	1		0	0	0	σ_E^2								
2	2		0	0	0	0	σ_E^2							
2	3		0	0	0	0	0	σ_E^2						
3	1		0	0	0	0	0	0	σ_E^2					
3	2		0	0	0	0	0	0	0	σ_E^2				
3	3		0	0	0	0	0	0	0	0	σ_E^2			
4	1		0	0	0	0	0	0	0	0	0	σ_E^2		
4	2		0	0	0	0	0	0	0	0	0	0	σ_E^2	
4	3		0	0	0	0	0	0	0	0	0	0	0	σ_E^2

where
i = the row in which the first observation in a particular pair-wise comparison is located,
j = the column in which the first observation is located,
i' = the row in which the second observation in the comparison is located and
j' = the column in which the second observation is located.

However, experience of natural variation in fields tells us that adjacent plots are likely to have similar values of E. This does not invalidate the simple analysis of a simple experimental design: the similarity between neighbouring observations is taken care of by the randomization step (see Section 2.6). However, recognition of this similarity does open up the way to more precise estimation of treatment effects. The knowledge that adjacent plots are likely to have similar values of E can be expressed by the idea that there is a positive correlation between the values of E for adjacent plots in the same column and a similar correlation – not necessarily of equal strength – between adjacent plots in the same row. The coefficients of these two correlations are designated ρ_1 and ρ_2, respectively. If

$$\rho_1 = \rho_2 = 0$$

then the values of E in adjacent plots are independent, but if

$$\rho_1 > 0 \text{ and } \rho_2 > 0$$

then plots that are close together will generally have more similar values of E than those that are further apart. The closer the plots, the greater the similarity between them, and the larger the values of ρ_1 and ρ_2, the more extensive the regions of similar plots (in a trial with many rows and columns). If the correlation coefficient between adjacent plots in the same column is ρ_1, it follows that between plots separated by an intervening plot is ρ_1^2, that between plots separated by two intervening plots is ρ_1^3 and so on. Thus the pattern of correlations between plots in the same column can be expressed by the following matrix:

$$
\begin{array}{c|cccc}
i\ i' & 1 & 2 & 3 & 4 \\
\hline
1 & 1 & & & \\
2 & \rho_1 & 1 & & \\
3 & \rho_1^2 & \rho_1 & 1 & \\
4 & \rho_1^3 & \rho_1^2 & \rho_1 & 1
\end{array}.
$$

Likewise, the following matrix expresses the pattern of correlations between plots in the same row:

$$
\begin{array}{c|ccc}
j\ j' & 1 & 2 & 3 \\
\hline
1 & 1 & & \\
2 & \rho_2 & 1 & \\
3 & \rho_2^2 & \rho_2 & 1
\end{array}.
$$

These results can be generalized to give the correlation coefficient between two plots that are neither in the same column nor in the same row. This is determined by the number of columns and rows that separate them: for plots separated by $|i - i'|$ rows and $|j - j'|$ columns it is $\rho_1^{|i-i'|}\rho_2^{|j-j'|}$. Thus the pattern of correlations among all plots in the experiment is

$$
\begin{pmatrix}
1 & & & \\
\rho_1 & 1 & & \\
\rho_1^2 & \rho_1 & 1 & \\
\rho_1^3 & \rho_1^2 & \rho_1 & 1
\end{pmatrix}
\otimes
\begin{pmatrix}
1 & & \\
\rho_2 & 1 & \\
\rho_2^2 & \rho_2 & 1
\end{pmatrix}
=
$$

i	j	i' j'	1 1	1 2	1 3	2 1	2 2	2 3	3 1	3 2	3 3	4 1	4 2	4 3
1	1		1											
1	2		ρ_2	1										
1	3		ρ_2^2	ρ_2	1									
2	1		ρ_1	$\rho_1\rho_2$	$\rho_1\rho_2^2$	1								
2	2		$\rho_1\rho_2$	ρ_1	$\rho_1\rho_2$	ρ_2	1							
2	3		$\rho_1\rho_2^2$	$\rho_1\rho_2$	ρ_1	ρ_2^2	ρ_2	1						
3	1		ρ_1^2	$\rho_1^2\rho_2$	$\rho_1^2\rho_2^2$	ρ_1	$\rho_1\rho_2$	$\rho_1\rho_2^2$	1					
3	2		$\rho_1^2\rho_2$	ρ_1^2	$\rho_1^2\rho_2$	$\rho_1\rho_2$	ρ_1	$\rho_1\rho_2$	ρ_2	1				
3	3		$\rho_1^2\rho_2^2$	$\rho_1^2\rho_2$	ρ_1^2	$\rho_1\rho_2^2$	$\rho_1\rho_2$	ρ_1	ρ_2^2	ρ_2	1			
4	1		ρ_1^3	$\rho_1^3\rho_2$	$\rho_1^3\rho_2^2$	ρ_1^2	$\rho_1^2\rho_2$	$\rho_1^2\rho_2^2$	ρ_1	$\rho_1\rho_2$	$\rho_1\rho_2^2$	1		
4	2		$\rho_1^3\rho_2$	ρ_1^3	$\rho_1^3\rho_2$	$\rho_1^2\rho_2$	$\rho_1^2\rho_2$	$\rho_1\rho_2$	ρ_1	$\rho_1\rho_2$	ρ_2	1		
4	3		$\rho_1^3\rho_2^2$	$\rho_1^3\rho_2$	ρ_1^3	$\rho_1^2\rho_2^2$	$\rho_1^2\rho_2$	ρ_1^2	$\rho_1\rho_2^2$	$\rho_1\rho_2$	ρ_1	ρ_2^2	ρ_2	1

For example, consider the shaded plots in the plan of this field experiment (Figure 10.10), namely

- the plot in Row 4, Column 2 ($i = 4, j = 2$) and
- the plot in Row 2, Column 3 ($i' = 2, j' = 3$).

These are separated by $|i - i'| = |4 - 2| = 2$ rows and $|j - j'| = |2 - 3| = 1$ column. Hence the correlation coefficient between the values of E for these plots is $\rho_1^2\rho_2$, as indicated in the shaded element in the matrix above. The combination of matrices in this way is called the *direct product*: note the operator $\otimes$, which specifies this combination.

The covariance matrix among the values of E is then given by

$$\text{covariance matrix} = \begin{pmatrix} 1 & & & \\ \rho_1 & 1 & & \\ \rho_1^2 & \rho_1 & 1 & \\ \rho_1^3 & \rho_1^2 & \rho_1 & 1 \end{pmatrix} \otimes \begin{pmatrix} 1 & & \\ \rho_2 & 1 & \\ \rho_2^2 & \rho_2 & 1 \end{pmatrix} \sigma_E^2. \tag{10.34}$$

In all the mixed models considered previously, the covariance matrix for each random-effect term has been fully specified by the design of the experiment (the specification of treatments and blocks, or of the pedigree structure), except for the variance component, which was estimated during the model-fitting process. However, in the present case, not only the variance component σ_E^2, but also the correlation coefficients ρ_1 and ρ_2 are to be estimated from the data. Provided that the true values of either ρ_1 or ρ_2, or both, are greater than zero, the treatment effects will consequently be estimated with greater precision.

These ideas can be illustrated in the analysis of data from a field trial of wheat breeding lines, conducted in South Australia in 1994. (Reproduced by kind permission of Gil Hollamby.) A total of 107 lines and varieties were sown in a randomized complete block design, with three replications (blocks). In each replication, 3 standard varieties were each sown in 2 plots, namely varieties 82 (Tincurran), 89 (VF655) and 104 (WW1477). The remaining 104 varieties were each sown in a single plot. Thus each replication comprised 110 plots, which were arranged in 22 rows and 5 columns. The layout and randomization of the experiment are shown in Figure 10.11. The variety sown in each plot is indicated.

An area of 4.2 m × 0.75 m in each plot was harvested and the grain yield (kg) was measured. The first and last few rows of a spreadsheet holding the results are shown in Table 10.9.

Block		1						2					3		
Column	1	2	3	4	5	6	7	8	9	10	11	12	13	14	15
Row															
1	4	14	79	59	89	104	62	4	74	15	5	2	27	79	47
2	10	30	80	60	90	18	92	67	71	81	4	42	89	9	29
3	17	15	97	61	86	70	91	19	100	106	104	81	28	66	100
4	16	18	102	62	91	22	36	103	27	61	46	91	56	82	93
5	21	19	40	63	92	56	57	35	42	69	63	57	37	82	89
6	32	20	42	64	93	41	68	14	98	66	3	25	99	59	107
7	33	23	43	65	94	37	23	107	12	38	48	26	97	101	39
8	34	24	44	66	95	53	102	11	77	9	98	72	68	83	32
9	72	25	45	67	96	86	33	73	72	24	60	80	64	62	40
10	74	26	46	68	7	40	47	30	13	55	8	73	49	90	58
11	75	27	47	69	99	50	105	6	64	60	51	45	74	106	13
12	81	28	48	70	100	16	2	29	59	46	71	52	24	104	34
13	83	35	49	88	101	80	104	75	8	93	15	53	95	65	61
14	106	36	50	103	29	78	28	97	5	31	22	102	21	31	20
15	107	37	51	104	104	26	90	82	85	63	14	17	103	76	43
16	3	38	52	98	105	95	32	10	45	43	86	54	75	78	36
17	5	39	53	2	22	96	21	25	101	54	18	19	94	50	44
18	6	71	54	1	31	17	39	7	89	48	12	96	55	70	41
19	8	73	55	89	41	82	76	52	65	1	69	92	11	88	105
20	9	76	56	87	12	87	84	3	99	49	87	7	67	77	6
21	11	77	57	84	82	44	79	83	34	20	84	10	30	23	1
22	13	78	58	85	82	58	89	88	51	94	35	38	85	33	16

Figure 10.11 Arrangement of a field trial of wheat breeding lines in South Australia. (Source: Reproduced by kind permission of Gil Hollamby.)

Table 10.9 Yields of wheat varieties in a field trial in South Australia, with a randomized block design and with information on the spatial location of each plot.

	A	B	C	D	E
1	replicate!	row!	column!	variety!	yield
2	1	1	1	4	483
3	1	1	2	14	400
4	1	1	3	79	569
5	1	1	4	59	734
6	1	1	5	89	571
7	2	1	6	104	642
8	2	1	7	62	665
9	2	1	8	4	738
.					
.					
.					
329	3	22	13	85	442
330	3	22	14	33	423
331	3	22	15	16	551

Source: Reproduced by kind permission of Gil Hollamby.

The model that corresponds to the design of this experiment is the standard randomized complete block, namely

Response variate: yield

Fixed-effect model (treatment model): variety

Random-effect model (block model): replication

These data are imported, and this model fitted using analysis of variance, by the following statements:

```
IMPORT 'IMM edn 2\\Ch 10\\SA wheat yield.xlsx'
BLOCKSTRUCTURE replicate
TREATMENTSTRUCTURE variety
ANOVA [FPROB = yes] yield
```

Alternatively, the same model can be fitted by mixed modelling methods, replacing the BLOCK-STRUCTURE, TREATMENTSTRUCTURE and ANOVA statements by the following statements:

```
VCOMPONENTS [FIXED = variety] RANDOM = replicate
REML [PRINT = model, components, deviance, Wald] yield
```

The output of the REML statement is as follows:

REML variance components analysis
Response variate: yield
Fixed model: Constant + variety
Random model: replicate
Number of units: 330

Residual term has been added to model

Sparse algorithm with AI optimization

Estimated variance components

Random term	Component	s.e.
replicate	12642	12764

Residual variance model

Term	Model(order)	Parameter	Estimate	s.e.
Residual	Identity	Sigma2	13432	1278

Deviance: −2*Log-Likelihood

Deviance	d.f.
2471.63	221

Note: deviance omits constants which depend on fixed model fitted.

Tests for fixed effects

Sequentially adding terms to fixed model

Fixed term	Wald statistic	n.d.f.	F statistic	d.d.f.	F pr.
variety	153.02	106	1.44	221.0	0.012

Dropping individual terms from full fixed model

Fixed term	Wald statistic	n.d.f.	F statistic	d.d.f.	F pr.
variety	153.02	106	1.44	221.0	0.012

Message: denominator degrees of freedom for approximate F tests are calculated using algebraic derivatives ignoring fixed/boundary/singular variance parameters.

The estimated variance component for the term 'replicate', $\hat{\sigma}^2_{\text{replicate}} = 12\,642$, is smaller than its own SE, suggesting that there may be no real variation among the replicates. However, the F value for this term in the anova (not shown) is highly significant ($F_{2,106} = 104.53, p < 0.001$). The reason for this wide discrepancy is not known to the author at the time of writing, but it does provide a salutary warning against over-reliance on the SE of a variance component estimate. The F test for 'variety' indicates that there is also highly significant variation among the varieties.

A model that allows for the possibility that there are correlations between the residual values in neighbouring plots, both within rows and within columns, is specified by the following statements:

```
VCOMPONENTS [FIXED = variety] RANDOM = row.column
VSTRUCTURE [TERMS = row.column] FACTOR = row, column; \
    MODEL = AR, AR; ORDER = 1, 1
REML \
    [PRINT = model, components, deviance, Wald, covariancemodels] \
    yield
```

The random-effects model 'replicate' has been replaced by 'row.column'. It might be thought that the term 'replicate' would be retained, so that the new model would be 'replicate + row.column'. However, the term 'replicate' is redundant when the term 'row.column' is included: any variation among replicates can be absorbed into the variation among columns. As in the analysis of the pedigree data, the VSTRUCTURE statement specifies the structure of the covariance matrix among the random effects. The TERMS option indicates that such structure is to be specified for the term 'row.column'. (There might be other terms in the random-effect model for which a different structure, or none, was to be specified.) The parameter FACTOR indicates the factors within this model term for which a covariance matrix is to be specified, and over which the direct product is to be formed in order to specify the overall covariance matrix. (Some factors in the term might be excluded from this specification.) The parameter MODEL specifies that an *auto-regressive* (AR) model is to be fitted for both 'row' and 'column': that is, a model in which the value in each plot is correlated with that in the neighbouring plot, as described above. The parameter ORDER indicates that for both 'row' and 'column' the AR model is to be first-order: that is, only a correlation between immediately neighbouring plots is to be directly specified. Correlations between more distant neighbours arise only as a consequence of this. A first-order AR model for both rows and columns is designated as AR1 × AR1.

The output of the REML statement is as follows:

REML variance components analysis

Response variate: yield
Fixed model: Constant + variety
Random model: row.column
Number of units: 330

row.column used as residual term with covariance structure as below

Sparse algorithm with AI optimization

Covariance structures defined for random model

Covariance structures defined within terms:

Term	Factor	Model	Order	No. rows
row.column	row	Auto-regressive (+scalar)	1	22
	column	Auto-regressive	1	15

Residual variance model

Term	Factor	Model(order)	Parameter	Estimate	s.e.
row.column			Sigma2	19794	3819
	row	AR(1)	phi_1	0.9039	0.0185
	column	AR(1)	phi_1	0.4288	0.0640

Estimated covariance models

Variance of data estimated in form:

$V(y) = Sigma2.R$

whore: V(y) io varianco matrix of data,

Sigma2 is the residual variance,

R is the residual covariance matrix.

Residual term: row.column

Sigma2: 19794.

R uses direct product construction:

Factor: row
Model: Auto-regressive

Covariance matrix (first 10 rows only):

	1	2	3	4	5	6	7	8	9	10
1	1.000									
2	0.904	1.000								
3	0.817	0.904	1.000							
4	0.739	0.817	0.904	1.000						
5	0.668	0.739	0.817	0.904	1.000					
6	0.603	0.668	0.739	0.817	0.904	1.000				
7	0.545	0.603	0.668	0.739	0.817	0.904	1.000			
8	0.493	0.545	0.603	0.668	0.739	0.817	0.904	1.000		
9	0.446	0.493	0.545	0.603	0.668	0.739	0.817	0.904	1.000	
10	0.403	0.446	0.493	0.545	0.603	0.668	0.739	0.817	0.904	1.000
	1	2	3	4	5	6	7	8	9	10

Factor: column
Model: Auto-regressive

Covariance matrix (first 10 rows only):

	1	2	3	4	5	6	7	8	9	10
1	1.000									
2	0.429	1.000								
3	0.184	0.429	1.000							
4	0.079	0.184	0.429	1.000						
5	0.034	0.079	0.184	0.429	1.000					
6	0.014	0.034	0.079	0.184	0.429	1.000				
7	0.006	0.014	0.034	0.079	0.184	0.429	1.000			
8	0.003	0.006	0.014	0.034	0.079	0.184	0.429	1.000		
9	0.001	0.003	0.006	0.014	0.034	0.079	0.184	0.429	1.000	
10	0.000	0.001	0.003	0.006	0.014	0.034	0.079	0.184	0.429	1.000
	1	2	3	4	5	6	7	8	9	10

Deviance: -2*Log-Likelihood

Deviance	d.f.
2207.29	220

Note: deviance omits constants which depend on fixed model fitted.

Tests for fixed effects

Sequentially adding terms to fixed model

Fixed term	Wald statistic	n.d.f.	F statistic	d.d.f.	F pr.
variety	772.52	106	7.29	136.7	<0.001

Dropping individual terms from full fixed model

Fixed term	Wald statistic	n.d.f.	F statistic	d.d.f.	F pr.
variety	772.52	106	7.29	136.7	<0.001

Message: denominator degrees of freedom for approximate F tests are calculated using algebraic derivatives ignoring fixed/boundary/singular variance parameters.

The specification of the model at the beginning of the output now includes some information about the structure in the covariance matrix, including the fact that the experiment has 22 rows and 15 columns. The annotation '(+scalar)' indicates that the model relating the residual value in each row to that in the next includes a constant term, which is subtracted from each residual before using the correlation coefficient ρ_1 to specify the relationship between them. This constant is required for only one of the two terms that define the covariance structure. The residual variance model relates entirely to the term 'row.column', as this term uniquely specifies every observation in the experiment. The model indicates that the estimate of residual variance is now $\hat{\sigma}^2_{\text{Residual}} = 19\,794$, somewhat higher than the value of $\hat{\sigma}^2_{\text{Residual}} = 13\,432$ given by the randomized complete block model. However, the estimates for the parameter 'phi_1' for the factors 'row' and 'column' indicate that the AR1 × AR1 model gives a substantially better fit to the data. These parameters are the estimates of the correlation coefficients ρ_1 and ρ_2 defined earlier. Both are substantially greater than their SEs, indicating that the true correlation coefficient is not zero in either case. The estimated correlation between adjacent rows, $\hat{\rho}_1 = 0.9039$, is much stronger than that between adjacent columns, $\hat{\rho}_2 = 0.4288$, probably because each row was only 0.75 m wide, whereas each column was 6 m wide. (The plots were trimmed to give the 4.2 m section harvested.) The covariance matrix for factor 'row' indicates the correlations among the first 10 plots in any column. The correlation between Rows 1 and 2 is $\hat{\rho}_1 = 0.904$, that between Rows 1 and 3 is $\hat{\rho}_1^2 = 0.817$ and so on. Similarly, the covariance matrix for factor 'column' indicates the correlations among the first 10 plots in any row. Note that these plots occupy the first 10 columns of the field experiment: the note 'first 10 rows only' refers to the covariance matrix and is the same regardless of the name of the factor in question.

The residual deviances from these two models are as follows:

$$\text{Deviance}_{\text{randomized-complete-block model}} = 2471.63.$$

$$\text{Deviance}_{\text{AR1}\times\text{AR1 model}} = 2207.29.$$

These deviances cannot be compared by a likelihood ratio test as described in Section 3.12, because the models are not *nested*, that is, neither model is a reduced form of the other. However, other methods for comparing them are available. In general, a model with more parameters can be made to fit a given data set more closely than one with fewer parameters, so the choice of the preferred model involves a trade-off between goodness of fit, a closer fit being better, and parsimony, fewer parameters being better. The likelihood ratio test itself is based on such a trade-off. It assesses whether the reduction in deviance (i e the improvement in fit) achieved by adding a parameter to the model is greater than could be expected by chance, by comparing this reduction with the chi-square distribution. A similar trade-off can be performed to compare two mixed models in which the random-effect models are not nested, provided that they are applied to the same data and the fixed-effect models are the same, by obtaining the *Akaike Information Criterion* (AIC) for each model, namely

$$\text{AIC} = 2k - 2\text{Log-likelihood} = 2k + \text{Deviance} \tag{10.35}$$

where
$k =$ number of parameters in the model.

The model with the lowest AIC is to be preferred. In the present case, the number of parameters in each model can be obtained from the GenStat output as (No. of units − d.f. of deviance), so

$$\text{AIC}_{\text{randomized-complete-block model}} = 2 \times (330-221) + 2471.63 = 2689.63$$

$$\text{AIC}_{\text{AR1}\times\text{AR1 model}} = 2 \times (330-220) + 2207.29 = 2427.29.$$

The AIC from the AR1 × AR1 model is much lower, confirming that this model fits the data better, and suggesting that the variety means will be estimated with considerably more precision when this model is used. A correction to the AIC to take account of the finite sample size of the data set is given by

$$\text{AIC}_c = \text{AIC} + \frac{2k(k+1)}{n-k-1} \tag{10.36}$$

where
n = sample size.

In the present case,

$$n = 330$$

and for such a large sample the adjustment makes little difference to the values obtained. An alternative, closely related basis for comparing non-nested models is the Bayesian information criterion,

$$\text{BIC} = -2\text{Log-likelihood} + k\log_e(n) = \text{Deviance} + k\log_e(n)$$

The model giving the lower value of BIC (Bayesian Information Criterion) is to be preferred. In the present case

$$\text{BIC}_{\text{randomized-complete-block model}} = 2471.63 + (330-221) \times \log_e(330) = 3103.73$$

$$\text{BIC}_{\text{AR1×AR1 model}} = 2207.29 + (330-220) \times \log_e(330) = 2845.19$$

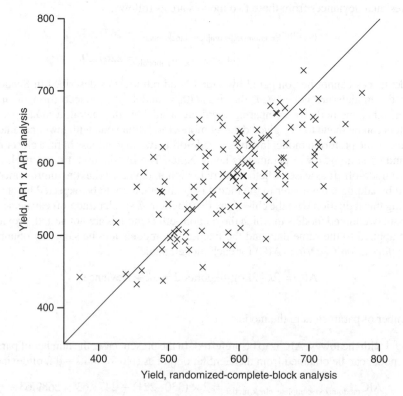

Figure 10.12   Comparison between estimates of mean yields of wheat varieties in South Australia from a randomized complete block analysis and an AR1 × AR1 analysis.

The BIC from the AR1 × AR1 model is much lower, again confirming that this model fits the data better. The two information criteria are thoroughly explained and compared by Burnham and Anderson (2004). Note that they cannot be confidently used to compare mixed models in cases where the fixed-effect models differ (S. Welham, personal communication).

The estimated variety means can be displayed by adding 'means' to the setting of the PRINT option in the REML statement. Those given by the randomized complete block model and the AR1 × AR1 model are compared in Figure 10.12: the correlation between them is $r = 0.656$. If the agreement between the two sets were perfect, the correlation would be $r = 1$ and the points in the figure would all lie on the diagonal line. The discrepancy is substantial, and would have a considerable effect on the decisions made by a plant breeder: the set of varieties chosen for further evaluation or commercial release would be very different, depending on which estimates were used to choose them. All the evidence from the analyses indicates that the estimates from the AR1 × AR1 model are closer to the true values and give a better indication of the potential of each variety.

There is a good deal more that can be done to improve the estimates of variety means from these data. Further terms can be added to the mixed model, representing additional aspects of spatial variation: for a full account, see Gilmour, Cullis and Verbyla (1997). In brief:

- The value of each plot may have a direct dependence on the value of non-adjacent plots, in addition to its indirect dependence via the values of adjacent plots – a second-order auto-regressive (AR2) model.

- A term can be added to the model to represent random variation among plots that is independent of the values in neighbouring plots: this is known as a *nugget* term.

- An irregular trend over the field may be represented by a fluctuating curve known as a *smoothing spline*, the precise shape of which is specified by a set of random effects (Verbyla *et al.*, 1999).

- Terms can be added to represent the effect of known husbandry factors, for example, the direction in which the tractor was driven when trimming each plot, which affects the area harvested.

10.12   Use of R to model spatial variation

The following commands import the data into R, convert 'replicate', 'row', 'column' and 'variety' to factors, and fit the standard randomized complete block model to the South Australian wheat data by analysis of variance:

```
rm(list = ls())
SA.wheat <- read.table(
    "IMM edn 2\\Ch 10\\SA wheat yield.txt",
    header = TRUE)
attach(SA.wheat)
#names(SA.wheat)
freplicate <- factor(replicate)
frow <- factor(row)
fcolumn <- factor(column)
fvariety <- factor(variety)
```

```
SA.wheat.model.1 <-
   aov(yield ~ fvariety + Error(freplicate))
summary(SA.wheat.model.1)
```

The following commands fit the same model by mixed modelling methods:

```
library(lme4)
SA.wheat.model.2 <- lmer(yield ~ fvariety + (1|freplicate))
summary(SA.wheat.model.2)
```

Part of the output of the function summary() is as follows:

```
Linear mixed model fit by REML
Formula: yield ~ fvariety + (1 | freplicate)
  AIC  BIC logLik deviance REMLdev
 3099 3514  -1441     3958    2881
Random effects:
 Groups      Name          Variance Std.Dev.
 freplicate (Intercept)   12642     112.44
 Residual                 13432     115.90
Number of obs: 330, groups: freplicate, 3
Fixed effects:
             Estimate Std. Error t value
(Intercept)   709.000     93.209   7.607
fvariety2      24.333     94.629   0.257
fvariety3     -93.333     94.629  -0.986
.
.
.
<effect estimates for other varieties>



.
.
.
fvariety106 -176.667     94.629  -1.867
fvariety107 -245.000     94.629  -2.589
```

The variance component estimates for 'replicate' and for the residual term agree with those obtained from GenStat. The variety effect estimates are calculated relative to the estimated mean for Variety 1, which is specified as the intercept. Thus, for example, the estimated mean for Variety 2 is

$$\text{Intercept} + \text{effect of Variety } 2 = 709.000 + 24.333 = 733.333.$$

At the time of writing, no straightforward method in R for fitting AR terms over rows and columns simultaneously is known to the author. However, the following statements fit a model with an AR term for the variation over the columns within each row:

```
library(nlme)
SA.wheat.model.3 <- lme(yield ~ fvariety,
    random = ~ 1|fcolumn, data = SA.wheat)
#summary(SA.wheat.model.3)$coefficients
#anova(SA.wheat.model.3)

SA.wheat.model.4 <- update(SA.wheat.model.3,
    correlation = corAR1(form = ~ row|fcolumn))
summary(SA.wheat.model.4)
```

The function lme() fits a model with 'row' as the random-effect term. The function update() adds the AR term to this model, and within it, the function corAR1() indicates that a first-order AR term is to be used.

Part of the ouput from the function summary() is as follows:

```
Linear mixed-effects model fit by REML
 Data: SA.wheat
       AIC       BIC      logLik
 2817.935 3192.724 -1298.968

Random effects:
 Formula: ~1 | fcolumn
         (Intercept) Residual
StdDev:    139.3432 59.49081

Correlation Structure: AR(1)
 Formula: ~row | fcolumn
 Parameter estimate(s):
      Phi
 0.424804
Fixed effects: yield ~ fvariety
               Value Std.Error  DF   t-value p-value
(Intercept)  677.8961  47.06486 209 14.403444  0.0000
fvariety2    -45.1606  40.22624 209  1.122666  0.2629
fvariety3    -70.1728  42.48693 209 -1.651632  0.1001
.
.

.
<effect estimates for other varieties>

.
.

.
fvariety106  -57.3876  42.66394 209 -1.345107  0.1800
fvariety107 -111.0300  42.49796 209 -2.612595  0.0096
```

The standard deviations for random effects indicate that the variance component estimate for columns is $139.3432^2 = 19416.53$, and for the residual term is $59.49081^2 = 3539.156$.

This is substantially smaller than the residual variances obtained from the models fitted in GenStat: the reason for this discrepancy is not known to the author. The estimated correlation between the residuals of plots in adjacent rows within each column is given by the parameter Phi = 0.424804. This is considerably lower than the corresponding value obtained from the model fitted in GenStat, phi_1 = 0.9093: the reason for this discrepancy also is not known to the author. The variety effect estimates are calculated in the same way as in the output from the randomized complete block model.

10.13   Use of SAS to model spatial variation

The following SAS statements import the South Australian wheat data and fit the standard randomized complete block model by analysis of variance:

```
PROC IMPORT OUT = wheat DBMS = EXCELCS REPLACE
    DATAFILE = "&pathname.\IMM edn 2\Ch 10\SA wheat yield.xlsx";
    SHEET = "for SAS";
RUN;
```

```
ODS RTF;
PROC ANOVA;
    CLASS replicate variety;
    MODEL yield = replicate variety;
    MEANS variety;
RUN;
ODS RTF CLOSE;
```

The following statements fit the same model by mixed modelling methods:

```
ODS RTF;
PROC MIXED ASYCOV NOBOUND;
    CLASS replicate variety;
    MODEL yield = variety /CHISQ DDFM = KR HTYPE = 1;
    RANDOM replicate;
RUN;
ODS RTF CLOSE;
```

Part of the output from PROC MIXED is as follows:

Convergence criteria met.

Covariance parameter estimates	
Cov Parm	**Estimate**
replicate	12642
Residual	13432

Asymptotic covariance matrix of estimates			
Row	Cov Parm	CovP1	CovP2
1	replicate	1.6292E8	−14843
2	Residual	−14843	1632715

Fit statistics	
−2 Res log likelihood	2881.5
AIC (smaller is better)	2885.5
AICC (smaller is better)	2885.5
BIC (smaller is better)	2883.7

Null model likelihood ratio test		
DF	Chi-square	Pr > ChiSq
1	137.16	<0.0001

Type 1 tests of fixed effects						
Effect	Num DF	Den DF	Chi-square	F value	Pr > ChiSq	Pr > F
variety	106	221	153.02	1.44	0.0019	0.0120

The variance component estimates, and the F statistic for 'variety' and its numerator and denominator d.f., agree with those obtained from GenStat.

The following statements fit the model including AR terms for both 'row' and 'column':

```
ODS RTF;
PROC MIXED ASYCOV NOBOUND DATA = wheat;
   CLASS replicate variety;
   MODEL yield = variety
      / CHISQ DDFM = KR HTYPE = 1;
   REPEATED /SUBJECT = intercept TYPE = SP(POWA) (row column);
   PARMS 0 0 1;
RUN;
ODS RTF CLOSE;
```

In PROC MIXED, the MODEL statement specifies the response variable and the fixed-effect model. The REPEATED statement specifies the random-effect model and indicates that this involves repeated measures over some set of entities – in the present case, the rows and columns of the experiment. The option setting 'SUBJECT = intercept' indicates that the random-effect terms include only the main effects of the factors specified, not the corresponding variation in slopes related to any variates. The option setting 'TYPE = SP(POWA) (row column)' indicates that the type of model to be fitted is a SPatial model of the Anisotropic POWer type – that is, a model in which the correlation between units declines exponentially as the distance between them increases, but the rate constant for the decline – the correlation coefficient between adjacent observations – is not the same in the two dimensions. The option setting also specifies the dimensions under consideration – in this case, 'row' and 'column'. It is necessary to specify initial values for the iterative process to estimate the three parameters of the SP(POWA) model, namely the correlation coefficients between rows and between columns and the residual variance, and this is done by the PARMS statement. Fortunately, in the present case, it turns out not to be necessary that these initial values should be very close to the final estimates.

Part of the output from PROC MIXED is as follows:

Convergence criteria met.

Covariance parameter estimates		
Cov Parm	**Subject**	**Estimate**
SP(POWA) row	Intercept	0.9039
SP(POWA) column	Intercept	0.4289
Residual		19795

Asymptotic covariance matrix of estimates					
Row	**Cov Parm**	**CovP1**	**CovP2**	**CovP3**
1	SP(POWA) row	0.000464	−0.00016	79.0117
2	SP(POWA) column	−0.00016	0.006113	64.7785
3	Residual	79.0117	64.7785	18369590

Fit statistics	
−2 Res log likelihood	2617.1
AIC (smaller is better)	2623.1
AICC (smaller is better)	2623.2
BIC (smaller is better)	2633.4

PARMS model likelihood ratio test		
DF	**Chi−square**	**Pr > ChiSq**
2	401.50	<0.0001

Type 1 tests of fixed effects						
Effect	**Num DF**	**Den DF**	**Chi-square**	**F value**	**Pr > ChiSq**	**Pr > F**
variety	106	117	731.14	6.90	<0.0001	<0.0001

The estimates of the residual variance and the correlation coefficients agree fairly closely with those obtained from GenStat, but the F statistic for 'variety' and its denominator d.f. are rather different, probably due to the use of the Kenward–Roger method for the determination of the denominator d.f. in SAS (see Section 8.6). The estimated variety means can be obtained by adding the statement

```
LSMEANS variety;
```

to the PROC MIXED code, both for the randomized complete block model and the AR model. They agree with those obtained from GenStat.

10.14 Summary

The basic concepts of mixed modelling can be extended in several directions. The following have been illustrated in this chapter:

- Non-normal distribution of residuals:

  - sigmoid curve, binomial distribution of the response variable, logistic link function applied to the response variable (Sections 10.2–10.4);

  - contingency table, Poisson distribution of the response variable, logarithmic link function applied to the response variable (Sections 10.5–10.8);

- Non-normal distribution of effects in terms other than the residual term:

  - contingency table, gamma distribution of factorial terms, logarithmic link function (Sections 10.6–10.7).

- Specification of the variance structure via a covariance matrix:

  - covariance between relatives in a pedigree structure (Section 10.10).

- Estimation of aspects of the variance structure not fully specified by the investigator

  - correlation between observations in a regular array (Sections 10.11–10.13).

10.15 Exercises

10.1 The seeds of some species of clover (*Trifolium* spp.) will not germinate immediately after ripening, but must undergo a period of 'softening', by exposure to fluctuating high and low temperatures, usually on the soil surface. In an investigation of this phenomenon, seeds of eight clover species were sown into strips of fine mesh cotton bags, 100 seeds of a single species in each bag. Each strip comprised one bag of each species. At the start of summer, the bags were laid on the soil surface in full sun at a site cleared of vegetation. At intervals during the summer and autumn, two or four strips were removed from the field, and the seeds in each bag were tested to determine how many were still 'hard' (i.e. would not germinate when moistened in favourable conditions). The first and last few rows of the spreadsheet holding the data are presented in Table 10.10: the full data set is held in the file 'clover seed softening.xlsx' on this book's website (see Preface). (Data reproduced by kind permission of Hayley Norman.)

Table 10.10　Numbers of 'hard' and 'soft' seeds of several species of clover after exposure on the soil surface for varying lengths of time.

Observations on the same replication at the same time come from a single strip of bags. Time = interval in days from the start of exposure on the soil surface. nhard = number of hard seeds in a single bag. nsoft = number of soft seeds in a single bag.

	A	B	C	D	E
1	species!	replication!	time	nhard	nsoft
2	T. glomeratum	1	0	98	2
3	T. glomeratum	2	0	99	1
4	T. glomeratum	3	0	99	1
5	T. glomeratum	4	0	99	1
6	T. spumosum	1	0	94	6
7	T. spumosum	2	0	93	7
8	T. spumosum	3	0	95	5
9	T. spumosum	4	0	96	4
.					
.					
.					
234	T. scutatum	1	174	4	96
235	T. scutatum	2	174	11	89
236	T. scutatum	3	174	5	95
237	T. scutatum	4	174	5	95
238	T. pilulare	1	174	17	83
239	T. pilulare	2	174	28	72
240	T. pilulare	3	174		
241	T. pilulare	4	174		

Source: Data reproduced by kind permission of Hayley Norman.

As the species studied all belong to the genus *Trifolium* they are expected to have some characteristics in common, but no prior information is available to us concerning the seed-softening behaviour of the individual species.

(a) Decide whether 'species' should be specified as a fixed-effect or a random-effect term in the model to be fitted to these data and explain your decision.

(b) Specify a regression model for this experiment. Following your decision concerning 'species', which term(s) in the model should be regarded as fixed and which as random? What is the response variable?

(c) Fit your mixed model to the data, specifying an appropriate error distribution for the response variable, and an appropriate link function to relate the response variable to the linear model.

(d) Consider whether there is evidence that any terms can be omitted from the model. If so, fit the modified model to the data.

(e) Make a graphical display, showing the fitted relationship between the proportion of hard seed remaining for each species and the time elapsed since the start of exposure, and showing the scatter of the observed values around this relationship.

10.2 Return to the data set concerning the relationship between the distance from bushland and the level of predation on seeds, introduced in Exercise 7.4 in Chapter 7. In the earlier analysis of these data, it was assumed that the percentage of predation could be regarded as a normally distributed variable.

(a) Using the information on the number of seeds of each species per cage, convert each value of the percentage of predation ('%pred') to the actual number of seeds removed by predation.

(b) Re-fit your mixed model to the data, using the number of seeds removed by predation as the response variable, and specifying an appropriate error distribution for this response variable, and an appropriate link function to relate the response variable to the linear model.

(c) Obtain diagnostic plots of the residuals. Do these plots indicate that the assumptions underlying the analysis are more nearly fulfilled as a result of the changes to the response variable, the error distribution and the link function?

(d) Make a graphical display, showing the fitted relationship between the proportion of seeds removed by predation and the distance from the bushland, taking into account any other model terms (i.e. residue, cage type, species and/or interaction terms) that your analysis indicates are important. Your plot should also show the scatter of the observed values around this relationship.

10.3 Return to the data on the efficacy of lithium as a treatment for amyotrophic lateral sclerosis (Section 7.5). Define

σ_1^2 = variance component for term 'id'

σ_2^2 = variance component for term 'id.visit_day' and

ρ_{12} = correlation coefficient between effects of 'id' and 'id.visit_day'.

Specify a single covariance matrix, with 168 rows, representing the 84 levels of 'id' and the 84 levels of 'id.visit_day'. Indicate the covariance in every element of the matrix, that is, every row.column combination.

10.4 Records of matings and of phenotypic variables were kept in a pedigree of the shrimp *Penaeus vannamei*. The variables recorded were as shown in Table 10.11, and the first and last few rows of the spreadsheet holding the data are presented in Table 10.12: the full data set is held in the file 'shrimp pedigree.xlsx' on this book's website (see Preface). (Data reproduced by kind permission of Dr Shaun Moss, Director, Shrimp Department, Oceanic Institute, Hawaii, USA.) The variable of economic interest is the weekly growth rate.

(a) Consider whether any of the variables studied should be modelled as fixed-effect terms when estimating the genetic and residual components of variance of the weekly growth rate.

(b) Fit a mixed model to the data, taking account of the pedigree structure. Obtain estimates of the genetic and residual components and estimate the heritability of the weekly growth rate.

Table 10.11 Variables recorded in a pedigree of the shrimp *Penaeus vannamei*.

Name of variable	Description
Kid	Identifier of the shrimp
Ksire	Identifier of the male parent
Kdam	Identifier of the female parent
SEX	M = male; F = female
F	Coefficient of inbreeding, relative to the founders of the pedigree
WeekGro	Weekly growth rate, post-larva to harvest, (grams)
Brood	Brood identifier, useful for the recognition of maternal effects when a female had more than one brood of full-sibs
CGU	Contemporaneous group
Seq	A database reference
DD	Degree days accumulated during the growth period. This will be the same for all members of a brood and contemporaneous group

Table 10.12 Pedigree and phenotype data from shrimps of the species *Penaeus vannamei*.

	A	B	C	D	E	F	G	H	I	J
	Kid	Ksire	Kdam	SEX	F	WeekGro	Brood	CGU	seq	DD
1										
2	17	1	6	M					5120	
3	18	2	7	M					5121	
4	19	3	9	M					5122	
5	28	4	10	M					5473	
268	68533	63411	10166	M	0.02	1.52	101723411	34	22904	972
269	68547	63411	10166	F	0.02	1.22	101723411	34	22917	972
270	68571	63411	10166	F	0.02	1.72	101723411	34	22939	972
271	69500	11966	63406	M	0.03	1.54	634071966	34	25533	972
272	69528	11966	63406	F	0.03	1.45	634071966	34	25559	972
273	69545	11966	63406	F	0.03	1.43	634071966	34	25574	972
274	69676	11966	63391	M	0.02	1.29	633921966	34	25691	972
275	69712	11966	63391	F	0.02	1.38	633921966	34	25724	972
276	69718	11966	63391	F	0.02	1.41	633921966	34	25730	972

Source: Data reproduced by kind permission of Dr Shaun Moss, Director, Shrimp Department, Oceanic Institute, Hawaii, USA.

(c) Compare graphically the phenotypic value of the weekly growth rate for each individual and the estimated genetic effect on this variable.

10.5 Return to the data on yields of wheat genotypes, investigated in an alphalpha design, presented in Table 9.5.

(a) Explore the possibility of analysing these data using an auto-regressive model, instead of the model based on the alphalpha design fitted in Sections 9.6–9.8.

(b) How well does your auto-regressive model fit the data, relative to the model fitted earlier?

(c) Compare graphically the estimates of the genotype mean yields obtained by the two methods. How much effect will the choice of method have on the decisions made by a breeder seeking genetic improvement?

10.6 Return to the data from a field trial of wheat breeding lines conducted in South Australia, analysed in Sections 10.11–10.13. In those sections, it was noted that the randomized complete block model and the AR1 × AR1 model fitted to these data cannot be compared formally by means of a chi-square test, as neither model is a reduced form of the other.

(a) Specify an auto-regressive model of which the randomized complete block model *is* a reduced form.

(b) Fit your new model to the data and interpret the results. Conduct a formal significance test to compare it with the randomized complete block model and explain the result of the test.

(c) Make a graphical comparison of the estimated mean yields of the breeding lines obtained from the two models. How much effect will the choice of models have on the decisions made by a plant breeder?

(d) Can your auto-regressive model be compared by a formal significance test with the auto-regressive model used in Sections 10.11 and 10.13? Compare the fit of the two models as well as possible and consider which should be preferred. Make a graphical comparison of the estimated mean yields of the breeding lines obtained from these two models and consider how much effect the choice between them will have on the decisions made by a plant breeder.

10.7 Return to the data from a meta-analysis of clinical trials to study the effect of aspirin when given to heart-attack patients, analysed in Exercise 8.2.

(a) Rearrange the data in a 'stacked' form, with a separate row for the aspirin-treated and placebo-treated patients in each trial, and a column headed 'treatment' that distinguishes the two types of row.

(b) Perform a fixed-effect meta-analysis on the 'stacked' data set, using a method that

- estimates the effect of treatment as a log odds ratio,
- assumes that the number of deaths in each treatment in each trial follows a binomial distribution and
- does not depend on an approximation based on the normal distribution.

Compare the results with those of the corresponding analysis based on the normal approximation, obtained previously. In what situations would you expect to find poor agreement between the results of the two analyses?

(c) Perform a random-effect meta-analysis that meets the same criteria. Compare the results with those of the corresponding analysis based on the normal approximation, and suggest reasons for the discrepancies that you find.

Simple log odds ratios cannot be calculated when the number of responders in a category (in the present case, patients dying on a particular treatment in a particular trial) is zero or is close to the sample size (in the present case, the number of patients), and log odds ratio methods may give unreliable results in 'sparse' data sets close to these limiting conditions. Methods for dealing with this problem in meta-analysis are discussed by Sutton *et al.* (2000, Section 4.3, pp. 63–69). However, it does not arise in the present data set.

References

Abecasis, G.R., Cardon, L.R. and Cookson, W.O. (2000a) A general test of association for quantitative traits in nuclear families. *American Journal of Human Genetics*, **66**, 279–292.

Abecasis, G.R., Cookson, W.O.C. and Cardon, L.R. (2000b) Pedigree tests of transmission disequilibrium. *European Journal of Human Genetics*, **8**, 545–551.

Astle, W.A. and Balding, D.J. (2009) Population structure and cryptic relatedness in genetic association studies. *Statistical Science*, **24**, 451–471.

Brown, H. and Prescott, R. (2006) *Applied Mixed Models in Medicine*, 2nd edn, John Wiley & Sons, Ltd, Chichester, UK, 476 pp.

Bulmer, M.G. (1979) *Principles of Statistics*, 2nd edn, Dover Publications, New York, 252 pp.

Burnham, K.P. and Anderson, D.R. (2004) Multimodel inference: understanding AIC and BIC in model selection. *Sociological Methods and Research*, **33**, 261–304.

Falconer, D.S. and Mackay, T.F.C. (1996) *Introduction to Quantitative Genetics*, 4th edn, Longman, Harlow, 464 pp.

Finney, D.J. (1971) *Probit Analysis*, 3rd edn, Cambridge University Press, Cambridge, 333 pp.

Gilmour, A.R., Cullis, B.R. and Verbyla, A.P. (1997) Accounting for natural and extraneous variation in the analysis of field experiments. *Journal of Agricultural, Biological and Environmental Statistics*, **2**, 269–293.

Goldin, L.R., Bailey-Wilson, J.E., Borecki, I.B. *et al.* (eds) (1997) Genetic analysis workshop 10: detection of genes for complex traits. *Genetic Epidemiology*, **14** (6), 549–1152.

Lee, Y. and Nelder, J.A. (1996) Hierarchical generalized linear models (with discussion). *Journal of the Royal Statistical Society, Series B*, **58**, 619–678.

Lee, Y. and Nelder, J.A. (2001) Hierarchical generalized linear models: a synthesis of generalised linear models, random-effect models and structured dispersions. *Biometrika*, **88**, 987–1006.

McCullagh, P. and Nelder, J.A. (1989) *Generalized Linear Models*, 2nd edn, Chapman & Hall, London, 511 pp.

Pinhero, J.C. and Bates, D.M. (2000) *Mixed-Effects Models in S and S-PLUS*, Springer, New York, 528 pp.

Robinson, G.K. (1991) That BLUP is a good thing: the estimation of random effects. *Statistical Science*, **6**, 15–51.

Rönnegård, L., Shen, X. and Alam, M. (2010) hglm: a package for fitting hierarchical generalized linear models. *The R Journal*, **2**, 20–28.

Schall, R. (1991) Estimation in generalized linear models with random effects. *Biometrika*, **78**, 719–727.

Sham, P. (1998) *Statistics in Human Genetics*, Arnold, London, 290 pp.

Snedecor, G.W. and Cochran, W.G. (1989) *Statistical Methods*, 8th edn, Iowa State University Press, Ames, IA, 503 pp.

Speed, D., Hemani, G., Johnson, M.R. and Balding, D.J. (2013) Improved heritability estimation from genome-wide SNPs. *The American Journal of Human Genetics*, **93**, 1151–1155.

Sutton, A.J., Abrams, K.R., Jones, D.R. *et al.* (2000) *Methods for Meta-analysis in Medical Research*, John Wiley & Sons, Ltd, Chichester, 317 pp.

Verbyla, A.P., Cullis, B.R., Kenward, M.G. and Welham, S.J. (1999) The analysis of designed experiments and longitudinal data by using smoothing splines. *Applied Statistics*, **48**, 269–311.

Wilkinson, G.N. and Rogers, C.E. (1973) Symbolic description of factorial models for analysis of variance. *Applied Statistics*, **22**, 392–399.

Yang, J., Benyamin, B., McEvoy, B.P. *et al.* (2010) Common SNPs explain a large proportion of the heritability for human height. *Nature Genetics*, **42**, 565–569.

11

Why is the criterion for fitting mixed models called REsidual Maximum Likelihood?

11.1 Maximum likelihood and residual maximum likelihood

In the preceding chapters, we have established the need for mixed models, that is, statistical models with more than one random-effect term, and have seen how to construct such models and how to interpret the results obtained when they are fitted to data. We have noted that the criterion used to fit a mixed model – that is, to obtain the best estimates of its parameters – is called *REsidual Maximum Likelihood* (*REML*), but we have not so far examined the meaning of this term. In this chapter, we will explore the concept of maximum likelihood and its use as a criterion for the estimation of model parameters. We will then show how the criterion for parameter estimation used in the earlier chapters can be viewed as *residual* or *restricted* maximum likelihood. The argument will proceed as follows:

- Consideration of a model comprising only the random-effect term E. The estimation of its variance σ^2 using the maximum-likelihood criterion.

- Consideration of the simplest linear model, comprising the fixed-effect term μ and the random-effect term E. The simultaneous estimation of μ and σ^2 using the maximum-likelihood criterion. An alternative estimate of σ^2 using the REML criterion.

- Extension of the REML criterion to the general linear model.

- Extension to models with more than one random-effect term.

Introduction to Mixed Modelling: Beyond Regression and Analysis of Variance, Second Edition. N. W. Galwey.
© 2014 John Wiley & Sons, Ltd. Published 2014 by John Wiley & Sons, Ltd.
Companion website: http://www.wiley.com/go/beyond_regression

11.2   Estimation of the variance σ^2 from a single observation using the maximum-likelihood criterion

In the analysis of real data, it is usually necessary to estimate fixed-effect parameters, such as the mean, and random-effect parameters, such as variance components, simultaneously. However, we will start with a simplified, hypothetical situation in which the mean of a variable is known, and only its variance has to be estimated. Consider a variable E such that

$$E \sim N(0, \sigma^2). \tag{11.1}$$

Suppose that a sample of values of E, namely ε_i, $i = 1 \ldots n$, is to be taken, and that the variance σ^2 is to be estimated from this sample. We define the *maximum-likelihood estimate* of the standard deviation σ as the value of σ that maximizes the probability of the data. (For a full discussion of the merits of maximum likelihood as an estimation criterion, see Edwards, 1972.) We will denote this maximum-likelihood estimate by $\hat{\sigma}$. It follows that the square of this estimate, $\hat{\sigma}^2$, is the maximum-likelihood estimate of σ^2.

What value of σ has the required property? We will first address this question for a sample comprising a single observation of E, denoted by ε: that is, the case where

$$n = 1.$$

We will not obtain a formal algebraic solution to the problem (for such an account see, for example, Bulmer, 1979, Chapter 11, pp. 197–200), but will explore graphically the probability distributions of E that are produced by a range of possible values that might be proposed for σ, designating such a proposed value by the symbol $\hat{\sigma}$, and we will note the probability of the observed value ε for each of these values (Figure 11.1). These probabilities are infinitesimal: they are probability densities, designated by $f(E)$ (see Section 1.2). Nevertheless, their relative magnitudes can be compared. In this case, the value that maximizes $f(E)$ is

$$\hat{\sigma} = \varepsilon. \tag{11.2}$$

The observed value of E is itself the maximum-likelihood estimate of the standard deviation: any other value of $\hat{\sigma}$ makes the observed value less probable. A smaller value of $\hat{\sigma}$ gives a probability distribution that is too narrow, so that ε lies relatively far from its maximum and $f(E)$ is low. A larger value of $\hat{\sigma}$ gives a distribution that is too flat, so that although ε lies relatively close to its maximum, $f(E)$ is again low.

11.3   Estimation of σ^2 from more than one observation

Note that probability density is represented in Figure 11.1 by two methods, namely by the height on the vertical axis and by contour shading. The second method is useful when we move on to consider samples comprising two observations of E, denoted by ε_1 and ε_2: that is, the case where

$$n = 2.$$

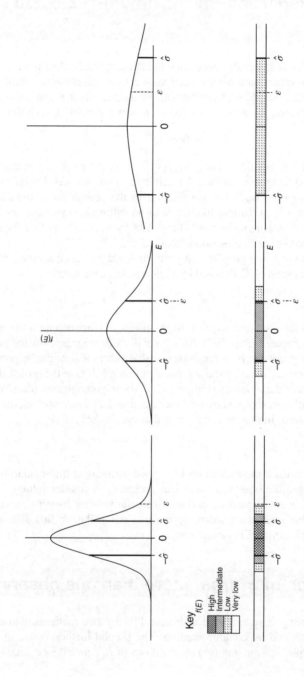

Figure 11.1 Comparison of possible estimates of the standard deviation of a variable E, when a single observation of E has been made. The upper part of the figure shows the probability density of E, $f(E)$, on the vertical axis: the lower part shows this value by contour shading. The heavy lines indicate the values of E at a distance $\pm\hat{\sigma}$ from the origin.

Figure 11.2 extends the graphical representation of probability distributions to this *bivariate* case. It is necessary to regard ε_1 and ε_2 as observations of *two different variables*, designated E_1 and E_2. These are represented by two dimensions in the figure, and the probability density of every possible combination of values of E_1 and E_2 is represented in the third dimension. In the upper part of the figure, $f(E_1, E_2)$ is represented on the vertical axis, in the lower part, it is represented by contour shading. Once again, the figure explores a range of possible values for σ and shows the probability density of the observed values $(\varepsilon_1, \varepsilon_2)$ for each proposed value of σ. The figure shows that in this case, the value of σ that maximizes the value of $f(E_1, E_2)$, $\hat{\sigma}$, is such that

$$\sqrt{2}\hat{\sigma} = \sqrt{\varepsilon_1^2 + \varepsilon_2^2}$$

and hence

$$\hat{\sigma} = \sqrt{\frac{\varepsilon_1^2 + \varepsilon_2^2}{2}}. \tag{11.3}$$

The coefficient 2 in this equation reflects the sample size, n. To help justify its presence, consider a sample in which

$$(\varepsilon_1, \varepsilon_2) = (\hat{\sigma}, \hat{\sigma}).$$

Such a 'typical' observation lies at a distance

$$\sqrt{\hat{\sigma}^2 + \hat{\sigma}^2} = \sqrt{2}\hat{\sigma}$$

from the origin.

This argument can be extended to show that in general,

$$\hat{\sigma} = \sqrt{\frac{\sum_{i=1}^{n} \varepsilon_i^2}{n}}. \tag{11.4}$$

Because $\hat{\sigma}$ is the maximum-likelihood estimate of σ, it follows that

$$\hat{\sigma}^2 = \frac{\sum_{i=1}^{n} \varepsilon_i^2}{n} \tag{11.5}$$

is the maximum-likelihood estimate of σ^2: the estimate of the variance from the whole sample is simply the mean of the estimates from the individual observations. The value $\hat{\sigma}^2$ has the additional merit of being an *unbiased* estimate of σ^2: that is, if an infinite population of small samples is taken, and the value $\hat{\sigma}^2$ is obtained in each small sample, the mean of all these estimates will be the true value, σ^2. However, it does *not* follow that $\hat{\sigma}$ is an unbiased estimate of σ.

11.4   The μ-effect axis as a dimension within the sample space

Now consider the simplest linear model,

$$Y = \mu + E \tag{11.6}$$

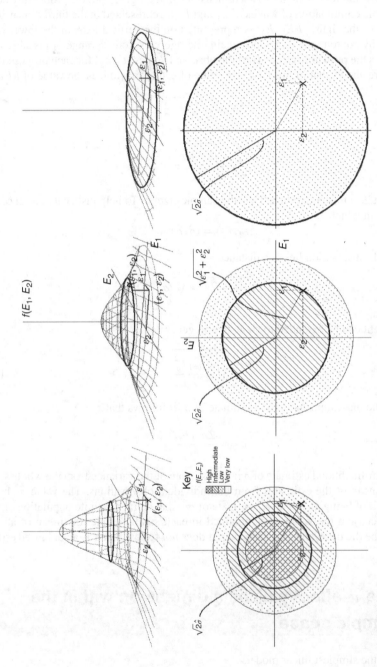

Figure 11.2 Comparison of possible estimates of the standard deviation of a variable E, when two observations of E have been made. The heavy circular line indicates the values of E_1 and E_2 at a distance $\sqrt{2}\hat{\sigma}$ from the origin.

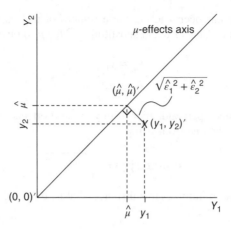

Figure 11.3   Graphical representation of the model $Y = \mu + E$, when two observations of Y have been made.

where Y is the observed variable. It is no longer possible to observe E, but estimates of its values can be obtained from a sample of observations y_i, $i = 1 \ldots n$. Figure 11.3 shows a graphical representation of this model in the case where $n = 2$ – the simplest case in which both μ and σ can be estimated. The observations y_1 and y_2 are regarded as values of two different variables, Y_1 and Y_2, and these variables provide axes that define a two-dimensional space with its origin at $\mathbf{0} = (0,0)'.$[1] This is known as the *sample space*. The observations, y_1 and y_2, define the point $\mathbf{y} = (y_1, y_2)'$. According to the model, the observations are related to μ and E by the following equations:

$$y_1 = \mu + \varepsilon_1 \tag{11.7}$$

$$y_2 = \mu + \varepsilon_2. \tag{11.8}$$

Any value of μ that may be proposed specifies a point, $\boldsymbol{\mu} = (\mu, \mu)'$, which represents the contribution of μ to y_1 and y_2. Any such value defines a line, passing through the origin and the point $\boldsymbol{\mu}$, which may be infinitely extended and which we will call the μ-effects axis. Any estimate of μ, which we will call $\hat{\mu}$, defines a point $\hat{\boldsymbol{\mu}} = (\hat{\mu}, \hat{\mu})'$ which lies on this axis. Because μ contributes equally to y_1 and y_2, the μ-effects axis makes the same angle ($45°$) with both the Y_1 axis and the Y_2 axis. Whatever value of $\hat{\mu}$ is chosen, Pythagoras's theorem can be used to show that

$$(\text{distance}(\mathbf{0}\hat{\boldsymbol{\mu}}))^2 = \hat{\mu}^2 + \hat{\mu}^2$$

and hence

$$\text{distance}(\mathbf{0}\hat{\boldsymbol{\mu}}) = \sqrt{2}\hat{\mu}. \tag{11.9}$$

---

[1] Any point in n-dimensional space can be represented by a *vector* of values – a vertical column, which can in turn be represented by a single value – in this case $\mathbf{0} = \begin{pmatrix} 0 \\ 0 \end{pmatrix}$. This vector can also be represented as $\mathbf{0} = (0,0)'$, using the prime symbol (') to indicate that a row of values is to be transposed to the corresponding column. For the rest of this chapter, the coordinates of a point in n-dimensional space will be represented in this way.

When the estimate $\hat{\mu}$ has been chosen, the estimates of the residual effects are given by rearranging Equations 11.7 and 11.8 and substituting $\hat{\mu}$ for μ to give

$$\hat{\varepsilon}_1 = y_1 - \hat{\mu} \tag{11.10}$$

$$\hat{\varepsilon}_2 = y_2 - \hat{\mu}. \tag{11.11}$$

If $\hat{\mu}$ is chosen by dropping a perpendicular from $\mathbf{y}$ to the μ-effects axis, Pythagoras's theorem shows that the length of the line from $\mathbf{y}$ to $\hat{\boldsymbol{\mu}}$ is

$$\text{distance}(\hat{\boldsymbol{\mu}}\mathbf{y}) = \sqrt{\hat{\varepsilon}_1^2 + \hat{\varepsilon}_2^2}, \tag{11.12}$$

and this is the choice that minimizes this distance: that is, this choice gives the *least squares estimate*.

11.5 Simultaneous estimation of μ and σ^2 using the maximum-likelihood criterion

A natural estimate of μ is obtained by dropping a perpendicular from the point $\mathbf{y}$ to the μ-effects axis, as has been done in Figure 11.3. Some more geometry (Figure 11.4) shows that in this case,

$$\text{distance}(\mathbf{0}\hat{\boldsymbol{\mu}}) = \frac{y_1 + y_2}{\sqrt{2}}. \tag{11.13}$$

Combining Equations 11.9 and 11.13, we obtain

$$\sqrt{2}\hat{\mu} = \frac{y_1 + y_2}{\sqrt{2}}$$

and hence

$$\hat{\mu} = \frac{y_1 + y_2}{2}. \tag{11.14}$$

Once again, the coefficient 2 in this formula reflects the sample size. Again the argument can be extended to show that in general

$$\hat{\mu} = \frac{\sum_{i=1}^n y_i}{n}. \tag{11.15}$$

The estimate of the population mean obtained by this geometrical argument is simply the familiar mean of the observations in the sample.

Figure 11.5 shows the bivariate probability distributions of Y_1 and Y_2 that are produced by a range of possible values that might be proposed for the parameters μ and σ, and shows the probability density of the observed values y_1 and y_2 for each proposed combination of parameter values. The central frame in the figure shows the relationship of the data to the probability distribution when the estimates

$$\hat{\mu} = \frac{y_1 + y_2}{2} \tag{11.16}$$

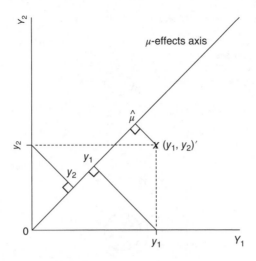

Figure 11.4   Demonstration of the relationship between y_1, y_2 and $\hat{\mu}$.

From Pythagoras's theorem,

$$\text{distance}(\mathbf{0y}_1)^2 + \text{distance}(\mathbf{y}_1(0, y_1)')^2 = y_1^2.$$

But

$$\text{distance}(\mathbf{y}_1(0, y_1)') = \text{distance}(\mathbf{0y}_1).$$

Substituting and rearranging,

$$\text{distance}(\mathbf{0y}_1) = \frac{y_1}{\sqrt{2}}.$$

Similarly,

$$\text{distance}(\mathbf{0y}_2) = \frac{y_2}{\sqrt{2}}.$$

But

$$\text{distance}(\mathbf{0y}_2) = \text{distance}(\mathbf{y}_1\hat{\boldsymbol{\mu}}).$$

(The demonstration of this is straightforward but is not given here.) Hence

$$\text{distance}(\mathbf{0}\hat{\boldsymbol{\mu}}) = \text{distance}(\mathbf{0y}_1) + \text{distance}(\mathbf{y}_1\hat{\boldsymbol{\mu}})$$

$$= \text{distance}(\mathbf{0y}_1) + \text{distance}(\mathbf{0y}_2)$$

$$= \frac{y_1}{\sqrt{2}} + \frac{y_2}{\sqrt{2}} = \frac{y_1 + y_2}{\sqrt{2}}.$$

and

$$\hat{\sigma} = \sqrt{\frac{\hat{\varepsilon}_1^2 + \hat{\varepsilon}_2^2}{2}} \tag{11.17}$$

are proposed. The four surrounding frames show that when a higher or lower estimate of μ is proposed, or a larger or smaller estimate of σ, the probability density of the data is reduced: that is, the values $\hat{\mu}$ and $\hat{\sigma}$ given by Equations 11.16 and 11.17 are the maximum-likelihood estimates of μ and σ, respectively.

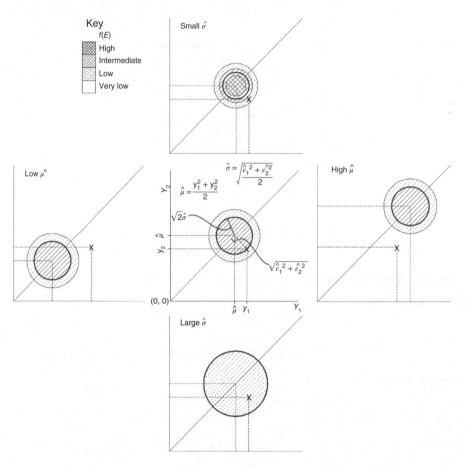

Figure 11.5   Comparison of possible estimates of the mean and standard deviation of a variable Y, when two observations of Y have been made. The heavy circular line indicates the values of Y_1 and Y_2 at a distance $\sqrt{2}\hat{\sigma}$ from the point $(\hat{\mu}, \hat{\mu})'$.

11.6   An alternative estimate of σ^2 using the REML criterion

As in the case where only the variance was to be estimated (Sections 11.2 and 11.3), if $\hat{\sigma}$ is the maximum-likelihood estimate of σ, it follows that $\hat{\sigma}^2$ is the maximum-likelihood estimate of σ^2. However, when the mean and the variance must both be estimated from the data, the maximum-likelihood estimate $\hat{\sigma}^2$ is no longer unbiased, as it was in the hypothetical case where the mean was known. The values $\hat{\varepsilon}_i$ are calculated as deviations from $\hat{\mu}$, and they cluster around this value rather more closely than the unknown values ε_i cluster around the true mean

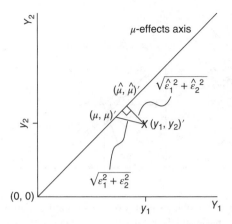

Figure 11.6 Demonstration that the observed values of a variable cluster more closely about the estimated mean than around the true mean.
The probability that the estimated mean coincides exactly with the true mean is infinitesimal. Unless this occurs, $\sqrt{(\hat{\epsilon}_1^2 + \hat{\epsilon}_2^2)/2}$ is always smaller than $\sqrt{(\epsilon_1^2 + \epsilon_2^2)/2}$.

μ (Figure 11.6). Therefore, in an infinite population of small samples, the values of $\hat{\sigma}^2$ will somewhat underestimate the true value, σ^2.

We usually make an adjustment to overcome this deficiency, replacing

$$\hat{\sigma} = \sqrt{\frac{\sum_{i=1}^{n} \hat{\epsilon}_i^2}{n}} \qquad (11.18)$$

by the familiar formula

$$\hat{\sigma} = \sqrt{\frac{\sum_{i=1}^{n} \hat{\epsilon}_i^2}{n-1}}. \qquad (11.19)$$

In the present case,

$$\hat{\sigma} = \sqrt{\frac{\hat{\epsilon}_1^2 + \hat{\epsilon}_2^2}{2}}$$

is replaced by

$$\hat{\sigma} = \sqrt{\frac{\hat{\epsilon}_1^2 + \hat{\epsilon}_2^2}{1}}. \qquad (11.20)$$

But this, of course, is not the maximum-likelihood estimate of σ. How can this adjusted formula be brought back within the framework of maximum likelihood?

If the estimate $\hat{\mu}$ is obtained first from Equation 11.15, and is then regarded as fixed, and if σ is subsequently estimated within this constraint, a new picture emerges. Figure 11.7 shows that when $\hat{\mu}$ is restricted to this particular function of the data, the observation **y** must lie on a line perpendicular to the μ-effects axis, passing through $\hat{\mu}$. Thus the estimation of σ is reduced

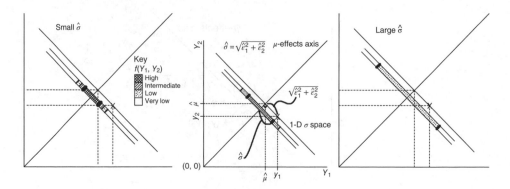

Figure 11.7 Comparison of possible estimates of the standard deviation of a variable Y, when two observations of Y have been made and the estimate of the mean is restricted to $\hat{\mu} = (y_1 + y_2)/2$. The heavy lines indicate the values of Y_1 and Y_2 at a distance $\hat{\sigma}$ from the point $(\hat{\mu}, \hat{\mu})'$.

from the two-dimensional problem illustrated in Figure 11.5 to the one-dimensional problem illustrated in Figure 11.7. This figure shows, using contour shading, the probability density of Y_1 and Y_2 that is produced by a range of possible values that might be proposed for σ, within the constraint imposed by Equation 11.15. Within this constraint, the value that maximizes $f(y_1, y_2)$ is

$$\hat{\sigma} = \sqrt{\frac{\hat{\varepsilon}_1^2 + \hat{\varepsilon}_2^2}{1}}.$$

This is known as the *residual*- or *restricted*-maximum-likelihood estimate and is the familiar adjusted formula given in Equations 11.19 and 11.20.

This approach to the estimation of μ and σ can be extended to higher values of n by defining additional dimensions, Y_3 to Y_n, the μ-effects axis making the same angle with each of them. The maximum-likelihood estimate $\hat{\mu}$ is obtained from Equation 11.15, and the point

$$\hat{\boldsymbol{\mu}} = (\hat{\mu}, \hat{\mu} \ \dots \ \hat{\mu})' \tag{11.21}$$

(a vector comprising n values) lies on the μ-effects axis. In the case of this simplest linear model, there is almost no distinction between obtaining the parameter estimate $\hat{\mu}$ and the fitted values $\hat{y}_1$ to $\hat{y}_n$: these are given by

$$\hat{\mathbf{y}} = (\hat{y}_1, \hat{y}_2 \ \dots \ \hat{y}_n)' = (\hat{\mu}, \hat{\mu} \ \dots \ \hat{\mu})' = \hat{\boldsymbol{\mu}}. \tag{11.22}$$

The estimated residual effects,

$$\hat{\boldsymbol{\varepsilon}} = (\hat{\varepsilon}_1, \hat{\varepsilon}_2 \ \dots \ \hat{\varepsilon}_n)',$$

must then lie in an $n - 1$ dimensional sub-space perpendicular to the μ-effects axis and passing through $\hat{\boldsymbol{\mu}}$, and the REML estimate of σ is given by the familiar formula

$$\hat{\sigma} = \sqrt{\frac{\sum_{i=1}^{n} \hat{\varepsilon}_i^2}{n - 1}}. \tag{11.23}$$

11.7 Bayesian justification of the REML criterion

There is an alternative, Bayesian justification for the REML criterion. Instead of considering the probability density of the observations for different values of σ conditional on the estimate $\hat{\mu}$, we can consider the probability density at every possible value of μ. If we assume that μ has a uniform prior probability distribution (i.e. that all values are judged equally probable prior to inspection of the data), we can then integrate over μ, from $-\infty$ to $+\infty$, to obtain the total probability density for any value of σ. We then choose as our estimate the value of σ that maximizes this probability density. This approach is illustrated in Figure 11.8 and leads to the same value of $\hat{\sigma}$ as is given by the REML criterion in Equation 11.23. To make the argument fully Bayesian, we can assume a prior probability distribution for σ, then combine it with the *likelihood function*, that is, the probability density given by every value of σ, to obtain a posterior distribution.

11.8 Extension to the general linear model: The fixed-effect axes as a sub-space of the sample space

The REML approach can be further extended to models that have more than one fixed-effect term, such as the general linear model (which encompasses all ordinary regression models),

$$Y = \beta_1 X_1 + \beta_2 X_2 + \cdots + \beta_p X_p + E, \tag{11.24}$$

fitted to the data

$$y_i, x_{ij}, \quad i = 1 \ldots n; \quad j = 1 \ldots p.$$

The observations of the response variable are represented by a point in the n-dimensional sample space $Y_1 \ldots Y_n$, and the explanatory variables, $X_1 \ldots X_p$, define axes that specify a p-dimensional sub-space of this sample space, just as the model term μ was represented by the single dimension of the μ-effects axis. We will call this sub-space the *fixed-effects sub-space*. The orientation of these *fixed-effects-model axes* in relation to the dimensions $Y_1 \ldots Y_n$ is determined by the values x_{ij}. We will first consider a very simple example of this type, in which there is a single explanatory variable and just two data points (Table 11.1). Figure 11.9 shows that these two points lie exactly on the line

$$Y = 6 + 0.6X, \tag{11.25}$$

that is, there can be no residual variation when a model with two parameters (the constants 6 and 0.6) is fitted to two data points.

In order to apply the methods outlined above to these data, we first re-express the relationship between the variables in the form of the general linear model, thus

$$Y = \beta_1 X_1 + \beta_2 X_2 \tag{11.26}$$

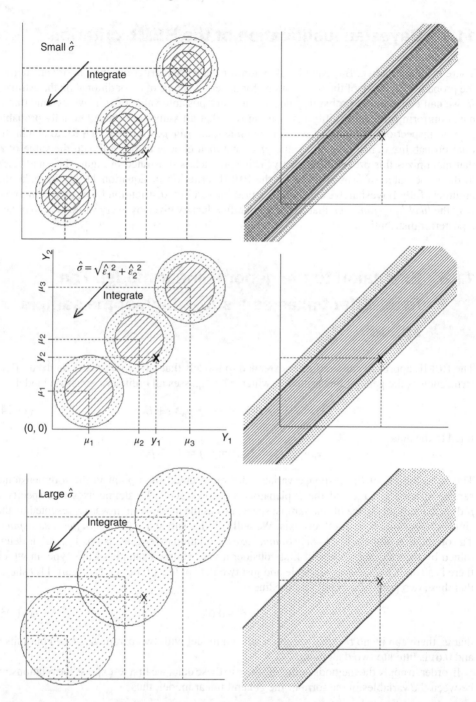

Figure 11.8   (a–c) Bayesian justification of the REML criterion for the estimation of σ.
The conventions are the same as in Figures 11.5 and 11.7. The 2-D probability distribution of possible observations (Y_1, Y_1) is shown for three suggested estimates of σ (in separate left-hand frames) and for three possible values of μ (in each left-hand frame). Integration of the probability density over all possible values of μ gives the distribution in the corresponding right-hand frame. This integration can be envisaged as dragging or smearing the disc representing the bivariate probability distribution in the direction of the diagonal arrow. The REML estimate of σ maximizes the probability of the data on this basis.

Table 11.1 Two observations of an explanatory variable X and a response variable Y.

i	X	Y
1	$x_1 = 10$	$y_1 = 12$
2	$x_2 = 20$	$y_2 = 18$

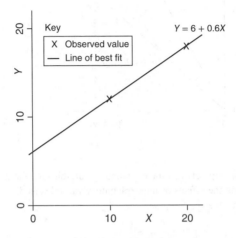

Figure 11.9 The line of best fit for the relationship between the explanatory variable X and the response variable Y when two pairs of observations have been made.

where

$$\hat{\beta}_1 = 6,$$
$$\hat{\beta}_2 = 0.6$$

and the values of X_1, X_2 and Y are as given in Table 11.2. Note that X_1 is a *dummy variable*, all the values of which are 1, used to express the constant term in general-linear model form.

The relationships between the sample space Y_1, Y_2, the dimensions represented by X_1 and X_2 and the data points in this model are represented graphically in Figure 11.10. The directions

Table 11.2 The two observations of an explanatory variable and a response variable presented in Table 11.1, expressed in the notation of the general linear model.

i	X_1	X_2	Y
1	$x_{11} = 1$	$x_{21} = 10$	$y_1 = 12$
2	$x_{12} = 1$	$x_{22} = 20$	$y_2 = 18$

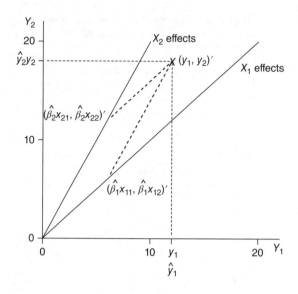

Figure 11.10   The relationship between the explanatory variables X_1 and X_2 and the response variable Y, expressed by representing the effects of the explanatory variables as dimensions in the sample space defined by two observations of Y.

of the X-effects axes, relative to the Y_1 and Y_2 axes, are determined as follows. The fitted values of Y are given by the following equations:

$$\hat{y}_1 = \hat{\beta}_1 x_{11} + \hat{\beta}_2 x_{21} \qquad (11.27)$$

$$\hat{y}_2 = \hat{\beta}_1 x_{12} + \hat{\beta}_2 x_{22}. \qquad (11.28)$$

(Because the two observations lie exactly on the line of best fit, the fitted values are, in this case, the same as the observed values.) These equations show that the contribution of X_1 to $\hat{y}_1$ is $\hat{\beta}_1 x_{11}$, and the contribution of X_1 to $\hat{y}_2$ is $\hat{\beta}_1 x_{12}$. Therefore, the X_1-effects axis is a line connecting the origin to the point $(\hat{\beta}_1 x_{11}, \hat{\beta}_1 x_{12})$, projected indefinitely. Similarly, the contribution of X_2 to $\hat{y}_1$ is $\hat{\beta}_2 x_{21}$ and its contribution to $\hat{y}_2$ is $\hat{\beta}_2 x_{22}$, so the X_2-effects axis is a line connecting the origin to the point $(\hat{\beta}_2 x_{21}, \hat{\beta}_2 x_{22})$, again projected indefinitely.

Because there are two data points and two parameters, there are two Y axes and two X-effect axes. Therefore, in this case, the X-effect axes define not just a sub-space of the sample space, but the whole sample space – the two-dimensional plane of the paper. For the same reason, there is no distinction between the observed and fitted values of Y: just as the observed points lie exactly on the line of best fit in Figure 11.9, the point representing the observed values of Y, lying in the plane defined by the Y axes, is the same as the point representing the fitted values of Y, lying in the plane defined by the X-effect axes. However, the distinction between observed and fitted values becomes apparent when the same model is fitted to a data set comprising three observations (Table 11.3). The relationship between these points and the fitted line is shown in Figure 11.11. Each observation has a residual component now, represented by its distance from the line of best fit.

Table 11.3 Three observations of an explanatory variable X and a response variable Y.

i	X	Y
1	$x_1 = 10$	$y_1 = 15$
2	$x_2 = 20$	$y_2 = 21$
3	$x_3 = 15$	$y_3 = 9$

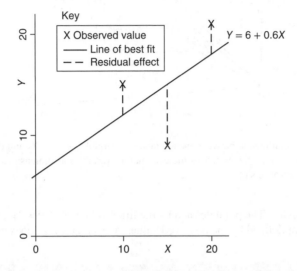

Figure 11.11 The line of best fit for the relationship between the explanatory variable X and the response variable Y when three pairs of observations have been made.

Once again, we re-express the relationship between the variables in the form of a general linear model (Table 11.4). The relationships between the sample space Y_1, Y_2, Y_3, the dimensions represented by X_1 and X_2 and the data points are now as shown in Figure 11.12. The axes representing the X effects again define a plane, which is now a two-dimensional sub-space of the three-dimensional sample space defined by the axes Y_1, Y_2 and Y_3 – we will call this the

Table 11.4 The three observations of an explanatory variable and a response variable presented in Table 11.3, expressed in the notation of the general linear model.

i	X_1	X_2	Y
1	$x_{11} = 1$	$x_{21} = 10$	$y_1 = 15$
2	$x_{12} = 1$	$x_{22} = 20$	$y_2 = 21$
3	$x_{13} = 1$	$x_{23} = 15$	$y_3 = 9$

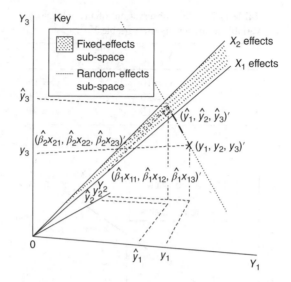

Figure 11.12   The relationship between the explanatory variables X_1 and X_2 and the response variable Y, expressed by representing the effects of the explanatory variables as dimensions in the sample space defined by three observations of Y.

fixed-effects sub-space. The point defined by the fitted values of Y, $\hat{\mathbf{y}} = (\hat{y}_1, \hat{y}_2, \hat{y}_3)'$, lies in this plane, but the point defined by the observed values, $\mathbf{y} = (y_1, y_2, y_3)'$, does not.

11.9   Application of the REML criterion to the general linear model

The distance between the points representing the fitted and the observed values in Figure 11.10 is related to the residual values. It is the square root of their sum of squares, namely

$$\text{distance}(\hat{\mathbf{y}}\mathbf{y}) = \sqrt{\sum_{i=1}^{n} \hat{\varepsilon}_i^2}, \tag{11.29}$$

and the maximum-likelihood estimate of σ is given by

$$\hat{\sigma} = \frac{\text{distance}(\hat{\mathbf{y}}\mathbf{y})}{\sqrt{n}} = \sqrt{\frac{\sum_{i=1}^{n} \hat{\varepsilon}_i^2}{n}}. \tag{11.30}$$

This is the same formula as was obtained in the simpler case of the variation of a sample of observations about the mean value (Equation 11.18). In the present case, the residual values are as shown in Table 11.5. Hence

$$\text{distance}(\hat{\mathbf{y}}\mathbf{y}) = \sqrt{3^2 + 3^2 + (-6)^2} = \sqrt{54} = 7.348,$$

Table 11.5 Fitted values and residuals from the
line of best fit to the data presented in Table 11.3.

i	y_i	$\hat{y}_i$	$\hat{\varepsilon}_i$
1	15	12	3
2	21	18	3
3	9	15	-6

the maximum-likelihood estimate of σ is

$$\hat{\sigma} = \sqrt{\frac{54}{3}} = 4.243,$$

and the maximum-likelihood estimate of σ^2 is

$$\hat{\sigma}^2 = \frac{54}{3} = 18.$$

However, as in the simpler case (Section 11.6, Figure 11.7), the point defined by the observed values is constrained to lie in an $(n-p)$-dimensional sub-space of the sample space, orthogonal (perpendicular) to the model sub-space, which we will call the *random-effects sub-space*. Hence the maximum-likelihood estimate of σ can again be adjusted to give a REML estimate, replacing the divisor n by $(n-p)$ to give the familiar formulae

$$\hat{\sigma} = \frac{(\text{distance}(\hat{\mathbf{y}}\mathbf{y}))}{\sqrt{n-p}} = \sqrt{\frac{\sum_{i=1}^{n}\hat{\varepsilon}_i^2}{n-p}}. \tag{11.31}$$

and

$$\hat{\sigma}^2 = \frac{\sum_{i=1}^{n}\hat{\varepsilon}_i^2}{n-p}. \tag{11.32}$$

In the present case, the point $\mathbf{y}$ lies in a $(3-2)$-dimensional space, that is, a line, perpendicular to the plane defined by the X_1-effects and X_2-effects axes, as indicated in Figure 11.12. The REML estimate of σ^2 in this case is

$$\hat{\sigma}^2 = \frac{54}{3-2} = 54.$$

In Figure 11.13 contour shading is added to the plot, showing the probability density of Y_1, Y_2 and Y_3 produced by proposing that $\sigma = \hat{\sigma}$, subject to the constraint imposed by the fixed-effects sub-space, that is, by the equation

$$\hat{Y} = \hat{\beta}_1 X_1 + \hat{\beta}_2 X_2. \tag{11.33}$$

This unbiased estimate of the residual variance is substantially larger than the maximum-likelihood estimate. When the number of parameters in the model (p) is close to the number of observations (n), the residual degrees of freedom are few, and the adjustment from maximum likelihood to REML is an important change (Patterson and Thompson, 1971).

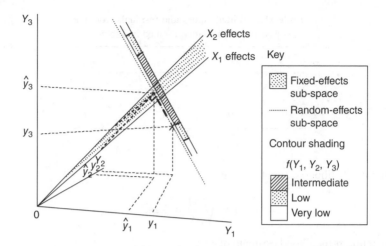

Figure 11.13 The sample space defined by three observations of Y, showing the probability density subject to the constraint imposed by the fixed-effects sub-space combined with the proposal that $\sigma = \hat{\sigma}$.

11.10 Extension to models with more than one random-effect term

Finally, this approach can be extended to models that have more than one random-effects term – mixed models. In the case of an ordinary general linear model with only one random-effect term, the fixed and random effects can be estimated by the exact, analytic solution of a set of simultaneous equations: the model-fitting process is simply the dropping of a perpendicular from the observed values of the response variable to the fixed-effects sub-space of the sample space, as we have seen in Sections 11.5 and 11.8. However, in a model with more than one random-effect term it is not generally possible to estimate the fixed and random effects analytically: REML algorithms proceed by obtaining initial estimates of the variance components for the random-effect terms and then using these as the basis for estimation of the fixed effects. These fixed-effects estimates are used to improve the estimates of variance components and so on until convergence is achieved (Payne, 2013, Section 5.3.9, pp. 627–629).

We have seen that the iterative process of fitting a mixed model using a REML algorithm is not guaranteed to succeed (Section 10.5), and it may also require large amounts of computer time and data space. A straightforward approach to the application of the REML criterion requires the inversion of a matrix in which

$$\text{number of rows} = \text{number of columns}$$

$$= \text{number of fixed effects} + \text{number of random effects}.$$

This is a serious computational challenge in the case of a model with a large number of effects. Consequently, in the more advanced applications of mixed modelling methods, much effort is devoted to finding techniques and specifications that will reduce this computational burden. These include the choice of realistic initial estimates for the variance components

(Section 10.5), the choice of the REML algorithm used, and other specifications that sometimes simplify the application of the algorithm chosen.

Two algorithms are available for applying the REML criterion, namely Fisher scoring and the average information (AI) algorithm. Their relative merits are described by Payne (2013, Section 5.3.1, pp. 596–600 and Section 5.3.9, pp. 627–629). The choice between the algorithms and the choice of other specifications within the Fisher scoring algorithm are largely concerned with the reduction of the matrix-inversion burden. The AI algorithm, which uses *sparse matrix* methods for some of the model terms, is usually faster and requires less data space, but it does not permit the estimation of standard errors for differences between random effects. If Fisher scoring is used, one of the model terms may be specified as an *absorbing factor*. Matrix-inversion operations for the terms that involve the absorbing factor and those that do not are then performed separately, substantially reducing the size of the matrices that must be inverted. Again, there is a price to be paid in lost information, this time concerning the estimates of covariance between parameters from the two sets.

This brief account will be sufficient to indicate that the serious application of mixed modelling methods requires experience, experimentation and careful thought. Mixed modelling is not a process that is easily automated, and this limits its application to routine statistical analyses. However, any researcher engaged in the statistical analysis of data should consider whether mixed models would increase the realism of their work. The alternative is to remain in a world in which only residual variation is recognized as random – in which the multiple strata at which researchers sample reality, and/or deal with exchangeable sets of levels, are denied.

11.11 Summary

The maximum-likelihood estimate of a parameter is the value that maximizes the probability of the data.

The use of the maximum-likelihood criterion for the estimation of the mean and variance of a normally distributed variable is illustrated.

The maximum-likelihood estimate of the variance differs from the usual, unbiased estimate. However, it is shown that the criterion for obtaining the usual estimate can be viewed as REstricted Maximum Likelihood (REML).

In order to view the variance estimation criterion in this way, a sub-space of sample space that represents the fixed effects, the fixed-effects sub-space, is first defined. The remaining sub-space represents the residual effects, and the REML estimate is the maximum-likelihood estimate within this random-effects sub-space.

The REML criterion can be extended to the general linear regression model, which encompasses all ordinary regression models.

The REML criterion can be further extended to models with more than one random-effect term (mixed models). However, in this case, it is not generally possible to estimate the fixed and random effects analytically.

REML algorithms obtain initial estimates of the variance components for the random-effect terms, then use these as the basis for the estimation of the fixed effects. These fixed-effect estimates are used to improve the estimates of the variance components, and so on until convergence is achieved.

The fitting of a mixed model using a REML algorithm is not guaranteed to succeed, and may require large amounts of computer time and data space.

The techniques that can be used to reduce this computational burden include realistic initial estimates of the variance components, and methods to reduce the size of the matrices that must be inverted.

Two REML algorithms are available, namely Fisher scoring and Average Information (AI). The AI algorithm is usually faster and requires less data space, but does not permit the estimation of standard errors for differences between random effects.

11.12 Exercises

11.1 Three observations were made of a random variable Y, namely

$$y_1 = 51, \quad y_2 = 35, \quad y_3 = 31.$$

(a) Assume that

$$Y \sim N(40, \sigma^2),$$

that is, the mean is known but the variance must be estimated. Obtain the maximum-likelihood estimate of σ^2 from this sample. Is this estimate unbiased?

(b) Now assume that

$$Y \sim N(\mu, \sigma^2),$$

that is, both parameters must be estimated. Obtain the maximum-likelihood estimates of μ and σ^2. Are these estimates unbiased? Obtain the residual-maximum-likelihood (REML) estimate of σ^2. Is this estimate unbiased?

(c) Make a sketch of the data space defined by this sample, corresponding to that presented in Figure 11.3 for a sample of two observations. Show the Y_1, Y_2 and Y_3 axes, the μ-effects axis and the observed values.

(d) Describe briefly how this geometrical representation can be used to obtain estimates of μ and σ^2, designated $\hat{\mu}$ and $\hat{\sigma}^2$. What is the distance from the point $\mathbf{y} = (y_1, y_2, y_3)'$ to the point $\hat{\boldsymbol{\mu}} = (\hat{\mu}, \hat{\mu}, \hat{\mu})'$?

When this graphical approach is used to represent a sample of two observations, and when any particular values, $\hat{\mu}$ and $\hat{\sigma}^2$, are postulated for μ and σ^2, the contours of the resulting probability distribution are circles, as shown in Figure 11.5.

(e) What is the shape of the corresponding contours for this sample of three observations?

(f) Sketch the contour at a distance $\sqrt{3}\hat{\sigma}$ from the point $(\hat{\mu}, \hat{\mu}, \hat{\mu})'$:

　(i) when the postulated values $\hat{\mu}$ and $\hat{\sigma}^2$ are the maximum-likelihood estimates and

　(ii) when the postulated value $\hat{\mu}$ is above the maximum-likelihood estimate and the postulated value $\hat{\sigma}^2$ is below the maximum-likelihood estimate.

Now suppose that $\hat{\mu}$ is restricted to the maximum-likelihood estimate.

(g) In what sub-space of the data space must the point $\mathbf{y}$ then lie?

(h) Within this random-effects sub-space, what will be the shape of the contours of the probability distribution? Mark the sub-space and a representative contour on your sketch.

11.2 Seven observations on two explanatory variates, X_1 and X_2, and a response variate, Y, are presented in Table 11.6.

Table 11.6 Observations of two explanatory variates and a response variate.

X_1	X_2	Y
42	7.3	128.0
58	9.2	145.0
93	3.9	101.6
70	4.1	90.9
35	8.4	135.6
61	3.7	85.0
29	8.0	108.7

(a) Fit the model

$$Y = \beta_0 + \beta_1 X_1 + \beta_2 X_2 + E$$

to these data, and obtain estimates of β_0, β_1 and β_2.
(b) When this model is fitted to these data, how many dimensions does each of the following have:

  (i) the data space?
  (ii) the fixed-effects sub-space?
  (iii) the random-effects sub-space?

(c) Obtain the estimated value of Y, and the estimate of the residual effect, for each
observation.

For Observation 5, the estimated value of $Y = 128.2$.
(d) What is the contribution to this value of

  (i) the constant effect?
  (ii) the effect of X_1?
  (iii) the effect of X_2?

It is assumed that

$$E \sim N(0, \sigma^2).$$

(e) Obtain the maximum-likelihood estimate of σ^2 and the REML estimate of σ^2.
(f) What is the minimum number of observations required to obtain estimates of β_0, β_1,
β_2 and σ^2? If the number of observations available is one less than this minimum,
what estimates can be obtained? What is then the relationship between the estimated
and observed values of Y?

11.3 Consider the final model fitted to the osteoporosis data in Section 7.2.

(a) When this model is fitted to these data, how many dimensions does each of the following have

  (i) the data space?
  (ii) the fixed-effects sub-space?
  (iii) the random-effects sub-space?

(b) What is the relationship between the number of dimensions of the random-effects
sub-space and the degrees of freedom of the deviance from this model?

References

Bulmer, M.G. (1979) *Principles of Statistics*, 2nd edn, Dover Publications, New York, 252 pp.

Edwards, A.W.F. (1972) *Likelihood. An Account of the Statistical Concept of Likelihood and Its Application to Scientific Inference*, Cambridge University Press, Cambridge, 235 pp.

Patterson, H.D. and Thompson, R. (1971) Recovery of inter-block information when block sizes are unequal. *Biometrika*, **58**, 545–554.

Payne, R.W. (ed) (2013) *The Guide to the GenStat Command Language (Release 16). Part 2: Statistics*, VSN International, Hemel Hempstead, 1021 pp.

Index

absorbing factor, 473

accumulated analysis of deviance. *See* analysis of deviance, accumulated

accumulated anova, 12–14, 19, 35, 71

AI. *See* algorithm, average information

Akaike Information Criterion (AIC), 31, 38, 73, 99, 124, 125, 127, 128, 177, 181, 202, 234, 235, 237, 238, 240, 241, 247, 248, 266, 268, 269, 271, 273, 275, 279, 310, 312, 313, 315, 317, 324, 325, 359, 361, 373, 375, 389, 411, 439, 440, 442, 443, 445, 446

algorithm

  average information, 66, 94, 154, 223, 298, 355, 425, 473

  Fisher scoring, 66, 473, 474

  sparse, 18, 19, 24, 65, 94, 111, 154, 223, 225, 244, 251, 252, 257, 259, 298, 299, 301, 304, 355, 371, 425, 428, 435, 437

aliasing

  partial, 11, 267

allele, 430

analysis of deviance, 396, 416

  accumulated, 396, 416

analysis of variance (anova)

anisotropic power model. *See* model, anisotropic power

anova. *See* analysis of variance (anova)

AR model. *See* model, autoregressive

array, 244, 363, 368, 377, 431, 447

assay, 87–91, 97–102, 130, 132, 133

association mapping, 430

asterisk, 11, 29, 106, 126, 370

average information, 19, 66, 94, 154, 223, 298, 355, 425, 473

balance, 58, 71, 91, 106–111, 123 , 126, 132, 149, 151, 195, 207, 214, 350–354, 362–374, 376, 377, 380

baseline value, 218, 242, 243, 283, 296

Bayesian approach. *See also* Bayesian statistics

  empirical, 185, 188, 304, 333, 338, 343

Bayesian Information Criterion (BIC), 31, 38, 73, 99, 124, 125, 127, 128, 177, 181, 202, 234, 235, 237, 238, 240, 241, 247, 248, 266, 268, 269, 271, 273, 275, 279, 310, 312, 313, 315, 317, 324, 325, 359, 361, 373, 375, 389, 411, 440, 441, 442, 443, 445, 446

Bayesian interpretation of BLUP. *See* interpretation of BLUP, Bayesian

Bayesian statistics, 183

belief

  posterior (*see* posterior belief)

  prior (*see* prior belief)

Best Linear Unbiased Estimate (BLUE), 168, 170–175, 187, 188, 190, 328, 380

Best Linear Unbiased Predictor (BLUP), 165–190, 328, 380, 430

bias, 67, 296, 303, 342, 430

  in meta-analysis, 296, 303, 342

BIC. *See* Bayesian Information Criterion (BIC)

binary outcome. *See* binary response

binary response, 322

Introduction to Mixed Modelling: Beyond Regression and Analysis of Variance, Second Edition. N. W. Galwey.
© 2014 John Wiley & Sons, Ltd. Published 2014 by John Wiley & Sons, Ltd.
Companion website: http://www.wiley.com/go/beyond_regression